Agent-Based and Individual-Based Modeling

Agent-Based and Individual-Based Modeling

A PRACTICAL INTRODUCTION   Steven F. Railsback and Volker Grimm

PRINCETON UNIVERSITY PRESS

Princeton and Oxford

Published by Princeton University Press, 41 William Street, Princeton, New Jersey 08540
In the United Kingdom: Princeton University Press, 6 Oxford Street, Woodstock, Oxfordshire OX20 1TW
press.princeton.edu

Jacket Art: *Orange Haiku*, 2010, oil on linen, 24″ × 24″ © Astrid Preston.

Library of Congress Cataloging-in-Publication Data

Railsback, Steven F.
   Agent-based and individual-based modeling : a practical introduction / Steven F. Railsback and Volker Grimm.
       p. cm.
   Includes bibliographical references and index.
   ISBN 978-0-691-13673-8 (hardback) — ISBN 978-0-691-13674-5 (pbk.) 1. Multiagent systems—Textbooks.
2. Science—Mathematical models. I. Grimm, Volker, 1958– II. Title.
   QA76.76.I58R35 2011
   006.3—dc23

                                                                                        2011017473

British Library Cataloging-in-Publication Data is available

This book has been composed in Minion Pro, Myriad Pro, and Courier
Printed on acid-free paper. ∞
Printed in the United States of America
10 9 8

Contents

Contents

Preface

In 2005, we published *Individual-Based Modeling and Ecology*, which laid out our ideas on why and how individual-based models (IBMs) can be used in ecology and, by analogy, in many other fields. As we wrote that book we realized that it could not serve well by itself as a text for classes on individual- or agent-based modeling: there was too much general and conceptual material to cover to allow us to also provide the detail and examples needed for a textbook. Hence, we produced this book to fill the textbook role.

Our ultimate goal in writing this book is to address a fundamental limitation to the use of agent-based and individual-based models (A/IBMs): outside of computer science and engineering, students interested in using A/IBMs have little access to training and expertise. Few professors in the biological and social sciences are trained in simulation modeling and software development. This situation stands in stark contrast to statistical modeling: almost every university department in every science has faculty skilled in statistics and expects its students to develop some facility in statistical modeling. As A/IBMs become an ever-more-important tool, we expect more and more departments will want to offer classes even though they lack experienced instructors. We designed this book to help "bootstrap" the adoption of this new technology by allowing instructors of all experience levels to get students started doing science with A/IBMs. We also expect this book to help students learn individual-based modeling by themselves if there is no class available.

This book was designed so it can be used by itself, but we think many of its users will benefit from reading *Individual-Based Modeling and Ecology* first or at the same time. Our first book focused on conceptual aspects of how to design A/IBMs and analyze them to do science, while this book focuses more on the details of implementing and analyzing models. One difference is that this book is not specific to ecology; it still reflects our own backgrounds in ecology, but we intend it to be useful for the many fields in which a textbook on A/IBMs is needed.

In disciplines other than ecology, IBMs are more often referred to as ABMs, so we use the term "agent-based" in this book more than "individual-based." There have been historical differences between individual- and agent-based models: IBMs focused on individual variability and local interactions, whereas ABMs focused on decision-making and adaptive behavior. But these differences are fading away so that we use both terms interchangeably, as we did in our first book. Likewise, we also implicitly include and address "multi-agent systems," which are just a branch of agent-based modeling that originated from computer science and research on artificial intelligence and artificial life.

Book Objectives

This book is designed to support introductory classes—or independent study—in agent-based modeling for scientists, including courses with instructors new to simulation modeling and computer programming. The course is targeted at graduate students and advanced undergraduates who are starting their research careers, but it is also appropriate for experienced scientists who want to add agent-based modeling to their toolkit. Students can expect to learn about both the conceptual and theoretical aspects of using ABMs and the details of implementing models on the computer using NetLogo software. Among the topics covered are:

- When and why to use ABMs, and how they are different from other models;
- How to design an ABM for a particular system and problem;
- A conceptual foundation for designing and describing models;
- Programming models and conducting simulation experiments in NetLogo; and
- How to analyze a model to solve scientific problems, including development of theory for complex systems.

Throughout the course we emphasize several themes about doing science with ABMs:

- *Using models for solving research problems.* The primary characteristic of scientific models is that they are designed to solve a specific problem about a system or class of systems. These problems might include predicting how the system responds to novel conditions, or just understanding the mechanisms that drive the system.
- *Basing models on theory, and using models to develop theory.* By theory, in the context of agent-based complex systems, we mean models of the individual characteristics and behaviors from which system behaviors emerge.
- *Learning and following the conventions and theory of scientific modeling.* Modeling is not simply an intuitive process that lacks standard procedures and theory. There is, in fact, much that modelers need to learn from our predecessors. Examples include knowing the importance of appropriate space and time scales and how to conduct standard kinds of model analysis.
- *Documenting models and testing software.* These tasks are often treated by novices as tedious distractions but they are in fact essential—and productive, and even sometimes fun—parts of scientific modeling.
- *Standardization.* One of the historical difficulties with ABMs is that the standard "languages" we have for thinking about and describing other kinds of models (e.g., differential equations; statistics) are not sufficient for formulating ABMs. A great deal of recent work has gone into developing "standards" for ABMs, and we emphasize their use. Throughout this book we use a standard protocol (called *ODD*) for describing models and a set of standard concepts for thinking about and designing ABMs; and NetLogo itself is a standard language for programming ABMs.

Why NetLogo?

Choosing which software platform to use in this book was a critical decision for us. There are many platforms for agent-based modeling, and they vary in many ways. We learned to use the most popular platforms and tried teaching several of them. This experience led us to two conclusions. First, there is no single ideal platform; platforms are inevitably compromises that cannot be best for all applications. Second, though, NetLogo clearly stands apart as the best platform for beginners and even for many serious scientific models.

NetLogo provides a simplified programming language and graphical interface that lets users build, observe, and use ABMs without needing to learn the complex details of a standard programming language. Yet many publishable scientific models can be, and have been, implemented in NetLogo. Just as importantly, the NetLogo team at Northwestern University provides an extremely complete, helpful, and professional set of documentation and tutorial materials. NetLogo was originally designed as an educational tool, but its use in science has grown very rapidly, and NetLogo itself has changed to better meet the needs of scientists.

Despite NetLogo's capabilities, some students will eventually want more programming flexibility and power. Therefore, we also try to prepare interested students for additional training in programming and software development. Throughout the course we identify basic software concepts that should make it easier for those students who choose to move on to, for example, a class in Java and training in one of the ABM platforms (e.g., MASON, Repast) that use Java.

Overview and Suggested Course Structure

The book has four parts. The first provides the foundation that students need to get started: a basic understanding of what agent-based modeling is and of the modeling process, and basic skills in implementing ABMs in NetLogo. Part II introduces specific model design concepts and techniques while widening and deepening the student's NetLogo skills. In part III we move on broader scientific issues: using knowledge of the system being modeled to design models with the right level of complexity, develop theoretical understanding of the system, and calibrate models. Finally, part IV focuses on analyzing models: how to learn from and do science with ABMs. At the end we provide suggestions for moving on to become a successful modeler without losing momentum.

A course using this book is expected to include both lecture and hands-on computer lab time. We present many modeling and software concepts that are best explained in a lecture format. But most of the learning will take place in a computer laboratory as students work on exercises and projects of increasing complexity and independence. For a college class, we envision each week to include one or two lecture hours plus at least one computer lab of two to four hours.

At the start of the course, chapters 1, 3, and 6 are designed for lecture, while chapters 2, 4, and 5 are designed as introductory computer labs led by the instructor; chapter 6 also includes an important lab exercise. For the remainder of the course, each chapter includes modeling concepts and programming techniques to be introduced in lecture, followed by exercises that reinforce the concepts and techniques. In the computer labs for parts II–IV, students should be able to work more independently, with the instructor spending less time in front of the class and more time circulating to help individual students. Exercises started in lab can be completed as homework.

The exercises in parts I–II are generally short and focused, and it is probably best to have the entire class do the same exercises. Starting with part III, the exercises are more open-ended and demand more creativity and independence. It is natural to transition to more project-like work at this point, and it should be easy to develop independent projects that accomplish the same objectives as the exercises we provide. We strongly encourage instructors to replace our exercises with ones based on their own experience wherever they can.

Our experience has been that working on exercises in teams, usually of two students, is productive at least early in a class. It is great when students teach and help each other, and it is good to encourage them to do so, for example by providing a computer lab where they can work in the same room. Anyone who programs regularly knows that the best way to find your mistake is to explain it to someone else. However, it is also very common for beginning programmers to get stuck and need just a little help to get moving again. Programming instructors often provide some kind of "consulting service"—office hours, tutors, email, and the like—to students needing help with their assignments outside of regular class hours.

Even for a graduate-level class, it may be ambitious to try to work through the entire book. Especially in part II, instructors should be able to identify subsections, or perhaps even a whole chapter or two, to leave out, considering their particular interests. We certainly discourage instructors from spending so much time in parts I and II that the key later chapters (especially 18, 19, and 22) are neglected; that would be like teaching art students to mix pigments and stretch canvas but never to paint pictures.

What We Expect of Instructors

We wrote this book with the expectation that many instructors using it will have little or no prior experience with ABMs. This approach is unusual for a textbook, but we had little choice, given how few experienced agent-based modelers there are.

What do we expect of instructors? First, we expect it will help to digest *Individual-Based Modeling and Ecology* fairly thoroughly, especially because it presents many of the ideas in this book in more detail and provides more explanation for why we do things the way we do.

Second, instructors will need to develop, or find in people who can help, enough experience with the NetLogo platform to stay ahead of their students. This platform can be very easy to learn, but it has a distinct style that takes getting used to. Prior experience with programming may or may not be helpful: NetLogo uses some standard programming techniques (e.g., declaring variables, writing subroutine-like procedures), but if you try to write NetLogo code the way you write C or Java or MatLab, you will miss most of its advantages. Luckily, there is no reason to fear NetLogo. Its excellent documentation and example models will be a great help to you and your students. And NetLogo is fun: if you try to learn it completely by yourself, your primary problem (after an initial 1–2 hours of confusion and frustration) will be making yourself stop. Also, NetLogo is becoming quite popular and it is likely that you can find experienced users on campus, or perhaps even a computer science professor or student willing to learn and help. Finally, there is an on-line NetLogo users community and even a special forum for educators. If you start working your way through NetLogo's tutorial materials and parts I and II of this book with some focus a few weeks before starting to teach, you should be OK.

The third thing we hope for from instructors is help keeping students from developing bad habits that are unfortunately common among novice modelers. NetLogo encourages an experimental approach: it is very easy and fun to make little code changes and see what happens. This experimentation is fine, but students need to learn that they are not really modeling until they do three things. First is to stop changing and expanding their model even before it seems "realistic" or "finished," so they can proceed with learning something from it. Second is to write down what their model is to do and why; and third is to provide solid evidence that their software actually does what they wrote down. After building many ABMs, we know that software that has not been tested seriously is very unlikely to be mistake-free, even with a platform as simple as NetLogo. And a written description of the model is necessary for both scientific communication and software testing. But neither of those tasks can be completed until the student stops changing the model's design. We hope instructors will do what computer programming instructors do: treat assignments as incomplete unless software is accompanied with documentation of (a) its purpose and design and (b) a serious attempt to find mistakes in it. You will find your students making much more rapid progress with modeling after these steps become habit.

For the Independent Learner

As much as we hope that classes on agent-based modeling will be offered at every university, we realize that this will not happen immediately and some people will need to teach themselves. If you are one of these people, we kept you in mind as we designed the book.

How should you proceed? We recommend you read a chapter and then work through its exercises until you are comfortable moving on. If you are nervous about learning NetLogo by yourself, keep in mind that we will point you to many learning materials and sources of support provided with NetLogo and by its users. The on-line user community may be especially helpful for you, and you should be able to find other NetLogo users nearby to consult with.

Course Web Site

Materials supporting this book and its users are available through

http://press.princeton.edu/titles/9639.html.

Please check this site for news and updates as you use the book. The information available through the site includes:

- Any changes to the text, example code, and exercises that result from new versions of NetLogo (this book is current with NetLogo 5.0);
- Supplementary materials, referred to in the book, that we plan to update;
- Materials to help instructors; and
- Ways to communicate with us, provide feedback, and share ideas.

Acknowledgments

Many people helped us develop the material and experience that went into this book. In particular, our dear friend Uta Berger of the Institute of Forest Growth and Forest Computer Sciences, Technische Universität Dresden (Dresden University of Technology), Germany, organizes the TUD Summer School in Individual- and Agent-Based Modeling. Much of this course was developed with Professor Berger at the Summer School. We thank Uta, the other instructors, and the many students. We also need to thank all the people who read drafts and provided useful feedback; unfortunately we cannot name them all individually.

The data for figure 1.1 were generously contributed by Dr. Henri Weimerskirch, Centre d'Etudes Biologiques de Chizé, Centre National de la Recherche Scientifique, France, and provided to us by Dr. Maite Louzao. Guy Pe'er provided the topographic data for his hilltopping butterfly model field sites; the data are from the work of John K. Hall and provided by Adi Ben-Nun of the Hebrew University GIS center. Professors Steve Hackett and Howard Stauffer of Humboldt State University reviewed sections for technical accuracy. The R code for contour plotting in chapter 12 was provided by Damaris Zurell, Institute of Geoecology, University of Potsdam. Computer scientists who contributed, especially to chapter 6, include Steve Jackson, Steve Lytinen, and Phil Railsback. Marco Janssen and Margaret Lang provided many good suggestions for making a textbook effective. And we thank the fabulous painter Astrid Preston (www.astridpreston.com) for again letting us use one of her pieces on the cover.

We greatly appreciate the support of Volker Grimm's home institution. The Department of Ecological Modelling at the Helmholtz Center for Environmental Research–UFZ, Leipzig, is a unique center of expertise in the practice and philosophy of modelling, especially individual-based modeling of the natural and human environments.

We especially thank Uri Wilensky and the NetLogo team at Northwestern University's Center for Connected Learning for producing and maintaining the software that makes this course possible and fun. In addition, we thank Seth Tisue for his time reviewing our manuscript and the many improvements he suggested.

We dedicate this book to our fathers, Charles Railsback and Hans Grimm, both of whom died in 2010. They were both, by nature, devoted to philosophical and rational inquiry, education, and—especially—to making the world a better place.

Agent-Based Modeling and NetLogo Basics

Models, Agent-Based Models, and the Modeling Cycle

1.1 Introduction, Motivation, and Objectives

Welcome to a course in agent-based modeling (or "individual-based" modeling, as the approach is called in some fields). Why is it important to learn how to build and use agent-based models (ABMs)? Let's look at one real model and the difference it has made.

1.1.1 A Motivational Example: Rabies Control in Europe

Rabies is a viral disease that kills great numbers of wild mammals and can spread to domestic animals and people. In Europe, rabies is transmitted mainly by red fox. When an outbreak starts in a previously rabies-free region, it spreads in "traveling waves": alternating areas of high and low infection rates.

Rabies can be eradicated from large areas, and new outbreaks can be controlled, by immunizing foxes: European governments have eradicated rabies from central Europe by manufacturing rabies vaccine, injecting it into baits, and spreading the baits from aircraft. However, this program is extremely expensive and works only if new outbreaks are detected and contained. Key to its cost-effectiveness are these questions: What percentage of wild foxes need to be vaccinated to eliminate rabies from an area, and what is the best strategy for responding to outbreaks?

Models have long been applied to such epidemiological problems, for wildlife as well as people. Classical differential equation models of the European rabies problem predicted that 70% of the fox population must be vaccinated to eliminate rabies. Managers planned to respond to new outbreaks using a "belt vaccination" strategy (which has worked well for other epidemics, including smallpox): not vaccinating the outbreak location itself but a belt around it, the width of which was usually determined by the limited emergency supply of vaccine. The 70% vaccination strategy did succeed, but the rabies problem has several characteristics suggesting that an agent-based modeling approach could make important contributions: the spread of rabies has important patterns in space as well as time, and is driven by individual behavior (in this case, the use of stationary territories by most fox but long-distance migration by young foxes). Hence, Florian Jeltsch and colleagues developed a simple ABM that represented fox families in stationary home ranges and migration of young foxes (Jeltsch et al. 1997). This model accurately simulated the spread of rabies over both space and time.

Dirk Eisinger and Hans-Hermann Thulke then modified the ABM specifically to evaluate how the distribution of vaccination baits over space affects rabies control (Thulke and Eisinger 2008, Eisinger and Thulke 2008, Eisinger et al. 2005). Their ABM indicated that eradication could be achieved with a vaccination rate much lower than 70%, a result that could save millions of euros and was confirmed by the few case studies where actual vaccination coverage was monitored. The reason for the lower vaccination rate predicted by the ABM is that the "wave" spread of rabies emerges from local infectious contacts that actually facilitate eradication. The ABM of Eisinger and Thulke also indicated that the belt vaccination strategy for outbreaks would fail more often than an alternative: compact treatment of a circle around the initial outbreak. Because the ABM had reproduced many characteristics of real outbreaks and its predictions were easy to understand, rabies managers accepted this result and began—successfully—to apply the compact vaccination strategy.

The rabies example shows that agent-based modeling can find new, better solutions to many problems important to our environment, health, and economy—and has already done so. The common feature of these problems is that they occur in systems composed of autonomous "agents" that interact with each other and their environment, differ from each other and over space and time, and have behaviors that are often very important to how the system works.

1.1.2 Objectives of Chapter 1

This chapter is your introduction to modeling and agent-based modeling. We get started by clarifying some basic ideas about modeling. These lessons may seem trivial at first, but they are in fact the very foundation for everything else in this course. Learning objectives for chapter 1 are to develop a firm understanding of:

- What models are, and what modeling is—why do we build models anyway?
- The modeling cycle, the iterative process of designing, implementing, and analyzing models and using them to solve scientific problems.
- What agent-based models are—how are ABMs different from other kinds of model, and why would you use them?

1.2 What Is a Model?

A model is a purposeful representation of some real system (Starfield et al. 1990). We build and use models to solve problems or answer questions about a system or a class of systems. In science, we usually want to understand how things work, explain patterns that we have observed, and predict a system's behavior in response to some change. Real systems often are too complex or develop too slowly to be analyzed using experiments. For example, it would be extremely difficult and slow to understand how cities grow and land uses change just with experiments. Therefore, we try to formulate a simplified representation of the system using equations or a computer program that we can then manipulate and experiment on. (To *formulate* a model means to design its assumptions and algorithms.)

But there are many ways of representing a real system (a city or landscape, for example) in a simplified way. How can we know which aspects of the real system to include in the model and which to ignore? To answer this question, the model's purpose is decisive. The question we want to answer with the model serves as a filter: all those aspects of the real system considered irrelevant or insufficiently important *for answering this question* are filtered out. They are ignored in the model, or represented only in a very simplified way.

Let us consider a simple, but not trivial, example: Did you ever search for mushrooms in a forest? Did you ask yourself what the best search strategy might be? If you are a mushroom expert, you would know how to recognize good mushroom habitat, but let us assume you are a neophyte. And even the mushroom expert needs a smaller-scale search strategy because mushrooms are so hard to see—you often almost step on them before seeing them.

You might think of several intuitive strategies, such as scanning an area in wide sweeps but, upon finding a mushroom, turning to smaller-scale sweeps because you know that mushrooms occur in clusters. But what does "large" and "small" and "sweeps" mean, and how long should you search in smaller sweeps until you turn back to larger ones?

Many animal species face similar problems, so it is likely that evolution has equipped them with good adaptive search strategies. (The same is likely true of human organizations searching for prizes such as profit and peace with neighbors.) Albatross, for example, behave like mushroom hunters: they alternate more or less linear long-distance movements with small-scale searching (figure 1.1).

The common feature of the mushroom hunter and the albatross is that their sensing radius is limited—they can only detect what they seek when they are close to it—so they must move. And, often the items searched for are not distributed randomly or regularly but in clusters, so search behavior should be adaptive: it should change once an item is found.

Why would we want to develop a model of this problem? Because even for this simple problem we are not able to develop quantitative mental models. Intuitively we find a search strategy which works quite well, but then we see others who use different strategies and find more mushrooms. Are they just luckier, or are their strategies better?

Now we understand that we need a clearly formulated purpose before we can formulate a model. Imagine that someone simply asked you: "Please, model mushroom hunting in the

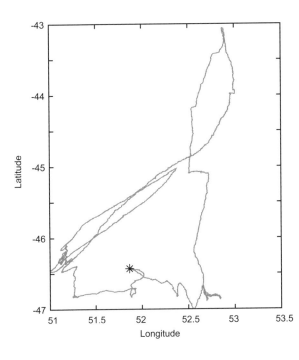

Figure 1.1
Flight path of a female wandering albatross (*Diomedea exulans*) feeding in the southern Indian Ocean. The flight begins and ends at a breeding colony (indicated by the star) in the Crozet Islands. Data recorded by H. Weimerskirch and colleagues for studies of adaptive search behavior in albatross (e.g., Weimerskirch et al. 2007).

forest." What should you focus on? On different mushroom species, different forests, identification of good and bad habitats, effects of hunting on mushroom populations, etc.? However, with the purpose "What search strategy maximizes the number of mushrooms found in a certain time?" we know that

- We can ignore trees and vegetation; we only need to take into account that mushrooms are distributed in clusters. Also, we can ignore any other heterogeneity in the forest, such as topography or soil type—they might affect searching a little, but not enough to affect the general answer to our question.
- It will be sufficient to represent the mushroom hunter in a very simplified way: just a moving "point" that has a certain sensing radius and keeps track of how many mushrooms it has found and perhaps how long it has been since it found the last one.

So, now we can formulate a model that includes clusters of items and an individual "agent" that searches for the items in the model world. If it finds a search item, it switches to smaller-scale movement, but if the time since it found the last item exceeds a threshold, it switches back to more straight movement to increase its chance of detecting another cluster of items. If we assume that the ability to detect items does not change with movement speed, we can even ignore speed.

Figure 1.2 shows an example run of such a model, our simple Mushroom Hunt model. In chapter 2 you will start learning NetLogo, the software platform we use in this book, by programming this little model.

This searching problem is so simple that we have good idea of what processes and behaviors are important for modeling it. But how in general can we know whether certain factors are important with regard to the question addressed with a model? The answer is: we can't! That is, exactly, why we have to formulate, implement (program in the computer), and analyze a model: because then we can use mathematics and computer logic to rigorously explore the consequences of our simplifying assumptions.

Our first formulation of a model must be based on our preliminary understanding of how the system works, what the important elements and processes are, and so on. These preliminary ideas might be based on empirical knowledge of the system's behavior, on earlier models addressing similar questions, on theory, or just on . . . imagination (as in the mushroom

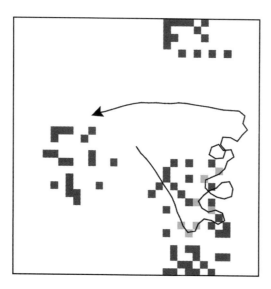

Figure 1.2
Path of a model agent searching for items that are distributed in clusters.

hunting example). However, if we have no idea whatsoever of how the system works, we cannot formulate a model! For example, even though scientists are happy to model almost everything, so far there seems to be no explicit model of human consciousness, simply because we have no clue what consciousness really is and how it emerges.

Because the assumptions in the first version of a model are experimental, we have to test whether they are appropriate and useful. For this, we need criteria for whether the model can be considered a good representation of the real system. These criteria are based on patterns or regularities that let us identify and characterize the real system in the first place. Stock market models, for example, should produce the kinds of volatility and trends in prices we see in real markets. Often we find that the first version of a model is too simple, lacks important processes and structures, or is simply inconsistent. We thus go back and revise our simplifying assumptions.

1.3 The Modeling Cycle

When thinking about a model of a mushroom hunter (or albatross), we intuitively went through a series of tasks. Scientific modeling means to go through these tasks in a systematic way and to use mathematics and computer algorithms to rigorously determine the consequences of the simplifying assumptions that make up our models.

Being scientific always means iterating through the tasks of modeling several times, because our first models can always be improved in some way: they are too simple or too complex, or they made us realize that we were asking the wrong questions. It is therefore useful to view modeling as iterating through the "modeling cycle" (figure 1.3). *Iterating* does not mean that we always go through the full cycle; rather, we often go through smaller loops, for example between problem formulation and verbal formulation of the model. The modeling cycle consists of the following tasks:

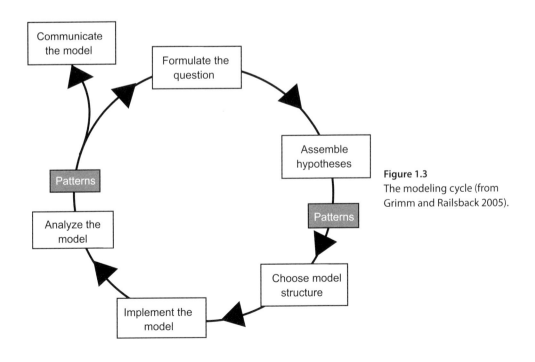

Figure 1.3
The modeling cycle (from Grimm and Railsback 2005).

1. *Formulate the question.* We need to start with a very clear research question because this question then serves as the primary compass and filter for designing a model. Often, formulating a clear and productive question is by itself a major task because a clear question requires a clear focus. For complex systems, getting focused can be difficult. Very often, even our questions are only experimental and later we might need to reformulate the question, perhaps because it turned out to be not clear enough, or too simple, or too complex.

   The question in our Mushroom Hunt model is: What search strategy maximizes the rate of finding items if they are distributed in clusters?

2. *Assemble hypotheses for essential processes and structures.* Agent-based modeling is "naive" (DeAngelis et al. 1994) in the sense that we are not trying to aggregate agents and what they are doing in some abstract variables like abundance, biomass, overall wealth, demographic rates, or nutrient fluxes. Instead, we naively and directly represent the agents and their behavior. We create these agents, put them in a virtual environment, then let the virtual world run and see what we can learn from it. (It is important, though, to ask ourselves: is it possible to answer our question using a more aggregated and thus simpler model?)

   Usually we have to formulate many hypotheses for what processes and structures are essential to the question or problem we address. We can start top-down and ask ourselves questions such as: What factors have a strong influence on the phenomena of interest? Are these factors independent or interacting? Are they affected by other important factors? We might draw so-called influence diagrams, or flow charts, or just caricatures of our system and question. But whatever technique we prefer, this task has to combine existing knowledge and understanding, a "brainstorming" phase in which we wildly hypothesize, and, most importantly, a simplification phase.

   We have to force ourselves to simplify as much as we can, or even more. The modeling cycle must be started with the most simple model possible, because we want to develop understanding gradually, while iterating through the cycle. A common mistake of beginners is to throw too much into the first model version—usually arguing that all these factors are well known and can't possibly be ignored. Then, the answer of the modeling expert is: yes, you might be right, but—let us focus on the absolute minimum number of factors first. Put all the other elements that you think might need to be in the model on your "wish list" and check their importance later.

   The reason for this advice is this: just our preliminary understanding of a system is not sufficient for deciding whether things are more or less important for a model. It is the very purpose of the model to teach us what is important. So, it is wise to have a model implemented as soon as possible, even if it is ridiculously simple. But the simpler the model is, the easier it is to implement and analyze, and the sooner we are productive. The real productive phase in a modeling project starts when we get the modeling cycle running: assumptions—implementation—analyses—interpretation—revised assumptions, and so on.

   It is difficult to formalize this task of the modeling cycle. One important help is heuristics for modeling: rules of thumb that are often, but not always, useful for designing models. We point out these heuristics throughout this book; use the index to find them. Compilations of modeling heuristics can be found in Starfield et al. (1990) and Grimm and Railsback (2005, chapter 2). And—in part III of this book we present *pattern-oriented modeling*, a very important strategy for formalizing both this and the next step in the modeling cycle.

   For the Mushroom Hunt model we assume that the essential process is switching between relatively straight large-scale "scanning" movement and small-scale searching, depending on how long it has been since the hunter last found an item.

3. *Choose scales, entities, state variables, processes, and parameters.* Once we choose some simplifying assumptions and hypotheses to represent our system of interest, it is time to sit down and think through our model in detail. We thus produce a written formulation of the model. Producing and updating this formulation is essential for the entire modeling process, including delivery to our "clients" (our thesis committee, journal reviewers, research sponsors, etc.). In chapter 3, we will start using a very helpful protocol for doing this.

   This step, for the Mushroom Hunt model, includes specifying how the space that hunters move through is represented (as square grids with size equal to the area the hunter can search in one time step), what kinds of objects are in the model (one hunter and the items it searches for), the state variables or characteristics of the hunter (the time it has hunted and the number of items it has found, and the time since last finding an item), and exactly how the hunter searches. (Full details are provided when we implement the model in chapter 2.)

4. *Implement the model.* This is the most technical part of the modeling cycle, where we use mathematics and computer programs to translate our verbal model description into an "animated" object (Lotka 1925). Why "animated"? Because, in a way, the implemented model has its own, independent dynamics (or "life"), driven by the internal logic of the model. Our assumptions may be wrong or incomplete, but the implementation itself is— barring software mistakes—always right: it allows us to explore, in a logical and rigorous way, the consequences of our assumptions and see whether our initial model looks useful.

   This task often is the most daunting one for neophytes in modeling, because they usually have no training in how to build software. Thus, our claim that the implementation always is "right" might sound ironic to beginners. They might struggle for months to get the implementation right—but only if they don't take advantage of existing software platforms for agent-based modeling. With the platform that we use in this book, NetLogo, you can often implement simple ABMs within a day or two, including the time to test your code and show that it is accurate. So please don't panic!

5. *Analyze, test, and revise the model.* While new modelers might think that designing a model and implementing it on the computer takes the most work, this task—analyzing a model and learning from it—is the most time-consuming and demanding one. With tools like NetLogo you will learn to quickly implement your own ABMs. But doing science with ABMs requires much more. Much of this book will be devoted to this task: how can we learn from our models? In particular, we will try to put forward the research program of "individual-based ecology" (Grimm and Railsback 2005) and apply it to other sciences. This program is dedicated to learning about the real world: we do not just want to see what happens when we create some agents and make up their behaviors—we want to see what agent behaviors can explain and predict important characteristics of real systems.

   To answer the mushroom hunting question, we could analyze the model by trying a variety of search algorithms and parameter values to see which produces the highest rate of finding items.

1.4 What Is Agent-Based Modeling? How Is It Different?

Historically, the complexity of scientific models was often limited by mathematical tractability: when differential calculus was the only approach we had for modeling, we had to keep models simple enough to "solve" mathematically and so, unfortunately, we were often limited to modeling quite simple problems.

With computer simulation, the limitation of mathematical tractability is removed so we can start addressing problems that require models that are less simplified and include more characteristics of the real systems. ABMs are less simplified in one specific and important way: they represent a system's individual components and their behaviors. Instead of describing a system only with variables representing the state of the whole system, we model its individual agents.

ABMs are thus models where individuals or agents are described as unique and autonomous entities that usually interact with each other and their environment locally. Agents may be organisms, humans, businesses, institutions, and any other entity that pursues a certain goal. Being unique implies that agents usually are different from each other in such characteristics as size, location, resource reserves, and history. Interacting locally means that agents usually do not interact with all other agents but only with their neighbors—in geographic space or in some other kind of "space" such as a network. Being autonomous implies that agents act independently of each other and pursue their own objectives. Organisms strive to survive and reproduce; traders in the stock market try to make money; businesses have goals such as meeting profit targets and staying in business; regulatory authorities want to enforce laws and provide public well-being. Agents therefore use *adaptive behavior*: they adjust their behavior to the current states of themselves, of other agents, and of their environment.

Using ABMs lets us address problems that concern *emergence*: system dynamics that arise from how the system's individual components interact with and respond to each other and their environment. Hence, with ABMs we can study questions of how a system's behavior arises from, and is linked to, the characteristics and behaviors of its individual components. What kinds of questions are these? Here are some examples:

- How can we manage tropical forests in a sustainable way, maintaining both economic uses and biodiversity levels critical for forests' stability properties (Huth et al. 2004)?
- What causes the complex and seemingly unpredictable dynamics of a stock market? Are market fluctuations caused by dynamic behavior of traders, variation in stock value, or simply the market's trading rules (LeBaron 2001, Duffy 2006)?
- How is development of human tissue regulated by signals from the genome and the extracellular environment and by cellular behaviors such as migration, proliferation, differentiation, and cell death? How do diseases result from abnormalities in this system (Peirce et al. 2004)?
- How do shorebird populations respond to loss of the mudflats they feed in, and how can the effects be mitigated cost-effectively (Goss-Custard et al. 2006)?
- What drives patterns of land use change during urban sprawl, and how are they affected by the physical environment and by management policies (Brown et al. 2004, Parker et al. 2003)?

ABMs are useful for problems of emergence because they are *across-level* models. Traditionally, some scientists have studied only systems, modeling them using approaches such as differential equations that represent how the whole system changes. Other scientists have studied only what we call agents: how plants and animals, people, organizations, etc. change and adapt to external conditions. ABMs are different because they are concerned with two (and sometimes more) levels and their interactions: we use them to both look at what happens to the *system* because of what its *individuals* do and what happens to the *individuals* because of what the *system* does. So throughout this course there will be a focus on modeling behavior of agents and, at the same time, observing and understanding the behavior of the system made up by the agents.

ABMs are also often different from traditional models in being "unsimplified" in other ways, such as representing how individuals, and the environmental variables that affect them, vary over space, time, or other dimensions. ABMs often include processes that we know to be important but are too complex to include in simpler models.

The ability of ABMs to address complex, multilevel problems comes at a cost, of course. Traditional modeling requires mathematical skills, especially differential calculus and statistics. But to use simulation modeling we need additional skills. This course is designed to give you three very important skills for using ABMs:

- A new "language" for thinking about and describing models. Because we cannot define ABMs concisely or accurately in the languages of differential equations or statistics, we need a standard set of concepts (e.g., emergence, adaptive behavior, interaction, sensing) that describe the important elements of ABMs.
- The software skills to implement models on computers and to observe, test, control, and analyze the models. Producing useful software is more complex for ABMs than for most other kinds of models.
- Strategies for designing and analyzing models. There is almost no limit to how complex a computer simulation model can be, but if a model is too complex it quickly becomes too hard to parameterize, validate, or analyze. We need a way to determine what entities, variables, and processes should and should not be in a model, and we need methods for analyzing a model, after it is built, to learn about the real system.

Full-fledged ABMs assume that agents are different from each other; that they interact with only some, not all other agents; that they change over time; that they can have different "life cycles" or stages they progress through, possibly including birth and death; and that they make autonomous adaptive decisions to pursue their objectives. However, as with any model assumption, assuming that these individual-level characteristics are important is experimental. It might turn out that for many questions we do not explicitly need all, or even any, of these characteristics. And, in fact, full-fledged ABMs are quite rare. In ecology, for example, many useful ABMs include only one individual-level characteristic, local interactions. Thus, although ABMs are defined by the assumption that agents are represented in some way, we still have to make many choices about what type of agents to represent and in what detail.

Because most model assumptions are experimental, we need to test our model: we must implement the model and analyze its assumptions. For the complex systems we usually deal with in science, just thinking is not sufficient to rigorously deduce the consequences of our simplifying assumptions: we have to let the computer show us what happens. We thus have to iterate through the modeling cycle.

1.5 Summary and Conclusions

Agent-based modeling is no longer a completely new approach, but it still offers many exciting new ways to look at old problems and lets us study many new problems. In fact, the use of ABMs is even more exciting now that the approach has matured: the worst mistakes have been made and corrected, agent-based approaches are no longer considered radical and suspicious, and we have convenient tools for building models. People like you are positioned to take advantage of what the pioneers have learned and the tools they built, and to get directly to work on interesting problems.

In this first chapter our goal is to provide some extremely fundamental and important ideas about modeling and agent-based modeling. Whenever you find yourself frustrated with either your own model or someone else's, in "big-picture" ways (What exactly does this model do? Is it a good model or not? Should I add this or that process to my model? Is my model "done"?), it could be useful to review these fundamental ideas. They are, in summary:

- A model is a purposeful *simplification* of a system for solving a particular *problem* (or category of problems).
- We use ABMs when we think it is important for a model to include the system's individuals and what they do.
- Modeling is a cycle of: formulating a precise question; assembling hypotheses for key processes and structures; formulating the model by choosing appropriate scales, entities, state variables, processes, and parameters; implementing the model in a computer program; and analyzing, testing, and revising.

Understanding this modeling cycle is so important that a recent review of modeling practice (Schmolke et al. 2010) concluded that explicitly thinking about and documenting each step in the cycle is the primary way we can improve how models are developed and used. Schmolke et al. then proposed a very useful format ("TRACE") for documenting the entire cycle of developing, implementing, and analyzing a model.

It is very important that you have a very basic understanding of these ideas from the start, but for the rest of part I we will focus on obtaining a basic understanding of how to implement models on the computer. In the rest of this course, however, we will come back to modeling ideas. As soon as you have some ability to program and analyze your own models *and* some understanding of how to use these modeling concepts, you will rapidly become a real modeler.

1.6 Exercises

1. One famous example of how different models must be used to solve different problems in the same system is grocery store checkout queues. If you are a customer deciding which queue to enter, how would you model the problem? What exact question would your model address? What entities and processes would be in the model? Now, if instead you are a store manager deciding how to operate the queues for the next hour or so, what questions would your model address and what would it look like? Finally, if you are a store designer and the question is how to design the checkout area so that 100 customers can check out per hour with the fewest employees, what things would you model? (Hint: think about queues in places other than stores.)

2. For the following questions, what should be in a model? What kinds of things should be represented, what variables should those things have to represent their essential characteristics, and what processes that change things should be in the model? Should the model be agent-based? If the question is not clear enough to decide, then reformulate the question to produce one that is sufficiently clear.

   a) How closely together should a farmer plant the trees in a fruit orchard?

   b) How much of her savings should an employee put in each of the five investment funds in her retirement program?

   c) Should a new road have one, two, or three lanes in each direction?

   d) Is it acceptable to allow a small legal harvest of whales?

   e) To complete a bachelor's degree in physics as soon as possible, what classes should a student register for this semester?

   f) How many trees per year should a timber company harvest?

   g) Banks make money by investing the money that their customers deposit, but they must also keep some money available as cash instead of invested. A bank can fail if its customers withdraw more cash than the bank has available, or if their investments do not make enough money to meet expenses. Government regulators require banks to keep a minimum percent of total deposits as cash that is not invested. To minimize bank failures, what should this minimum percent be?

h) To maximize profit, how many flights per day should Saxon Airlines schedule between Frankfurt (their international hub) and Leipzig?

i) To minimize system-wide delays and risk of accidents, how many flights per day should the European Aviation Administration allow between Frankfurt and Leipzig?

j) Two new movies will open in theaters next weekend. One is based on a comic book series and features a superhero, special effects, car chases, and violence. The other is a romantic comedy starring a beautiful actress and a goofy but lovable actor. Which movie will make the most money next weekend? Over the next four weeks? Over the next five years?

k) (Any other problems or questions from your studies, research, or experience in general, that might be the basis of a model or agent-based model.)

Getting Started with NetLogo

2.1 Introduction and Objectives

Now it is time to start playing with and learning about NetLogo, the software package we use in this course to implement ABMs. One of the great things about NetLogo is that you can install it and start exploring its built-in models in a few minutes, and then use its tutorials to learn the basics of programming models.

We introduce NetLogo first by explaining its most important elements and then via learning by doing: programming a very simple model that uses the most basic elements of NetLogo. We expect you to learn the basic elements of NetLogo mainly by using its excellent User Manual, which includes tutorials, detailed guides to the interface and programming, and the all-important NetLogo Dictionary. You should therefore plan, for this chapter, to spend as much time in the NetLogo materials as in this book.

At the end of this chapter we include a checklist of NetLogo elements that are very important to understand. Many of these elements are introduced in this chapter, and most of the others will be explained in the User Manual's tutorials. We recommend you come back to this checklist from time to time to check your knowledge and understanding of NetLogo. And please keep in mind that this book is about agent-based modeling, not just NetLogo. We are using NetLogo as a tool, so we have to focus on this tool at the start. But we return our focus to modeling as soon as possible.

The primary goal of this chapter is to familiarize you with the basic elements of NetLogo. Learning objectives for chapter 2 are to:

- Be familiar with the NetLogo User Manual and its parts, including the Interface Guide and Programming Guide and the NetLogo Dictionary.
- Complete the three tutorials in NetLogo's User Manual.
- Know what the Interface, Information, and Procedures tabs are. (The first release of NetLogo version 5.0 tentatively renamed these tabs to Run, Info, and Code; in this book we have chosen to use their original names. These tab names may therefore differ between our text and figures and your version of NetLogo.)
- Understand how to add buttons to the Interface and to use the World and change its settings.

- Know NetLogo's four built-in types of agent (the observer, patches, turtles, and links).
- Be familiar with NetLogo's basic code elements: procedures, primitives, commands, and reporters; and the basic organization of a NetLogo program, including the `setup` and `go` procedures. (By "code" we mean the programming statements that modelers write to implement their model.)
- Have programmed, by following our example, a first, very simple model.

2.2 A Quick Tour of NetLogo

To get started, let's install NetLogo and look around at the basics.

◼ Go to the NetLogo web site (http://ccl.northwestern.edu/netlogo/) and download and install NetLogo. This takes only a few minutes. You will see that versions are available for all the common computer operating systems.

This book is current with the beta release of NetLogo version 5.0, which was released just as we finalized the book in 2011. Any major new releases since then could make some of the code examples in this book obsolete. Please look at the web site for this book (described in the Preface) to check whether there are any major new releases since NetLogo 5.0 or revisions to our code examples.

◼ Start NetLogo, click on "Help" and then "NetLogo User Manual." This opens NetLogo's extensive documentation, which appears in your web browser. Parts of the User Manual that are especially important for you now are the "Introduction" section and the tutorials in the "Learning NetLogo" section. The "Interface Guide" and "Programming Guide" will be essential as soon as you start writing your own programs, so you should also become familiar with them.

◼ Before proceeding any further, work through "Tutorial #1: Models" in the NetLogo User Manual. Make sure you understand the `setup` and `go` buttons, sliders and switches, plots and monitors, what the View is and how to adjust its settings, how the View's coordinate system and `max-pxcor` and `max-pycor` work, and how to open and play with the Models Library.

◼ Look at and try some of the models in the Models Library. Be sure to look at the Information tab to see a written description of the model and what you can use it for and learn from it. We postpone discussing the Information tab until the next chapter, but it is very important!

The Models Library contains many intriguing models, grouped by discipline. However, many of these example models explain quite simple scientific problems, probably because they were designed to be easily understood and to illustrate NetLogo's excellent tools for modeling and displaying space and motion. Playing with the example models may therefore lead to the impression that NetLogo is good mainly just for producing animated output. You will soon learn, though, that this is not the case.

One part of the models library is especially important: the "Code examples." This section includes many well-documented examples of how to do specific things (e.g., control the order in which agents execute, read input files and write output files, import a picture, histogram your results, control colors) in NetLogo. Whenever you are not sure how to program something, you should look in the code examples for ideas, or even code you can simply copy.

By now you will have seen the basics of NetLogo terminology. There are four types of *agent* in NetLogo:

- *Mobile agents*, which in NetLogo are referred to as *turtles*. (Later, we will learn how to create our own *breeds*, or kinds, of turtles.)
- *Patches*, which are the square cells that represent space. The patches exist in the "World," the rectangular grid of patches displayed on the Interface tab.
- *Links*, which each connect two turtles and provide a way to represent relationships among turtles such as networks.
- The *observer*, which can be thought of as an overall controller of a model and its displays. The observer does things such as create the other agents and contain global variables.

(This use of the word "agent" is a little different from how it is typically used by agent-based modelers. When talking about ABMs instead of NetLogo, we use "agent" to refer to the individuals in a model that make up the population or system we are modeling, not to things like patches, links, and the observer.)

Each of these types of agent has certain variables and commands that it uses. The NetLogo Dictionary (in the "Reference" part of the User Manual) is where all these variables and commands are defined. There are several important built-in variables for each agent type, which you can find in the NetLogo Dictionary by clicking on the special category "Variables." These built-in variables represent things like location and color, which are used in almost all models. When you write your own program, you define the additional variables that your model needs for each agent type. Variables belonging to the observer are automatically "global" variables, which means that all agents can read and change their value. Global variables are often used to represent general environmental characteristics that affect all agents and to represent model parameters.

A *primitive* is one of NetLogo's built-in procedures or commands for telling agents what to do; these primitives are extremely important because they do much of the programming work for you. For example, the primitive `move-to` tells a turtle to move to the location of a patch or other turtle. The more powerful primitive `uphill` tells a turtle to move to the patch in its neighborhood that has the highest value of some variable.

These primitives are all defined in the NetLogo Dictionary; one of the main tricks to writing NetLogo programs is constantly looking through the dictionary to find primitives that already do what you want your program to do, to minimize how much code you have to write yourself. The sheer number of primitives makes this difficult at first, but it becomes easier after you learn the small subset of primitives that are used most often (see our NetLogo checklist at the end of this chapter).

NetLogo's primitives are in two categories: *commands* and *reporters*. Commands simply tell an agent to do something (e.g., `move-to`). Reporters calculate a value and report it back for use in the program (e.g., `mean` reports the mean of a list of numbers; `neighbors` reports a list of the surrounding patches).

When you look up some of these primitives in the NetLogo Dictionary, you will see an icon of a turtle, patch, link, or observer (the observer icon is an eye). These icons mean that the primitive can *only* be used by that kind of agent. The NetLogo term *context* is very important and a little confusing at first. Each piece of code in a program is "in the context" of one kind (or, occasionally, several kinds) of agent: turtles, patches, links, or the observer; and it can only be executed by that kind of agent. If you write a block of statements telling turtles to do something, you cannot include a primitive that can be used only by patches. This error is called "using a patch command in a turtle context."

How do you know what context you are in, when writing code? If you are writing a procedure that the observer executes (e.g., go and setup; any procedure executed by a button on the Interface), then you are in observer context. If you are writing a procedure executed by turtles (or patches), you are in turtle (or patch) context. But—the context can change within a procedure because primitives such as ask, create-turtles, and hatch begin new contexts within their brackets that contain code for other agents to execute. If the go procedure includes the statement ask turtles [move] then the procedure move is in turtle context. See the NetLogo Programming Guide on ask for more explanation.

Similarly, agents can only directly access the variables defined for their own type; patches cannot do calculations using a variable defined for turtles, the observer cannot use patch variables, and so on (but a patch can obtain the value of a turtle's variable using the primitive of). There are two very important exceptions to this restriction: observer variables are global and can be used by all agents, and turtles automatically can use the variables of the patch they are currently on.

▪ To understand agents, variables, and commands and reporters, read the "Agents,"
"Procedures," and "Variables" sections of the Programming Guide carefully.

At this point you probably feel confused and overwhelmed, but don't panic! All the material thrown at you here will be repeated, reinforced, and explained in much greater detail as we move through parts I and II of this course. Perhaps slowly at first, but then rapidly, NetLogo will become intuitive and easy. For now, just follow along and learn by doing, teaching yourself further by continually referring to the NetLogo documentation.

2.3 A Demonstration Program: Mushroom Hunt

Now we are ready to program our first model, the mushroom hunting model introduced in section 1.2. We will describe the model as we go.

▪ Create, via "File/New" in the NetLogo menu, a new NetLogo program. Save it, via "File/Save as", in a directory of your choice and with the file name "MushroomHunt.nlogo".

▪ Click the Settings button.

From the Interface tab, click the Settings button to open the Model Settings dialog, where you can check the World's geometry. For now we will use the default geometry with the origin (patch 0,0) at the center of the World, and both the maximum x-coordinate (max-pxcor) and maximum y-coordinate (max-pycor) set to 16. This makes a square of 33 × 33 patches. Even though we use the default geometry now, it is an important habit to always check these settings. Click OK to close the Model Settings dialog.

For Mushroom Hunt, we want to create a World of black patches, with a few clusters of red patches (representing mushrooms) sprinkled in it; create two turtles—our hunters; and then let the hunters search for the red patches. We therefore need to program how to initialize (create) the World and the hunters, and then the actions the hunters perform when the model runs.

We always use the NetLogo convention of using the name setup for the procedure that initializes the World and turtles, and the name go for the procedure that contains the actions to be performed repeatedly as the model runs. By now you know from reading the Programming

Guide that a procedure starts and ends with the keywords to and end, and contains code for one discrete part of a model.

To program the setup and go procedures, we first create buttons on the Interface tab that will be used to start these procedures.

▨ On the Interface tab, there is a drop-down menu, usually labeled "Button," that lets you select one of the Interface elements provided by NetLogo (Button, Slider, Switch, etc.). Click on this selector, which opens a list of these elements, then on "Button." Place the mouse cursor in the blank white part of the Interface, left of the black World window, and click. This puts a new button on the Interface and opens a window that lets you put in the characteristics of this new button (figure 2.1).

Figure 2.1
Adding a button to the Interface.

▨ In the "Commands" field, type "setup" and click OK.

Now we have created a button that executes a procedure called setup. The button's label is in red, which indicates that there is a problem: no corresponding command or procedure named setup exists yet on the Procedures tab. We will fix that in a minute.

▨ Create a second button, assign it the command go, click the "forever" checkbox in the Button window, and click OK.

This second button, go, now has a pair of circular arrows that indicate it is a "forever" button. "Forever" means that when the button is clicked, it will execute its procedure over and over until the button is clicked again.

Now, let us write the setup procedure.

■ Click on the Procedures tab, where you will find a blank white space to enter your code. To start (figure 2.2), type:

```
to setup
end
```

Figure 2.2
Starting to write code in the Procedures tab.

■ Click the Check button.

This button runs NetLogo's "syntax checker," which looks for mistakes in the format of your code: are there missing brackets, do primitives have the right number of inputs, did you call a procedure that does not exist, etc.? Using the syntax checker very often—after every statement you write—is essential for writing code quickly: you find mistakes immediately and know exactly where they are. However, this checker does not find all mistakes; we will discuss it and other ways to find mistakes in chapter 6.

Now, there should be no error message. If you switch back to the Interface tab, the label of the setup button is now black, indicating that it now finds a setup procedure to execute—even though this procedure is still empty and does nothing.

■ Go back to the Procedures tab and delete the word end. Click the Check button again. You will obtain an error message, saying that a procedure does not end with end.

■ Undo your deletion of end by typing CTRL-Z several times.

▧ Just as we created the `setup` procedure (with two lines saying `to go` and `end`), write the "skeleton" of the `go` procedure.

Now, we will tell the `setup` procedure to change the color of the patches.

▧ Change the `setup` procedure to say

```
to setup
  ask patches
  [
      set pcolor red
  ]
end
```

The `ask` primitive is the most important and powerful primitive of NetLogo. It makes the selected agents (here, all patches) perform all the actions specified within the brackets that follow it. These square brackets always delineate a block of actions that are executed together; you should read [as "begin block" and] as "end block." (Notice that NetLogo automatically indents your code so it is easy to see which code is within a set of brackets or the beginning and end of a procedure.)

   The statements we just wrote "ask" all patches to use a new value of their built-in color variable `pcolor`. Patches by default have `pcolor` set to black (which is why the World looks like a big black square), but in our new `setup` procedure we use the primitive `set` to change it to red. The primitive `set` is called an "assignment" operator: it assigns a value (red) to a variable (`pcolor`). (In many programming languages, this statement would be written `pcolor = red`, which absolutely does not work in NetLogo!)

▧ Test the new `setup` procedure by going to the Interface tab and clicking the `setup` button. All the patches should turn red.

However, we only want only a few clusters of patches to turn red. Let's have NetLogo select four random patches, then ask those four patches to turn 20 nearby patches red. Modify the `setup` procedure to be like this:

```
ask n-of 4 patches
[
  ask n-of 20 patches in-radius 5
```

```
      [
        set pcolor red
      ]
    ]
```

The primitives n-of and in-radius are new to us.

▪ Look up n-of and in-radius in the NetLogo Dictionary. You can find them by going to the dictionary and clicking on the "Agentset" category (a good place to look for primitives that seem to deal with groups of agents). But—an important trick—you can go straight to the definition of any primitive from your code by clicking within the word and pressing F1.

The dictionary explains that n-of reports a random subset of a group of agents, and in-radius reports all the agents (here, patches) within a specified radius (here, five patch-widths). Thus, our new setup code identifies four random patches, and asks each of them to randomly identify 20 other patches within a radius of 5, which are then turned red.

▪ Go to the Interface tab and press the setup button.

If all your patches were already red, nothing happens! Is there is mistake? Not really, because you only told the program to turn some patches red—but all patches were already red, so you see no change. To prevent this kind of problem, let's make sure that the model world is always reset to a blank and default state before we start initializing things. This is done by the primitive clear-all (or, abbreviated, ca), which you should add as the second line of the setup procedure:

```
to setup
  ca
  ask n-of 4 patches
  [
    ask n-of 20 patches in-radius 5
      [
        set pcolor red
      ]
  ]
end
```

Programming note: Clear-all

Forgetting to put ca (clear-all) at the beginning of the setup procedure is a common error for beginners (and, still, one author of this book). Almost always, we need to begin setup's initialization process by erasing everything left over from the last time we ran the model. As you build and use a NetLogo program, you hit the setup and go buttons over and over again. Clear-all removes all turtles and links, sets patch variables to their default values, and clears any graphs on your Interface. If you forget ca in setup, there will be junk (agents, colors, graph data) left from your previous run mixed in with the new stuff that setup creates. Learn to make ca automatically next after to setup.

If you click the `setup` button now, you will see that a World consisting of black and red patches appears. If you click the `setup` button several times, you see that four random clusters of red patches are indeed created.

Sometimes it looks like fewer, larger clusters are created, because clusters overlap. Other times, it looks like part of a cluster is created at the edge of the World; to understand why, go to the NetLogo Interface Guide section on the Views and read about "world wrapping," and see the Programming Guide's section called "Topology." Understanding NetLogo's world wrapping is extremely important!

■ From the main NetLogo menu, hit "File/Save." Save your work often! NetLogo will not recover your unsaved work if something bad happens. (And, from here on, your code should match the full program we provide at the end of this section; use that full program we provide if you get confused.)

Now let's make the number of clusters of red patches a *model parameter*. In our models and programs, we usually try not to write numbers that control important things directly into the deep code—which is called "hardwiring" them. Instead, we make them parameters that are easy to find and change. To do this, we need to create a global variable named `num-clusters` and give it a value of 4, then use `num-clusters` in the `ask` statement that creates the clusters. You can give a variable a value by using the primitive `set`, but first you need to define that variable (as you learned from the Variables section of the Programming Guide).

■ In the Procedure tab, go the top of all your code and insert

```
globals
[
  num-clusters
]
```

■ In the `setup` procedure, add this near the beginning, just after `clear-all`:

```
set num-clusters 4
```

and replace 4 in the `ask n-of 4 patches` statement with `num-clusters`.

Now, to change the number of clusters, we know exactly where to find and change the parameter's value, without digging through the program. (Later, we will use *sliders* to change parameters from the Interface.)

Now that we have set up the world of our Mushroom Hunt program, we need to create the agents, in our case two hunters. This is done by the primitive `create-turtles`, or `crt` for short. The statement `crt 2 []` creates two turtles and then makes them execute the code within the brackets.

■ Add the following code at the end of the `setup` procedure. Click Check to look for mistakes, then try it using the `setup` button on Interface:

```
crt 2
[
  set size 2
```

```
    set color yellow
  ]
```

Size and color are built-in turtle variables, so we can set them without defining them. If you click the setup button several times, you see that the hunters are placed, by default, in the center of the World, but their heading varies randomly.

Figure 2.3 shows what the Interface should look like.

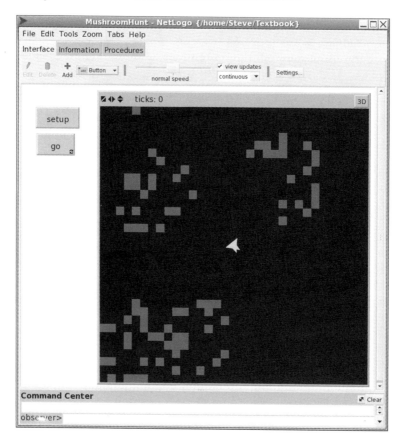

Figure 2.3
Mushroom Hunt interface with the model initialized.

Now we are going to program the go procedure. The go procedure defines a model's schedule: the processes (procedures) that will be performed over and over again, and their order. To keep the go procedure simple and easy to understand—which is very important because it is the heart of the program—we almost never program actual processes or actions within it. Instead, go simply calls other procedures where each action is programmed.

In Mushroom Hunt, we have only one action, the hunter's search. Thus, the go procedure needs to include only one statement, which tells all the turtles (our two hunters) to execute a turtle procedure where searching is programmed:

```
  ask turtles [search]
```

For this to work, of course, we need to write a procedure called search for the hunters to execute.

■ Add the above `ask turtles` statement to the `go` procedure. Then create an empty procedure so you can check the code so far:

```
to search
end
```

Now let's program how the hunters search. Each time `search` executes, we want the hunters to turn, either a little bit if they have found nothing recently (e.g., in the past 20 patches searched), or sharply to keep searching the same area if they did find a mushroom recently. Then they should step ahead one patch.

■ Add this code to the `search` procedure, looking up any parts you do not understand in the NetLogo Dictionary:

```
ifelse time-since-last-found <= 20
  [right (random 181) - 90]
  [right (random 21) - 10]

forward 1
```

Can you figure out from the NetLogo Dictionary's entry for `random` how the statement `right (random 181) - 90` causes a turtle to turn a random angle between -90 and +90 degrees?

Note how we use the `ifelse` primitive (or, sometimes, the related `if` and `ifelse-value`) to model decisions. If the *boolean condition* following `ifelse` is true, then the code in the first set of brackets is executed; if false, the code in the second set of brackets is executed. A boolean condition is a statement that is either true or false. Here, the boolean condition `time-since-last-found <= 20` consists of a comparison: it is true if the value of `time-since-last-found` is less than or equal to 20.

But when you now hit the Check button, there is a problem: what is `time-since-last-found`? We want it to be a variable that records how long it has been since each hunter found a mushroom. That means three things. First, each hunter must have its own unique value of this variable, so it must be a turtle variable. We need to define `time-since-last-found` as a turtle variable.

■ Just after the `globals` statement at the start of your program, add

```
turtles-own
[
  time-since-last-found
]
```

Second, we need to set the initial value of this variable when we create the hunters. We will set it assuming the hunters have not found a mushroom yet, by using a value larger than 20.

■ In the `setup` procedure, change the `create-turtles` statement to this:

```
crt 2
[
  set size 2
```

```
        set color yellow
        set time-since-last-found 999
    ]
```

Finally, the hunters must update `time-since-last-found` each step. If they find a mush-room, they need to reset `time-since-last-found` to zero (and pick the mushroom!); otherwise, they need to add one to `time-since-last-found`. We define finding a mush-room as landing on a red patch.

■ Add these statements to the end of the `search` procedure:

```
ifelse pcolor = red
  [
    set time-since-last-found 0
    set pcolor yellow
  ]
  [
    set time-since-last-found time-since-last-found + 1
  ]
```

Note that these statements only work because NetLogo lets turtles read and change variables of the patch they are on; here, the turtles use `pcolor`, their patch's color. Also note that to add one to the value of `time-since-last-found`, we had to use `set` to assign its new value to its old value plus one. (Be aware that in NetLogo you must use spaces around arithmetic opera-tors like "+"; otherwise NetLogo thinks they are part of the variable name.)

Make sure you understand that patch coordinates are discrete variables: they only have in-teger values, in units of patch-widths. In contrast, turtle coordinates are continuous variables: they can take any value within a patch, for example 13.11 units. It is therefore important to distinguish between primitives that refer to turtle coordinates and those that refer to patch coordinates. For example, `move-to` moves a turtle to the center of a patch, but `forward` can place a turtle anywhere.

If you now test the program by clicking the `go` button, you may not be able to see very much because the hunters move too fast. If so, adjust the execution speed controller on the Interface. (Click `go` a second time to stop execution.)

Voila! Here we have our first complete NetLogo program! (Make sure you save it!) Despite its simplicity, the program contains the most important elements of any NetLogo program. But before we stop, let us look at three more important NetLogo tools: Agent Monitors, the com-mand center, and `ticks`.

Agent Monitors are tools to see, and change, the variables of an agent (turtle, patch, or link).

■ Move the cursor over one of the hunters and right-click. A panel appears that, at the end, lists "turtle 0"; move the cursor to this entry and select "inspect turtle 0."

An Agent Monitor opens that includes two panels: a zoomed view on the inspected turtle and its environment, and a list of all variables of the turtle: its position, color, heading, etc. The turtle variable we created, `time-since-last-found`, is also listed. (Both panels can be closed and reopened by clicking the small black triangle at their upper left corner.) You can restart the model by clicking `go` and observe how the variables change.

Not only can you see an agent's variables with an Agent Monitor; you can also change their value whenever you want, just by entering a new value in the monitor.

▪ Stop the program and change the turtle's size to 5, and its `time-since-last-found` to -99. (Your changes become active when you hit the Enter key or move the cursor out of the dialog box where you enter them.)

There are more things you can do with Agent Monitors, including giving commands directly to the selected agent by typing in the same kinds of NetLogo statements you would use in your code. Read the "Agent Monitors" section of the User Manual's Interface Guide.

The Command Center appears at the bottom of the Interface tab. Here, you can send commands to the observer or to all patches, turtles, or links, and get the results of output primitives such as `show`. Read the "Command Center" section of the User Manual and try some commands such as telling the patches to `set pcolor blue`.

The ability to type in commands in Agent Monitors and the Command Center may not seem useful at first, but as you become a better NetLogo programmer you will find these tools extremely helpful. As a fun example, use the Command Center to do the following.

▪ After setting up the Mushroom Hunt program and letting it run for a second or two, stop it. Click on the text "observer>" in the Command Center. From the pop-up list, select "turtles>" so you can now issue a command to all turtles. In the adjacent window, enter `hatch 1 [right 160]`, which tells each hunter to create a second hunter, which then turns 160 degrees (figure 2.4).

Figure 2.4
Giving a command to all turtles via the Command Center.

▪ Run the program and see how the now four hunters search for red patches. You might repeat the `hatch` command to create even more hunters: just place the cursor in the command line and hit the Up arrow to retrieve the previous command, then Enter.

▪ After hatching some more turtles, have the Command Center tell you how many there are. Select the observer to send commands to, and enter `show count turtles`.

So far, our program does not deal with time explicitly: we do not keep track of how many times the `go` procedure has executed, so we cannot determine how much time has been simulated if we assume, for example, that each time the mushroom hunters move represents one minute. To track time, we call the primitive `tick` every time the `go` procedure is called. Read the Programming Guide section "Tick Counter"; then:

▪ At the end of the `setup` procedure, insert a line with only the statement `reset-ticks`. At the beginning of the `go` procedure, insert a line with only the statement `tick`.

If you now click `setup` and then `go`, you can follow the tick counter at the top of the World display; it shows how many times the `go` procedure has executed. (In previous versions of NetLogo, the `clear-all` statement reset the tick counter to zero, but now we need to do it by putting the `reset-ticks` statement at the end of `setup`.)

Finally, would it not be helpful to see the path of the hunter, as in figure 1.2? We can do this very easily by changing the turtles' built-in variable pen-mode from its default of "up" to "down."

▪ In the setup procedure, add the primitive pen-down to the block of statements following the crt statement.

Now your whole program should look like this:

```
globals
[
  num-clusters
]

turtles-own
[
  time-since-last-found
]

to setup

  ca
  set num-clusters 4

  ask n-of num-clusters patches
  [
    ask n-of 20 patches in-radius 5
    [
      set pcolor red
    ]
  ]

  crt 2
  [
    set size 2
    set color yellow
    set time-since-last-found 999
    pen-down
  ]

  reset-ticks

end


to go

  tick
```

```
  ask turtles [search]

end


to search

  ifelse time-since-last-found <= 20
    [right (random 181) - 90]
    [right (random 21) - 10]

  fd 1

  ifelse pcolor = red
  [
    set time-since-last-found 0
    set pcolor yellow
  ]
  [
    set time-since-last-found time-since-last-found + 1
  ]

end
```

2.4 Summary and Conclusions

2.4.1 Summary

Our goal in this chapter was to get you over the biggest hump in becoming a NetLogo user: learning the very basics such as what patches, turtles, links, and the observer are; how the graphical displays on the Interface tab work; what procedures, primitives, commands, and reporters are; and how programs are organized, with the `setup` procedure and button to initialize the model and the `go` procedure and button to make it run. And we hope you have started learning to program fearlessly by adding a statement or two, using the Check button, seeing what happens on the Interface, then adding some more.

To work successfully with NetLogo, you must use the User Manual all the time. All of us, no matter how experienced, use the Programming Guide and NetLogo Dictionary to remind ourselves how to do things and to learn how to use features for the first time. There is no need to remember how to use each primitive or (especially) to remember all the primitives and what they do; just look in the dictionary. If you prefer the manual as a paper book, a link at the bottom of the navigation panel lets you download the manual as a printable PDF file.

2.4.2 A Word to New Programmers

Learning a programming language is much more fun and efficient if you learn together with other people, and if you can consult experienced users when you are stuck. Often we figure out how to fix a problem when trying to explain it to others, and often we need just a tiny bit of simple help to start making rapid progress. So if you are learning NetLogo in a class, help each other out. If not, try to find NetLogo users in your community; there are many in many places. Another source of information and help is the NetLogo Users Forum (see exercise 6, below).

2.4.3 A Word to Experienced Programmers and Scientists

People who are experienced users of other programming languages are often very frustrated by NetLogo at first, and experienced scientists may find NetLogo's "personality" a little less serious than that of other tools. You may have a different hump to get over than complete beginners: getting used to NetLogo's style.

As an example, experienced programmers often quickly decide that they cannot program their models in NetLogo because they do not immediately see how to translate the logic of conventional languages ("do" and "while" loops, arrays, messaging objects, etc.) into NetLogo. But please trust us and persist: with just a little experience and understanding, you will find that you can do whatever you need to—just in a different way that often turns out to be easier. The main advantage of NetLogo over other platforms is that you very quickly can concentrate on doing science: after an hour or two of getting used to its style and programming basics, you can be programming, analyzing, and revising models extremely quickly instead of spending weeks or months just on programming.

NetLogo does tend to impose a certain way of thinking about problems: you try to formulate your problem so that the built-in features of NetLogo can be used. But this is true for any scientific tool, in particular analytical models that impose calculus or other mathematical frameworks as a way of thinking. The important thing is, for now, just to be aware of this interaction between tool and model structure!

2.4.4 NetLogo Checklist and Mini-reference

Here we provide a checklist of the elements of NetLogo that we think are essential for you to eventually understand. The elements that you have already learned about are underlined. You don't need to remember all these characteristics of NetLogo in detail, but be aware of them and be able to look them up as you need.

Main menu bar

- File/Save—Remember to save your changes frequently as you build a model!
- File/Models Library
    - Code Examples: Look here for examples of how to do things.

Interface tab

- World—use its settings to change or view
    - Location of origin: The location 0,0 must be inside the World!
    - World size
    - Wrapping
    - Tick counter
- Agent Monitors—In the World display, right-click on patches, turtles, or links to
        - See variables
        - Change variables

- Execute procedures
  - "Follow" and "watch" a turtle
- The Command Center
- Buttons—Execute procedures from the Interface:
  - `Setup`
  - `Go`: Understand the difference between "once-only" and "forever"
  - `Step`: A button that executes the `go` procedure just once
- Sliders—Define and initialize a global variable
- Switches—Define and initialize a boolean variable
- Other interface objects—choosers, inputs, monitors, outputs
- Plots

Procedures tab

- Contexts—Which agents can execute each piece of code:
  - Observer, patch, turtle (breed), and link contexts are possible
  - Each procedure has a context
  - Each `ask` statement starts a new context
  - Contexts are determined by:
    - `ask` statement: who is asked?
    - using primitives or variables specific to one context
    - global if there is no `ask`: setup, go
- Variable types—
  - number
  - boolean (true/false)
  - string (alphanumeric characters)
  - color
  - agent
  - agentset
  - list
- Built-in variables and agentsets (see "Variables" in the NetLogo Dictionary)—
  - Global agentsets and variables:
    - `patches` (agentset)
    - `turtles`, breeds (agentset)
    - `links` (agentset)
    - `ticks` (number of times the command `tick` has executed)
  - Patch variables:
    - `pcolor`
    - `pxcor`, `pycor` (integers)
    - `plabel`
  - Turtle variables:
    - `color`
    - `xcor`, `ycor` (real numbers, so location is continuous)
    - `label`
- Defining variables—
  - globals
  - patches-own
  - turtles-own
  - links-own
- The Check button (syntax checker)

Information tab

- ODD format

Help

- Programming Guide
- NetLogo Dictionary
- Interface Guide
- NetLogo forum email list

Creating output

- Summary statistics
- Plots
- Output files

BehaviorSpace

2.5 Exercises

1. From the NetLogo User Manual, read the three tutorials carefully. You do not need to actually build the model used in Tutorial 3 because we will do something similar in chapter 4. But look for the following concepts and make sure you understand them:

   - Finding and running models from the Models Library
   - `Setup` and `go` buttons, "once" and "forever" buttons
   - How the speed controller works—how can it make your computer appear to run faster?
   - Sliders and switches to control model parameters
   - Finding information about a model on its Information tab
   - Changing the World size—the number of patches
   - Patch coordinates
   - Using the Command Center
   - Colors; setting patch and turtle colors
   - Agent Monitors and commanders
   - Procedures: a section of code that programs one thing
   - `Setup` procedures to initialize a model
   - `Go` procedures to run a model
   - The commands `clear-all`, `ask`, `set`, `setxy`
   - "Calling" a procedure from another procedure; the `setup` and `go` procedures call several other procedures that each contain one behavior
   - Defining variables for turtles and patches
   - Agentsets
   - The tick counter and its primitives `reset-ticks` and `tick`

2. Become familiar with the Interface and Programming Guides and NetLogo Dictionary, so you can use them efficiently as references. (You can skip the Interface parts about the

3D View, which we do not use.) From now on, we will assume that you have completed the tutorials and learned how to use the User Manual and the NetLogo Dictionary, and how to look up examples in the Models Library and its Code Examples section.

3. In the Mushroom Hunt model, we create four clusters of red patches. For each of these clusters, we ask 20 patches to turn red. Are there always 80 red patches? You can check this very quickly using the Command Center, entering the statement `show count patches with [pcolor = red]`. Do this several times after hitting `setup` each time. Why do you get the answers you do?

4. Make the following small changes in the Mushroom Hunt model. Make them one at a time, check each, and save the model with a new name after each change.

   - There is only one hunter instead of two.
   - The hunter's initial heading is always 45 degrees.
   - Instead of four clusters that each have 20 mushrooms, there are eight clusters of 10 mushrooms. Each cluster has a radius of only three patches.
   - Whenever the hunter finds a mushroom, it writes "I found one!" to the Command Center (see the primitives `print`, `write`, `type`, and `show`).
   - When the hunter has not recently found a mushroom, it turns by a random angle between -45 and +45 degrees instead of between -10 and +10.
   - The hunter starts in the lower left corner instead of in the middle (see the primitives `min-pxcor`, `min-pycor`, and `setxy`).
   - The hunter counts how many mushrooms it catches. This requires a new turtle variable that must be incremented each time a mushroom is found. Instead of writing "I found one!" to the command center, the hunter now writes out how many items it has found so far.

5. Open and explore at least two other models in the NetLogo library. Try to understand the model's purpose from the Information tab, and its code.

6. Join the on-line NetLogo Users Group by following the links on the NetLogo web site. The users group is an email forum where users ask and answer questions about NetLogo programming and modeling. Belonging to the users group is valuable because it gives you an idea of the kinds of projects people use NetLogo for, the problems they encounter, and the solutions. (Note the option to receive only one summary email per day.) As you learn NetLogo you should participate in this group by asking and answering questions. However, it is very important to follow some user group etiquette so you do not abuse the extremely gracious people who answer many email questions! Especially when you are new to NetLogo, emailing the user group should be your last, not first, resort when you have a problem or question. Try everything else you possibly can first: scour the documentation and ask anyone you can for help. Especially: thoroughly search the archives of the users group (all their old email messages; you can find out how to search them when you join the group) to see if your question has already been answered.

Describing and Formulating ABMs: The ODD Protocol

3.1 Introduction and Objectives

Formulating an ABM means progressing from the heuristic part of modeling, in which we first think about the problem, data, ideas, and hypotheses, to the first formal and rigorous representation of the model. To formulate a model, we try to write it down in words, diagrams, equations, etc., which requires us to make a series of decisions about the model's structure. Beginners often hesitate to write down their first model version, but it is important to realize that a model simply does not exist before it has been formulated explicitly so people can understand it. Why?

The first purpose of the full model formulation is to make ourselves, the model's authors, think explicitly about all parts of the model. Until we try to write the model down, it is unlikely that we've identified all the decisions we must make in designing it. The second purpose is to communicate the model to our colleagues or supervisors, which usually leads to further discussions and modifications of the formulation. Then, third, we use the formulation as the basis for the model's implementation—the computer program that executes it. To be translated unambiguously into computer code, the formulation has to be complete and explicit regarding each and every element and aspect of the model. And of course when we are finally ready to publish results of our work, we need to include a complete and accurate description of the model.

In this chapter we step back from learning NetLogo and work on the problem of how to describe, and even how to think about, the design of an ABM. What characteristics of an ABM do we need to describe in its formulation, and how do we describe them concisely yet clearly? When we use traditional equation-based models, we know exactly how to answer these questions: we write down the equations and their parameter values. But when we use ABMs, these questions are difficult because the models can be more complex and because we do not have a traditional notation such as differential equations. However, there is a standard protocol for describing ABMs that is now widely used by agent-based modelers, which we introduce in this chapter. This protocol has proven useful not just for describing ABMs but also as a framework for thinking about the models as we formulate them.

Why do we introduce this model description and formulation protocol so soon, before we've even learned to program models, much less design them or publish results? One reason is simply that we use the protocol throughout the rest of this book. In fact, the example model description in this chapter is the basis for your next lessons in NetLogo, which come in the following two chapters. Another important reason, though, is so you learn from the very start to think about ABMs in a systematic and organized way. Without such a systematic approach to formulating and describing models, it is much harder to produce science using them!

Throughout the rest of this course you will gain a great deal of experience using ODD and learn more about each of its parts. Learning objectives for this chapter are to:

- Develop a firm understanding of the "Overview" and "Details" elements of ODD (we use these elements throughout the course but explain them only here).
- Develop an introductory understanding of the "Design concepts" element (we explore the design concepts much more in part II).
- Understand, from its ODD description, the model we will program and use in chapters 4 and 5.

3.2 What Is ODD and Why Use It?

With ABMs, it can be difficult to keep all of the model's characteristics in mind. In fact, many descriptions of ABMs in the literature are incomplete, which makes it impossible to reimplement the models and replicate their results. Replication, however, is key to science: models that cannot be reproduced are unscientific. Moreover, ABM descriptions are often a wordy mixture of factual description and lengthy justifications, explanations, and discussions of all kinds. Often, to quickly assess what the model itself really is, we have to read pages and pages even when the model itself turns out to be quite simple. How can we describe ABMs in a way that is easy to understand yet complete? One well-known way to deal with such problems is standardization: it is much easier to understand written material if we know in advance what kinds of information will be presented and what order it will appear in. If we consistently use a protocol that is effective for ABMs, then it becomes much easier to both write and read model formulations.

To bring the benefits of standardization to ABMs, a large group of experienced modelers (Grimm et al. 2006; see also Grimm and Railsback 2005, chapter 10) developed the ODD protocol for describing ABMs. ODD is designed to create factual model descriptions that are complete, quick and easy to grasp, and organized to present information in a consistent order. ODD is now gaining widespread acceptance in the ecological and social science literature (Polhill et al. 2008), and there is a newly updated guide for using ODD (Grimm et al. 2010; a version is available through this book's web site).

One lesson from using ODD so far is that the protocol is very useful for formulating ABMs as well as for its original purpose of just describing them. Once modelers have learned ODD, they find themselves using it from the very start to help make all the basic formulation decisions such as what kinds of things should be in the ABM, what behaviors agents should have, and what outputs are needed. Just as differential equations provide a way to think about (as well as describe) mathematical modeling problems, and frequentist and Bayesian theory provide ways to think about statistical modeling problems, ODD provides a way to both think about and describe agent-based modeling problems.

"ODD" stands for "Overview, Design concepts, and Details": the protocol starts with three elements that provide an overview of what the model is about and how it is designed, followed by an element of design concepts that depict the ABM's essential characteristics, and it ends with three elements that provide the details necessary to make the description complete. We now describe the seven elements of the ODD protocol (figure 3.1), which are exactly the seven elements of an ABM that its developer must design.

Purpose

The first element is a clear and concise statement of the question or problem addressed by the model: what system we are modeling, and what we are trying to learn about it. Why does ODD start with stating the model's purpose when the purpose is not really part of the model? Because it is impossible to make any decisions about the model—what should and should not be in it, whether its structure makes sense—if we do not first know what the model is for. And knowing a model's purpose is like having a roadmap to the rest of the model description: when we know what problem the model is to solve, we have a much better expectation of what it should look like.

Entities, State Variables, and Scales

Next we provide an outline of the model: what are its *entities*—the kinds of things represented in the model—and what variables are used to characterize them? ABMs usually have

	Elements of the ODD protocol
Overview	1. Purpose
	2. Entities, state variables, and scales
	3. Process overview and scheduling
Design concepts	4. Design concepts • Basic principles • Emergence • Adaptation • Objectives • Learning • Prediction • Sensing • Interaction • Stochasticity • Collectives • Observation
Details	5. Initialization
	6. Input data
	7. Submodels

Figure 3.1
Overview of the ODD protocol for describing ABMs (as described by Grimm et al. 2010).

the following types of entities: one or more types of agents; the environment in which the agents live and interact, which is often broken into local units or patches; and the "global" environment that affects all agents. These model entities are characterized by their *state variables*, which is how the model specifies their state at any time. An agent's state is defined by its properties, or attributes (size, age, savings, opinion, memory), and often also by its behavioral strategy (searching behavior, bidding strategy, learning algorithm). Some state variables are static and do not change, for example the sex and species of an animal and the location of immobile agents such as plants and cities. Still, these variables are in fact "variable" because they are different among agents. If all agents are identical in some variable, there is no need for it: if only females are represented in an animal population model (common for species in which males contribute little to reproductive success) there is no need to include "sex" as a state variable.

State variables do not, however, include quantities that can be deduced or calculated from the states of the agent and its environment. Consider, for example, agents that represent landowners, which are not mobile. Then, the distance to a certain service center in the landscape can be calculated from the respective positions of the landowner and service center. This distance is thus not a state variable characterizing the landowner per se, but characterizes the landowner plus its environment. As a rule of thumb, state variables of an ABM's entities are variables that cannot be deduced immediately from other variables.

Many ABMs are spatially explicit: they represent where agents are in a space or environment, and the space is often heterogeneous. The environment can be represented continuously (each point has a different value of the environmental variables), but much more often space is represented "discretely" by cells or patches, usually on a square grid. The advantage of this discrete structure is that spatial effects within the patches are ignored, greatly reducing the amount of data and calculation required to simulate spatial processes. Patches may be characterized by one to many state variables, including the coordinates defining where they are in space.

The global environment, finally, refers to variables that vary in time, but not necessarily in space. Examples might be weather variables such as temperature, the frequency with which some kind of system-wide disturbance occurs, or tax rates. These environmental variables are typically external: provided by data or submodels that are not affected by the ABM's entities.

The remaining thing to be specified here are the model's *temporal* and *spatial scales*. Temporal scales refer to how time is represented: how long a time is simulated (the "temporal extent"), and how the passage of time is simulated (these topics are covered more thoroughly in chapter 14). Most ABMs represent time using discrete time steps of, for example, a day, week, or year (the "temporal resolution" or "time step size"). The use of time steps means that all the processes and changes happening at times shorter than a time step are only summarized and represented by how they make state variables jump from one time step to the next. An ecological model might, for example, consider mortality of a certain species in yearly time steps, so it does not distinguish whether individuals died early or late in the year. A stock market model that uses a daily time step would represent all the trades during each day only as the daily mean asking and selling prices and number of trades, whereas a model of the same market with a shorter time step could keep track of individual trades.

The temporal extent of a model refers to the typical length of a simulation: how much time (number of time steps, years, minutes . . .) is simulated? This question also depends on what observations or model outputs we are interested in and the time scales at which they occur. The temporal extent is usually determined by system-level phenomena produced by the model, whereas temporal resolution is usually determined by the agent-level phenomena driving the model internally.

If the model is spatially explicit, then the total size, or extent, of the space must be described. If grid cells or patches are used to represent variation over space, then their size—the spatial resolution—must also be specified. As with the temporal resolution, the right spatial resolution of an ABM depends on key behaviors, interactions, and phenomena. Spatial relationships and effects within a grid cell are ignored; only spatial effects among cells are represented. For example, in models describing urban dynamics, a single household often is represented by a grid cell or patch. This means that the spatial relations *within* a household (e.g., who sleeps in which bedroom) are ignored because they are considered unessential for explaining urban patterns.

Process Overview and Scheduling

Whereas the previous element of ODD is about the structure of a model, here we deal with dynamics: the processes that change the state variables of model entities. Now, it turns out to be very useful that we first specified the model's entities, because every process—with one very important exception—describes the behavior or dynamics of the model's entities. Thus, to provide an overview of a model's processes we simply have to ask ourselves: what are the model entities doing? What behaviors do the agents execute as simulated time proceeds? What updates and changes happen in their environment? At this point in the protocol we write down a succinct description of each of the processes executed by the model's entities. If the process is extremely simple, it can be described completely here. But often we treat processes as "submodels," representing them here only by a self-explanatory name—mating, selling, buying, etc.—and provide the details in the final part of the ODD description.

The only processes that are not linked to one of the model's entities are *observer* processes: we, the creators of the model, want to observe and record what the model entities do, and why and when they do it. Therefore, we need to specify observation processes that do things like display the model's status on graphical displays and plots and write statistical summaries to output files. The ODD protocol includes description of observer processes because the way we observe a model—the kind of "data" we collect from it and how we look at those data—can strongly affect how we interpret the model and what we learn from it.

At the same time we specify the processes that are explicitly represented in the model, we also specify the model's schedule: the order in which the processes are executed by the computer. An ABM's schedule, when described well, provides a concise yet complete outline of the whole model. When you want to see what some model implemented in NetLogo does, you will soon learn to to find and read its schedule and understand very quickly what entities are in the model, what processes or behaviors they execute, in what order.

The concept of an *action* is essential for understanding and describing an ABM's schedule. A model's schedule can be thought of as a sequence of actions; an action specifies (a) which model entities execute (b) which processes, in (c) what order. In NetLogo, for example, an action is typically coded with a statement such as

```
ask turtles [move]
```

This statement specifies which entities—all turtles—execute which process—their move procedure, which presumably represents how they move in their environment. And, if you read the NetLogo documentation carefully, you will see that the ask statement causes the turtles to execute move one at a time, and in randomized order: each time the ask statement is executed, the order in which turtles do move is randomly shuffled.

Some ABMs have schedules simple enough that they can be specified completely by simply listing the model processes in this part of ODD in the order they are executed, adding any needed detail on the order in which individual agents execute. In many ABMs, though, scheduling is rather complicated and hard to describe accurately with words. For such models, we recommend use of *pseudo-code* that describes the schedule (pseudo-code mimics the logical structure of computer programs, but using words for people, not computers, to read). You will see many schedules described in this book, from which you can learn how to describe them well.

Design Concepts

This section of the ODD protocol describes how the model implements a set of basic concepts that are important for designing ABMs (table 3.1; see also chapter 7). These design concepts provide a standardized way to think about very important characteristics of ABMs that cannot be described well using other conceptual frameworks such as differential equations. For example, what model outcomes emerge from what characteristics of the agents and their environment? (Conventional models typically produce only one outcome, which is calculated directly from equations, but ABMs can produce many kinds of results that arise in complex, unpredictable ways.) What adaptive decisions do the agents make? If they are assumed to make these decisions to increase some measure of their objectives (e.g., "fitness" for organisms, "utility" for economic agents), what is that measure? Does the model include *collectives*, aggregations of agents that have their own behaviors? Describing how these design concepts are used in a model, by answering the questions in table 3.1, is like using a checklist to ensure that important design decisions are made consciously.

In the ODD protocol there are eleven different design concepts (table 3.1). Do not worry about understanding them thoroughly now, as we will look at each concept more closely in part II of this course. Not all of the concepts are important for all ABMs; some ABMs are simple enough that we need to address only a few design concepts in detail. But even in such cases it is important to consider the full list of design concepts and show what common characteristics of ABMs we choose not to employ.

Initialization

We must describe how we set up the model World at the beginning of the simulation, because results of the model often depend on these *initial conditions*. Examples of initial conditions to describe are the number of agents created and the initial values given to their state variables (location, size, etc.); and how the initial values of environment variables are set. Sometimes we want a model's results to depend on its initial conditions, for example when we want to understand how the system responds to some new situation (for example, the introduction of a new idea or species). In other cases, we want to make model results independent of initial conditions, which can be achieved by running the model until any "memory" of the initial state is "forgotten" (we would get essentially the same results if we changed the initial conditions). In any case, to make a model and its results reproducible, we have to specify the initial state of all the state variables of all entities in the model.

Input Data

Models often include environmental variables like temperature or market price that change over time and are read into, instead of simulated within, the model. These inputs are usually read in from data files as the model executes. ("Input" here does *not* refer to parameter values or initialization data, which are also sometimes read in from files at the start of a simulation.)

Submodels

Up to this point, the ODD protocol has specified the "skeleton" of the model: its entities, state variables, and the names all of the processes and how they are scheduled. These are the basic ideas that give the model its shape. Here, we put flesh on the bones: all the major processes in the model are considered submodels. You can think of a submodel as a model of one process

Table 3.1 Design Concepts for ABMs, with Key Questions for an ODD Description to Answer

Concept	Key questions
Basic principles	1. What general concepts, theories, hypotheses, or modeling approaches underlie the model's design? How is the model related to previous thinking about the problem it addresses?
	2. How were these principles incorporated in the model's design? Does the model implement the principles in its design; or address them as a study topic, e.g., by evaluating and proposing alternatives to them?
Emergence	3. What are the model's important results and outputs? Which of them emerge from mechanistic representation of the adaptive behavior of individuals, and which are imposed by rules that force the model to produce certain results?
Adaptation	4. What adaptive behaviors do agents have? In what ways can they respond to changes in their environment and themselves? What decisions do they make?
	5. How are these behaviors modeled? Do *adaptive traits* (models of adaptive behavior) assume agents choose among alternatives by explicitly considering which is most likely to increase some specific objective (*direct objective-seeking*), or do traits simply force agents to reproduce behavior patterns observed in real systems (*indirect objective-seeking*)?
Objectives	6. For adaptive traits modeled as direct objective seeking, what measure of agent objectives (for example, "fitness" in ecology, "utility" in economics) is used to rate decision alternatives? This objective measure is the agent's internal model of how it would benefit from each choice it might make. What elements of future success are in the objective measure (e.g., survival to a future reproductive period; probability of staying in business for some period; profits at the next reporting period)? How does the objective measure represent processes that link adaptive behaviors to important variables of the agents and their environment?
	7. How were the variables and mechanisms in the objective measure (e.g., risks of mortality or going out of business, the conditions necessary for reproduction or profitability) chosen considering the model's purpose and the real system it represents? How is the agent's current internal state considered in modeling decisions? Does the objective measure change as the agent changes?
Learning	8. Do individuals change their adaptive traits over time as a consequence of their experience? If so, how?
Prediction	9. How do agents predict future conditions (environmental and internal) in their adaptive traits? What assumptions about, or mechanisms of, the real individuals being modeled were the basis for how prediction is modeled?
	10. How does simulated prediction make use of mechanisms such as memory, learning, or environmental cues? Or is prediction "tacit": only implied in simple adaptive traits?
Sensing	11. What variables of their environment and themselves are agents assumed to sense and therefore be able to consider in their behavior?
	12. What sensing mechanisms are modeled explicitly, and which sensed variables are agents instead assumed simply to "know"?
	13. With what accuracy or uncertainty are agents assumed to "know" or sense which variables? Over what distances (in geographic, network, or other space)?
Interaction	14. How do the model's agents interact? Do they interact directly with each other (e.g., does one agent directly change the state of others)? Or is interaction mediated, such as via competition for a resource?
	15. With which other agents does an agent interact?
	16. What real interaction mechanisms were the model's representation of interaction based on? At what spatial and temporal scales do they occur?
Stochasticity	17. How are stochastic processes (based on pseudorandom numbers) used in the model and why? Are stochastic processes used: to initialize the model? Because it is believed important for some processes to be variable but unimportant to represent the causes of variability? To reproduce observed behaviors using empirically determined probabilities?
Collectives	18. Are collectives—aggregations of agents that affect the state or behavior of member agents and are affected by their members—represented in the model?
	19. If so, how are collectives represented? Do they emerge from the traits of agents, or are agents given traits that impose the formation of collectives? Or are the collectives modeled as another type of agent with its own traits and state variables?
Observation	20. What outputs from the model are needed to observe its internal dynamics as well as its system-level behavior? What tools (graphics, file output, data on individuals, etc.) are needed to obtain these outputs?
	21. What outputs are needed to test the model and to solve the problem the model was designed for?

in the ABM; submodels are often almost completely independent of each other and can be designed and tested independently. The submodels are listed in the schedule, and now we must describe them in full detail. To make the ABM reproducible, we must describe all equations, logical rules, and algorithms that constitute the submodels. But no one will be impressed if all we do is write down these details; we also need to document *why* we formulated the submodels as we did. What literature did we use, what assumptions did we make, where did we get parameter values, and how did we test and calibrate the submodel? Under what conditions is the submodel more or less reliable? In publications, we might present only the description and leave the full justification of the model assumptions to an appendix or separate report.

Let us now see the ODD description of our first real ABM. The example we chose is extremely simple—but it was also the basis of three interesting journal articles. Therefore, we learn immediately that, while some ABMs take months or even years to become productive, quite often we can learn a lot, even about the real world, from very simple ABMs.

3.4 Our First Example: Virtual Corridors of Butterflies

Many animals *disperse*—leave their home location and move long distances for purposes such as mating—at some point in their life. Dispersing animals respond to the landscape, avoiding some features and being attracted to others. These behavioral responses to the landscape can channel their movement into pathways referred to as *corridors*. Corridors are conceived of as linear elements in the landscape that facilitate dispersal; examples include hedgerows, fences, and vegetation along roads. However, our perception of corridors certainly is limited because we can't see the landscape through the animals' eyes. Could it be that some places where we see high numbers of dispersing animals are "virtual corridors"—they do not have especially beneficial features (unlike, for example, a hedgerow, which provides concealment from predators) but instead they emerge from more subtle characteristics of the landscape and from how the animals decide where to go?

To demonstrate the concept of virtual corridors, Peer et al. (2005) chose an extremely simple system in which it is easy to observe real individuals: mate-finding by butterflies. In many butterfly species, males and unmated females apply the "hilltopping" strategy: they simply move uphill until they are concentrated on hilltops where they can meet and mate. The model developed by Peer et al. (2005) is a good example of how simple a model can be and still capture the essence of a system with regard to a certain question. Because the model is so simple, we skip the heuristic, brainstorming phase of formulating models and jump directly to the model formulation using the ODD protocol.

Purpose

The model was designed to explore questions about virtual corridors. Under what conditions do the interactions of butterfly hilltopping behavior and landscape topography lead to the emergence of virtual corridors, that is, relatively narrow paths along which many butterflies move? How does variability in the butterflies' tendency to move uphill affect the emergence of virtual corridors?

Entities, State Variables, and Scales

The model has two kinds of entities: butterflies and square patches of land. The patches make up a square grid landscape of 150×150 patches, and each patch has one state variable: its elevation. Butterflies are characterized only by their location, described as the patch they are on. Therefore, butterfly locations are in discrete units, the x- and y-coordinates of the center of their patch. Patch size and the length of one time step in the simulation are not specified because the model is generic, but when real landscapes are used, a patch corresponds to 25 × 25 m². Simulations last for 1000 time steps; the length of one time step is not specified but should be about the time it takes a butterfly to move 25–35 m (the distance from one cell to one of its neighbor cells).

Process Overview and Scheduling

There is only one process in the model: movement of the butterflies. On each time step, each butterfly moves once. The order in which the butterflies execute this action is unimportant because there are no interactions among the butterflies.

Design Concepts

The *basic principle* addressed by this model is the concept of virtual corridors—pathways used by many individuals when there is nothing particularly beneficial about the habitat in them. This concept is addressed by seeing when corridors *emerge* from two parts of the model: the adaptive movement behavior of butterflies and the landscape they move through. This *adaptive behavior* is modeled via a simple empirical rule that reproduces the behavior observed in real butterflies: moving uphill. This behavior is based on the understanding (not included in the model) that moving uphill leads to mating, which conveys fitness (success at passing on genes, the presumed ultimate objective of organisms). Because the hilltopping behavior is assumed *a priori* to be the objective of the butterflies, the concepts of *Objectives* and *Prediction* are not explicitly considered. There is no *learning* in the model.

Sensing is important in this model: butterflies are assumed able to identify which of the surrounding patches has the highest elevation, but to use no information about elevation at further distances. (The field studies of Pe'er 2003 addressed this question of how far butterflies sense elevation differences.)

The model does not include *interaction* among butterflies; in field studies, Pe'er (2003) found that real butterflies do interact (they sometimes stop to visit each other on the way uphill) but decided it is not important to include interaction in a model of virtual corridors.

Stochasticity is used to represent two sources of variability in movement that are too complex to represent mechanistically. Real butterflies do not always move directly uphill, likely because of (1) limits in the ability of the butterflies to sense the highest area in their neighborhood, and (2) factors other than topography (e.g., flowers that need investigation along the way) that influence movement direction. This variability is represented by assuming butterflies do not move uphill every time step; sometimes they move randomly instead. Whether a butterfly moves directly uphill or randomly at any time step is modeled stochastically, using a parameter q that is the probability of an individual moving directly uphill instead of randomly.

To allow *observation* of virtual corridors, we will define (later, in section 5.2) a specific "corridor width" measure that characterizes the width of a butterfly's path from its starting patch to a hilltop.

Initialization

The topography of the landscape (the elevation of each patch) is initialized when the model starts. Two kinds of landscapes are used in different versions of the model: (1) a simple artificial topography, and (2) the topography of a real study site, imported from a file containing elevation values for each patch. The butterflies are initialized by creating five hundred of them and setting their initial location to a single patch or small region.

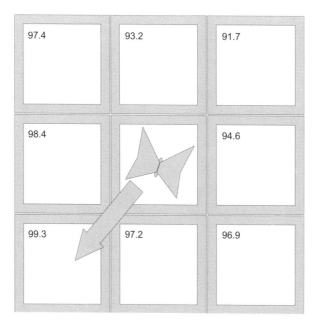

Figure 3.2
The butterfly movement action. A butterfly moves to one of its eight neighboring patches. Patches have an elevation variable, displayed here in their upper left corner. If a random number is less than q, the butterfly moves to the neighbor patch with highest elevation, indicated by the arrow. Otherwise, the butterfly chooses one of the eight neighbor patches randomly.

Input Data

The environment is assumed to be constant, so the model has no input data.

Submodels

The movement submodel defines exactly how butterflies decide whether to move uphill or randomly. First, to "move uphill" is defined specifically as moving to the neighbor patch that has the highest elevation; if two patches have the same elevation, one is chosen randomly. "Move randomly" is defined as moving to one of the neighboring patches, with equal probability of choosing any patch. "Neighbor patches" are the eight patches surrounding the butterfly's current patch. The decision of whether to move uphill or randomly is controlled by the parameter q, which ranges from 0.0 to 1.0 (q is a global variable: all butterflies use the same value). On each time step, each butterfly draws a random number from a uniform distribution between 0.0 and 1.0. If this random number is less than q, the butterfly moves uphill; otherwise, the butterfly moves randomly (figure 3.2).

This formulation in ODD format contains all the information needed to implement the model on a computer. As you learn NetLogo, you will see that the elements of ODD correspond to the typical elements of a NetLogo program. NetLogo has, for example, specific dialogs and primitives for setting the spatial scale and keeping track of time steps, defining and initializing the model's entities, scheduling actions, and observing results. Consequently, in chapter 4 we will be able to use this ODD description of the hilltopping butterfly model as the blueprint for your first real NetLogo program.

3.5 Summary and Conclusions

Describing a model on paper is perhaps the most important part of modeling: very few benefits of modeling can be achieved without it. For ABMs, it is especially important to use standard concepts and formats to describe and design models: these models are complex, so we need a clear, standard way to think and write about them. The standard "languages" of differential equations and statistical modeling cannot describe ABMs, so instead we use the ODD protocol and its design concepts. Without such a standard, ABMs are often too incompletely described to be replicated, which makes them unscientific. Establishing ODD as a standard language for communicating ABMs is therefore one of the main purposes of this book.

ODD is *generic* and *hierarchical*. *Generic* means that is can be used to describe any ABM in any field of application or discipline. *Hierarchical* means that it starts with an overview of a model's structure, scales, processes, and scheduling, so we can understand the model's basics, before presenting the details needed to understand how processes are actually represented. In between, important concepts underlying the model's design are explained, for example: what key behaviors in the model are emergent instead of being imposed? To what extent and why is stochasticity included? How do we observe model behavior to better understand how model output emerges?

In section 3.2 we mentioned one important benefit of ODD in addition to its original purpose as a standard way of describing ABMs: it also provides a very useful framework for thinking about and formulating ABMs. And in section 3.4 we mentioned another benefit that will become very important: once a model is written down in ODD format, it is very clear how to translate it into a NetLogo program. We will illustrate the close correspondence between ODD and NetLogo in the next chapter as you return to learning NetLogo. In fact, part II of this book, where you really learn modeling and programming rapidly, is organized around ways to implement ODD design concepts in NetLogo.

3.6 Exercises

1. On the web site for this book is a list of scientific publications that describe ABMs using the ODD protocol. To develop your understanding of the protocol, find several of these papers and read the model descriptions. (Some of the terms used in the protocol, especially for the design concepts, have changed as the protocol has evolved. For example, in ecology the term "fitness" was used, but the protocol now speaks of "objectives." For a more detailed discussion of the design concepts as applied to ecology, read chapter 5 of Grimm and Railsback 2005.)

2. Also on the web site for this book is a list of scientific publications that used ABMs but not the ODD protocol. Find one or more of these and read their model descriptions. Try to complete the ODD protocol for the models. What information is easy to extract from the publication, and what seems to be missing?

3. Think of a simple problem for which a model might be useful to obtain better understanding and for which it would be essential to represent agents explicitly. (You could use one of the problems listed in exercise 2 of chapter 1.) What factors (kinds of entities, variables, processes) do you consider important for this problem? Which of these factors seem absolutely essential, and which belong on the "wish list" for the first version of a model? Try to formulate an ODD description of a simple ABM that includes only the key entities and processes of your system. Discuss and revise the ODD description. Don't worry if you are unsure of many design decisions—formulating and analyzing ABMs requires skills and experience that you will gain from this book.

Implementing a First Agent-Based Model

4.1 Introduction and Objectives

In this chapter we continue your lessons in NetLogo, but from now on the focus will be on programming—and using—real ABMs that address real scientific questions. (The Mushroom Hunt model of chapter 2 was neither very agent-based nor scientific, in ways we discuss in this chapter.) And even though this chapter is mostly still about NetLogo programming, it starts addressing other modeling issues. It should prime you to start actually using an ABM to produce and analyze meaningful output and address scientific questions, which is what we do in chapter 5.

Learning objectives for chapter 4 are to:

- Understand how to translate a model from its written description in ODD format into NetLogo code.
- Understand how to define global, turtle, and patch variables.
- Become familiar with NetLogo's most important primitives, such as `ask`, `set`, `let`, `create-turtles`, `ifelse`, and `one-of`.
- Start learning good programming practices, such as making very small changes and constantly checking them, and writing comments in your code.
- Produce your own software for the Butterfly model described in chapter 3.

4.2 ODD and NetLogo

In chapter 3 we introduced the ODD protocol for describing an ABM, and as an example provided the ODD formulation of a butterfly hilltopping model. What do we do when it is time to make a model described in ODD actually run in NetLogo? It turns out to be quite straightforward because the organizations of ODD and NetLogo correspond closely. The major elements of an ODD formulation have corresponding elements in NetLogo.

Purpose

From now on, we will include the ODD descriptions of our ABMs on NetLogo's Information tab. These descriptions will then start with a short statement of the model's overall purpose.

Entities, State Variables, and Scales

Basic entities for ABMs are built into NetLogo: the World of square patches, turtles as mobile agents, and the observer. The state variables of the turtles and patches (and perhaps other types of agents) are defined via the `turtles-own []` and `patches-own []` statements, and the variables characterizing the global environment are defined in the `globals []` statement. In NetLogo, as in ODD, these variables are defined right at the start.

Process Overview and Scheduling

This, exactly, is represented in the `go` procedure. Because a well-designed `go` procedure simply calls other procedures that implement all the submodels, it provides an overview (but not the detailed implementation) of all processes, and specifies their schedule, that is, the sequence in which they are executed each tick.

Design Concepts

These concepts describe the decisions made in designing a model and so do not appear directly in the NetLogo code. However, NetLogo provides many primitives and interface tools to support these concepts; part II of this book explores how design concepts are implemented in NetLogo.

Initialization

This corresponds to an element of every NetLogo program, the `setup` procedure. Pushing the `setup` button should do everything described in the Initialization element of ODD.

Input Data

If the model uses a time series of data to describe the environment, the program can use NetLogo's input primitives to read the data from a file.

Submodels

The submodels of ODD correspond closely but not exactly to procedures in NetLogo. Each of a model's submodels should be coded in a separate NetLogo procedure that is then called from the `go` procedure. (Sometimes, though, it is convenient to break a complex submodel into several smaller procedures.)

These correspondences between ODD and NetLogo make writing a program from a model's ODD formulation easy and straightforward. The correspondence between ODD and NetLogo's design is of course not accidental: both ODD and NetLogo were designed to capture and organize the important characteristics of ABMs.

4.3 Butterfly Hilltopping: From ODD to NetLogo

Now, let us for the first time write a NetLogo program by translating an ODD formulation, for the Butterfly model. The way we are going to do this is hierarchical and step-by-step, with many interim tests. We strongly recommend you always develop programs in this way:

- Program the overall structure of a model first, before starting any of the details. This keeps you from getting lost in the details early. Once the overall structure is in place, add the details one at a time.
- Before adding each new element (a procedure, a variable, an algorithm requiring complex code), conduct some basic tests of the existing code and save the file. This way, you always

proceed from "firm ground": if a problem suddenly arises, it very likely (although not always) was caused by the last little change you made.

First, let us create a new NetLogo program, save it, and include the ODD description of the model on the Information tab.

▪ Start NetLogo and use File/New to create a new NetLogo program. Use File/Save to save the program under the name "Butterfly-1.nlogo" in an appropriate folder.

▪ Get the file containing the ODD description of the Butterfly model from the chapter 4 section of the web site for this book.

▪ Go the Information tab in NetLogo, click the Edit button, and paste in the model description at the top.

▪ Click the Edit button again.

You now see the ODD description of the Butterfly model at the beginning of the documentation. Now, let us start programming this model with the second part of its ODD description, the state variables and scales. First, define the model's state variables—the variables that patches, turtles, and the observer own.

▪ Go to the Procedure tab and insert

```
globals [ ]
patches-own [ ]
turtles-own [ ]
```

▪ Click the Check button.

There should be no error message, so the code syntax is correct so far. Now, from the ODD description we see that turtles have no state variables other than their location; because Net-Logo already has built-in turtle variables for location (xcor, ycor), we need to define no new turtle variables. But patches have a variable for elevation, which is not a built-in variable, so we must define it.

▪ In the program, insert elevation as a state variables of patches:

```
patches-own [ elevation ]
```

If you are familiar with other programming languages, you might wonder where we tell NetLogo what type the variable "elevation" is. The answer is: NetLogo figures out the type from the first value assigned to the variable via the set primitive.

Now we need to specify the model's spatial extent: how many patches it has.

▪ Go to the Interface tab, click the Settings button, and change "Location of origin" to Corner and Bottom Left; change the number of columns (max-pxcor) and rows (max-pycor) to 149. Now we have a world, or landscape, of 150 × 150 patches, with patch 0,0 at the lower left corner. Turn off the two world wrap tick boxes, so that our model world has closed boundaries. Click OK to save your changes.

You probably will see that the World display (the "View") is now extremely large, too big to see all at once. You can fix this:

■ Click the Settings button again and change "Patch size" to 3 or so, until the View is a nice size. (You can also change patch size by right-clicking on the View to select it, then dragging one of its corners.)

The next natural thing to do is to program the `setup` procedure, where all entities and state variables are created and initialized. Our guide of course is the Initialization part of the ODD description. Back in the Procedures tab, let us again start by writing a "skeleton."

■ At the end of the existing program, insert this:

```
to setup
  ca
  ask patches
  [

  ]
  reset-ticks
end
```

Click the Check button again to make sure the syntax of this code is correct. There is already some code in this `setup` procedure: `ca` to delete everything, which is almost always first in the `setup` procedure, and `reset-ticks`, which is almost always last. The `ask patches` statement will be needed to initialize the patches by giving them all a value for their elevation variable. The code to do so will go within the brackets of this statement.

Assigning elevations to the patches will create a topographical landscape for the butterflies to move in. What should the landscape look like? The ODD description is incomplete: it simply says we start with a simple artificial topography. We obviously need some hills because the model is about how butterflies find hilltops. We could represent a real landscape (and we will, in the next chapter), but to start it is a good idea to create scenarios so simple that we can easily predict what should happen. Creating just one hill would probably not be interesting enough, but two hills will do.

■ Add the following code to the `ask patches` command:

```
ask patches
[
  let elev1 100 - distancexy 30 30
  let elev2 50 - distancexy 120 100

  ifelse elev1 > elev2
    [set elevation elev1]
    [set elevation elev2]

  set pcolor scale-color green elevation 0 100
]
```

The idea is that there are two conical hills, with peaks at locations (x,y-coordinates) 30,30 and 120,100. The first hill has an elevation of 100 units (let's assume these elevations are in meters) and the second an elevation of 50 meters. First, we calculate two elevations (*elev1* and *elev2*) by assuming the patch's elevation decreases by one meter for every unit of horizontal distance between the patch and the hill's peak (remind yourself what the primitive `distancexy` does by putting your cursor over it and hitting the F1 key). Then, for each patch we assume the elevation is the greater of two potential elevations, *elev1* and *elev2*: using the `ifelse` primitive, each patch's elevation is set to the higher of the two elevations.

Finally, patch elevation is displayed on the View by setting patch color `pcolor` to a shade of the color `green` that is shaded by patch elevation over the range 0 and 100 (look up the extremely useful primitive `scale-color`).

Remember that the `ask patches` command establishes a patch context: all the code inside its brackets must be executable by patches. Hence, this code uses the state variable `elevation` that we defined for patches, and the built-in patch variable `pcolor`. (We could *not* have used the built-in variable `color` because `color` is a turtle, not a patch, variable.)

Now, following our philosophy of testing everything as early as possible, we want to see this model landscape to make sure it looks right. To do so, we need a way to execute our new `setup` procedure.

- On the Interface, press the Button button, select "Button," and place the cursor on the Interface left of the View window. Click the mouse button.

- A new button appears on the Interface and a dialog for its settings opens. In the Commands field, type "setup" and click OK.

Now we have linked this button to the `setup` procedure that we just programmed. If you press the button, the `setup` procedure is executed and the Interface should look like figure 4.1.

There is a landscape with two hills! Congratulations! (If your View does not look like this, figure out why by carefully going back through *all* the instructions. Remember that any mistakes you fix will not go away until you hit the `setup` button again.)

NetLogo brainteaser

When you click the new `setup` button, why does NetLogo seem to color the patches in spots all over the View, instead of simply starting at the top and working down? (You may have to turn the speed slider down to see this clearly.) Is this a fancy graphics rendering technique used by NetLogo? (Hint: start reading the Programming Guide section on Agentsets very carefully.)

Now we need to create some agents or, to use NetLogo terminology, turtles. We do this by using the `create-turtles` primitive (abbreviated `crt`). To create one turtle, we use `crt 1 [ ]`. To better see the turtle, we might wish to increase the size of its symbol, and we also want to specify its initial position.

- Enter this code in the `setup` procedure, after the `ask patches` statement:

```
crt 1
[
```

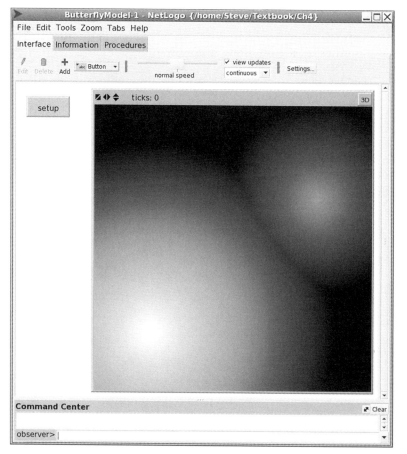

Figure 4.1
The Butterfly model's artificial landscape.

```
      set size 2
      setxy 85 95
    ]
```

■ Click the Check button to check syntax, then press the `setup` button on the Interface.

Now we have initialized the World (patches) and agents (turtles) sufficiently for a first minimal version of the model.

The next thing we need to program is the model's schedule, the `go` procedure. The ODD description tells us that this model's schedule includes only one action, performed by the turtles: movement. This is easy to implement.

■ Add the following procedure:

```
    to go
      ask turtles [move]
    end
```

■ At the end of the program, add the skeleton of the procedure `move`:

```
to move

end
```

Now you should be able to check syntax successfully.

There is one important piece of `go` still missing. This is where, in NetLogo, we control a model's "temporal extent": the number of time steps it executes before stopping. In the ODD description of model state variables and scales, we see that Butterfly model simulations last for 1000 time steps. In chapter 2 we learned how to use the built-in tick counter via the primitives `reset-ticks`, `tick`, and `ticks`. Read the documentation for the primitive `stop` and modify the `go` procedure to stop the model after 1000 ticks:

```
to go
  ask turtles [move]
  tick
  if ticks >= 1000 [stop]
end
```

■ Add a `go` button to the Interface. Do this just as you added the `setup` button, except: enter "go" as the command that the button executes, and activate the "Forever" tick box so hitting the button makes the observer repeat the `go` procedure over and over.

Programming note: Modifying Interface elements

To move, resize, or edit elements on the Interface (buttons, sliders, plots, etc.), you must first select them. Right-click the element and chose "Select" or "Edit." You can also select an element by dragging the mouse from next to the element to over it. Deselect an element by simply clicking somewhere else on the Interface.

If you click `go`, nothing happens except that the tick counter goes up to 1000—NetLogo is executing our `go` procedure 1000 times but the `move` procedure that it calls is still just a skeleton. Nevertheless, we verified that the overall structure of our program is still correct. And we added a technique very often used in the `go` procedure: using the `tick` primitive and `ticks` variable to keep track of simulation time and make a model stop when we want it to.

(Have you been saving your new NetLogo file frequently?)

Now we can step ahead again by implementing the `move` procedure. Look up the ODD element "Submodels" on the Information tab, where you find the details of how butterflies move. You will see that a butterfly should move, with a probability q, to the neighbor cell with the highest elevation. Otherwise (i.e., with probability $1 - q$) it moves to a randomly chosen neighbor cell. The butterflies keep following these rules even when they have arrived at a local hilltop.

■ To implement this movement process, add this code to the `move` procedure:

```
to move
  ifelse random-float 1 < q
```

```
      [ uphill elevation ]
      [ move-to one-of neighbors ]
  end
```

The `ifelse` statement should look familiar: we used a similar statement in the `search` procedure of the Mushroom Hunt model of chapter 2. If q has a value of 0.2, then in about 20% of cases the random number created by `random-float` will be smaller than q and the `ifelse` condition true; in the other 80% of cases the condition will be false. Consequently, the statement in the first pair of brackets (`uphill elevation`) is carried out with probability q and the alternative statement (`move-to one-of neighbors`) is done with probability $1-q$.

The primitive `uphill` is typical of NetLogo: following a gradient of a certain variable (here, *elevation*) is a task required in many ABMs, so `uphill` has been provided as a convenient shortcut. Having such primitives makes the code more compact and easier to understand, but it also makes the list of NetLogo primitives look quite scary at first.

The remaining two primitives in the `move` procedure (`move-to` and `one-of`) are self-explanatory, but it would be smart to look them up so you understand clearly what they do.

▪ Try to go to the Interface and click `go`.

NetLogo stops us from leaving the Procedures tab with an error message telling us that the variable q is unknown to the computer. (If you click on the Interface tab again, NetLogo relents and lets you go there, but turns the Procedures tab label an angry red to remind you that things are not all OK there.)

Because q is a parameter that all turtles will need to know, we define it as a global variable by modifying the `globals` statement at the top of the code:

```
  globals [ q ]
```

But defining the variable q is not enough: we also must give it a value, so we initialize it in the `setup` procedure. We can add this line at the end of `setup`, just before `end`:

```
  set q 0.4
```

The turtles will now move deterministically to the highest neighbor patch with a probability of 0.4, and to a randomly chosen neighbor patch with a probability of 0.6.

Now, the great moment comes: you can run and check your first complete ABM!

▪ Go to the Interface and click `setup` and then `go`.

Yes, the turtle does what we wanted it to do: it finds a hilltop (most often, the higher hill, but sometimes the smaller one), and its way is not straight but somewhat erratic.

Just as with the Mushroom Hunt model, it would help to put the turtle's "pen" down so we can see its entire path.

▪ In the `crt` block of the `setup` procedure (the code statements inside the brackets after `crt`), include a new line at the end with this command: `pen-down`.

You can now start playing with the model and its program, for example by modifying the model's only parameter, q. Just edit the `setup` procedure to change the value of q (or the initial location of the turtle, or the hill locations), then pop back to the Interface tab and hit

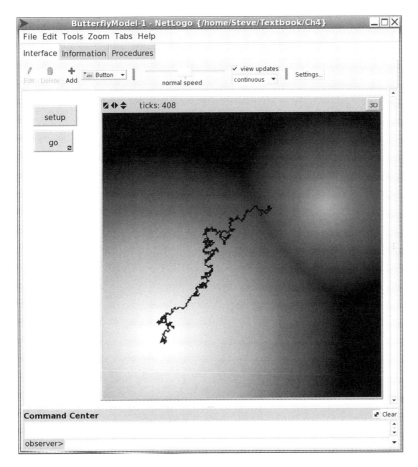

Figure 4.2
Interface of the Butterfly model's first version, with $q = 0.2$.

the `setup` and `go` buttons. A typical model run for $q = 0.2$, so butterflies make 80% of their movement decisions randomly, is shown in figure 4.2.

4.4 Comments and the Full Program

We have now implemented the core of the Butterfly model, but four important things are still missing: comments, observations, a realistic landscape, and analysis of the model. We will address comments here but postpone the other three things until chapter 5.

Comments are any text following a semicolon (;) on the same line in the code; such text is ignored by NetLogo and instead is for people. Comments are needed to make code easier for others to understand, but they are also very useful to ourselves: after a few days or weeks go by, you might not remember why you wrote some part of your program as you did instead of in some other way. (This surprisingly common problem usually happens where the code was hardest to write and easiest to mess up.) Putting a comment at the start of each procedure saying whether the procedure is in turtle, patch, or observer context helps you write the procedures by making you think about their context, and it makes revisions easier.

■ As a good example comment, change the first line of your `move` procedure to this, so you always remember that the code in this procedure is in turtle context:

```
to move ; A turtle procedure
```

The code examples in this book use few comments (to keep the text short, and to force readers to understand the code). The programs that you can download via the book's web site include comments, and you must make it a habit to comment your own code. In particular, use comments to:

- Briefly describe what each procedure or nontrivial piece of the code is supposed to do;
- Explain the meaning of variables;
- Document the context of each procedure;
- Keep track of what code block or procedure is ended by "]" or end (see the example below); and
- In long programs, visually separate procedures from each other by using comment lines like this:

```
; ————————————————
```

On the other hand, detailed and lengthy comments are no substitute for code that is clearly written and easy to read! Especially in NetLogo, you should strive to write your code so you do not need many comments to understand it. Use names for variables and procedures that are descriptive and make your code statements read like human language. Use tabs and blank lines to show the code's organization. We demonstrate this "self-commenting" code approach throughout the book.

Another important use of comments is to temporarily deactivate code statements. If you replace a few statements but think you might want the old code back later, just "comment out" the lines by putting a semicolon at the start of each. For example, we often add temporary test output statements, like this one that could be put at the end of the ask patches statement in the setup procedure to check the calculations setting elevation:

```
show (word elev1 " " elev2 " " elevation)
```

If you try this once, you will see that it gives you the information to verify that elevation is set correctly, but you will also find out that you do not want to repeat this check every time you use setup! So when you do not need this statement anymore, just comment it out by putting a semicolon in front of it. Leaving the statement in the code but commented out provides documentation of how you checked elevations, and lets you easily repeat the check when you want.

Here is our Butterfly model program, including comments.

```
globals [ q ] ; q is the probability that butterfly moves
              ; directly to the highest surrounding patch
patches-own [ elevation ]
turtles-own [ ]

to setup
  ca

  ; Assign an elevation to patches and color them by it
  ask patches
  [
    ; Elevation decreases linearly with distance from the
    ; center of hills. Hills are at (30, 30) and
    ; (120, 100). The first hill is 100 units high.
```

```
        ; The second hill is 50 units high.

        let elev1 100 - distancexy 30 30
        let elev2 50 - distancexy 120 100

        ifelse elev1 > elev2
            [set elevation elev1]
            [set elevation elev2]

        set pcolor scale-color green elevation 0 100
    ] ; end of "ask patches"

    ; Create just 1 butterfly for now
    crt 1
    [
      set size 2
      ; Set initial location of butterflies
      setxy 85 95
      pen-down
    ]

    ; Initialize the "q" parameter
    set q 0.4

    reset-ticks

end ; of setup procedure

to go ;  This is the master schedule
  ask turtles [move]
  tick
  if ticks >= 1000 [stop]
end

to move ; The butterfly move procedure, in turtle context
        ; Decide whether to move to the highest
        ; surrounding patch with probability q
  ifelse random-float 1 < q
    [ uphill elevation ] ; Move deterministically uphill
    [ move-to one-of neighbors ] ; Or move randomly
end ; of move procedure
```

■ Include comments in your program and save it.

The second thing that is missing from the model now is observation. So far, the model only produces visual output, which lets us look for obvious mistakes and see how the butterfly behaves. But to use the model for its scientific purpose—understanding the emergence of "virtual corridors"—we need additional outputs that quantify the width of the corridor used by a large number of butterflies.

Another element that we need to make more scientific is the model landscape. It is good to start programming and model testing and analysis with artificial scenarios as we did here, because it makes it easier to understand the interaction between landscape and agent behavior and to identify errors, but we certainly do not want to restrict our analysis to artificial landscapes like the one we now have.

Finally, we have not yet done any analysis on this model, for example to see how the parameter q affects butterfly movement and the appearance of virtual corridors. We will work on observation, a realistic landscape, and analysis in the next chapter, but now you should be eager to play with the model. "Playing"—creative tests and modification of the model and program ("What would happen if . . .")—is important in modeling. Often, we do not start with a clear idea of what observations we should use to analyze a model. We usually are also not sure whether our submodels are appropriate. NetLogo makes playing very easy. You now have the chance to heuristically analyze (that is: play with!) the model.

4.5 Summary and Conclusions

If this book were only about programming NetLogo, this chapter would simply have moved ahead with teaching you how to write procedures and use the Interface. But because this book is foremost about agent-based modeling in science, we began the chapter by revisiting the ODD protocol for describing ABMs and showing how we can quite directly translate an ODD description into a NetLogo program. In scientific modeling, we start by thinking about and writing down the model design; ODD provides a productive, standard way to do this. Then, when we think we have enough of a design to implement on the computer, we translate it into code so we can start testing and revising the model.

The ODD protocol was developed independently from NetLogo, but they have many similarities and correspond quite closely. This is not a coincidence, because both ODD and NetLogo were developed by looking for the key characteristics of ABMs in general and the basic ways that they are different from other kinds of model (and, therefore, ways that their software must be unique). These key characteristics were used to organize both ODD and NetLogo, so it is natural that they correspond with each other. In section 4.2 we show how each element of ODD corresponds to a specific part of a NetLogo program.

This chapter illustrates several techniques for getting started modeling easily and productively. First, start modeling with the simplest version of the model conceivable—ignore many, if not most, of the components and processes you expect to include later.

The second technique is to develop programs in a hierarchical way: start with the skeletons of structures (procedures, `ask` commands, `ifelse` switches, etc.); test these skeletons for syntax errors; and only then, step by step, add "flesh to the bones" of the skeleton. ODD starts with the model's most general characteristics; once these have been programmed, you have a strong "skeleton"—the model's purpose is documented in the Information tab, the View's dimensions are set, state variables are defined, and you know what procedures should be called in the `go` procedure. Then you can fill in detail, first by turning the ODD Initialization element into the `setup` procedure, and then by programming each procedure in full. Each step of the way, you can add skeletons such as empty procedures and `ask` statements, then check for errors before moving on.

Third, if your model will eventually include a complex or realistic environment, start with a simplified artificial one. This simplification can help you get started sooner and, more importantly, will make it easier for you identify errors (either in the programming or in the logic) in how the agents interact with their environment.

Finally, formatting your code nicely and providing appropriate comments is well worth the tiny bit of time it takes. Your comments should have two goals: to make it easier to understand and navigate the code with little reminders about what context is being used, what code blocks and procedures are ended by an] or end, and so on; and to document how and why you programmed nontrivial things as you did. But remember that your code should be easy to read and understand even with minimal comments. These are not just aesthetic concerns! If your model is successful at all, you and other people will be reading and revising the code many times, so take a few seconds as you write it to make future work easy.

Now, let's step back from programming and consider what we have achieved so far. We wanted to develop a model that helps us understand how and where in a landscape "virtual corridors" of butterfly movement appear. The hypothesis is that corridors are not necessarily linked to landscape features that are especially suitable for migration, but can also just emerge from the interaction between topography and the movement decisions of the butterflies. We represented these decisions in a most simple way: by telling the butterflies to move uphill, but with their variability in movement represented by the parameter q. The first results from a highly artificial landscape indicate that indeed our movement rule has the potential to produce virtual corridors, but we obviously have more to do. In the next chapter we will add things needed to make a real ABM for the virtual corridor problem: output that quantifies the "corridors" that model butterflies use, a realistic landscape for them to fly through, and a way to systematically analyze corridors.

4.6 Exercises

1. Change the number of butterflies to 50. Now, try some different values of the parameter q. Predict what will happen when q is 0.0 and 1.0 and then run the model; was your prediction correct?

2. Modify your model so that the turtles do not all have the same initial location but instead have their initial location set randomly. (Hint: look for primitives that provide random X and Y locations.)

3. From the first two exercises, you should have seen that the paths of the butterflies seem unexpectedly artificial. Do you have an explanation for this? How could you determine what causes it? Try analyzing a smaller World and add labels to the patches that indicate, numerically, their elevation. (Hint: patches have a built-in variable `plabel`. And see exercise 4.)

4. Try adding "noise" to the landscape, by adding a random number to patch elevation. You can add this statement to `setup`, just after patch elevation is set: `set elevation elevation + random 10`. How does this affect movement? Did this experiment help you answer the questions in exercise 3?

From Animations to Science

5.1 Introduction and Objectives

Beginners often believe that modeling is mainly about formulating and implementing models. This is not the case: the real work starts *after* a model has first been implemented. Then we use the model to find answers and solutions to the questions and problems we started our modeling project with, which almost always requires modifying the model formulation and software. This iterative process of model analysis and refinement—the modeling cycle—usually is not documented in the publications produced at the end. Instead, models are typically presented as static entities that were just produced and used. In fact, every model description is only a snapshot of a process.

The Models Library of NetLogo has a similar problem: it presents the models and gives (on the Information tab) some hints of how to analyze them, but it cannot demonstrate how to do science with them. These models are very good at animation: letting us see what happens as their assumptions and equations are executed. But they do not show you how to explore ideas and concepts, develop and test hypotheses, and look for parsimonious and general explanations of observed phenomena.

In this chapter we illustrate how to make a NetLogo program into a scientific model instead of just a simulator, by taking the Butterfly model and adding the things needed to do science. Remember that the purpose of this model was to explore the emergence of "virtual corridors": places where butterflies move in high concentrations even though there is nothing there that attracts butterflies. Our model so far simulates butterfly movement, but does not tell us anything about corridors and when and how strongly they emerge. Therefore, we will now produce quantitative output that can be analyzed, instead of just the visual display of butterfly movement. We will also replace our very simple and artificial landscape with a real one read in from a topography file. In the exercises we suggest some scientific analyses for you to conduct on the Butterfly model.

We assume that by now you have learned enough about NetLogo to write code yourself while frequently looking things up in the User Manual and getting help from others. Thus, we no longer provide complete code as we walk you through changes in the model, except when particularly complicated primitives or approaches are involved. Those of you working through this book without the help of instructors and classmates can consult our web site for assistance if you get stuck. And be aware that part II of this book should reinforce and deepen your understanding of all the programming techniques used here. If you feel like you have just been thrown into deep water without knowing how to swim, do your best and don't worry: things will become easier soon.

Learning objectives for this chapter are to:

- Learn the importance of version control: saving documented versions of your programs whenever you start making substantial changes.
- Understand the concept that using an ABM for science requires producing quantitative output and conducting simulation experiments; and execute your first simulation experiment.
- Learn to define and initialize a global variable by creating a slider or switch on the Interface.
- Develop an understanding of what reporters are and how to write them.
- Start learning to find and fix mistakes in your code.
- Learn to create output by writing to an output window, creating a time-series plot, and exporting plot results to a file.
- Try a simple way to import data into NetLogo, creating a version of the Butterfly model that uses real topographic data.

5.2 Observation of Corridors

Our Butterfly model is not ready to address its scientific purpose in part because it lacks a way to quantitatively observe the extent to which virtual corridors emerge. But how would we characterize a corridor? Obviously, if all individuals followed the same path (as when they all start at the same place and q is 1.0) the corridor would be very narrow; or if movement were completely random we would not expect to identify any corridor-like feature. But we need to quantify how the width of movement paths changes as we vary things such as q or the landscape topography. Let's do something about this problem.

First, because we are about to make a major change in the program:

■ Create and save a new version of your butterfly software.

> **Programming note:** Version control
>
> Each time you make a new version of a model, make a new copy of the program by saving it with a descriptive name, perhaps in a separate directory. (Do this first, before you change anything!) Then, as you change the model, use comments in the Procedures tab to document the changes you made to the program, and update the Information tab to describe the new version. Otherwise you will be very sorry! It is very easy to accidentally overwrite an older model version that you will need later, or to forget how one version is different from the others. It is even a good idea to keep a table showing exactly what changes were made in each copy of a model.
>
> Even when you just play around with a model, make sure you save it first as a temporary version so you don't mess up the permanent file.
>
> Programmers use the term "version control" for this practice of keeping track of which changes were made in which copy of the software. Throughout this course we will be making many modifications to many programs; get into the habit of careful version control from the start and spare yourself from lots of trouble and wasted time.

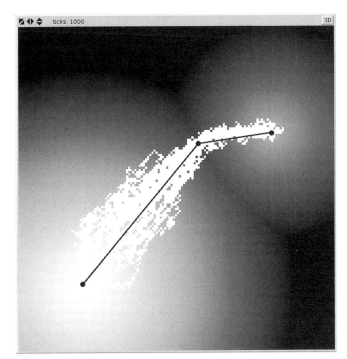

Figure 5.1
The measure of corridor width for a group of uphill-migrating butterflies (here, 50 butterflies that started at location 85,95 with $q = 0.4$). Corridor width is (a) the number of patches used by any butterflies (the white patches here), divided by (b) the average straight-line distance (in units of patch lengths) between butterflies' starting patch and ending hilltop. Here, most butterflies went up the large hill to the lower left, but some went up the smaller hill to the right. The number of white patches is 1956 and the mean distance between butterfly starting and ending points (the black dots) is 79.2 patches; hence, corridor width is 24.7 patches.

Pe'er et al. (2005) quantified corridor width by dividing the number of patches visited by all individuals during 1000 time steps by the distance between the start patch and the hill's summit. In our version of the model, different butterflies can start and end in different places, so we slightly modify this measure. First, we will assume each butterfly stops when it reaches a local hilltop (a patch higher than all its eight neighbor patches). Then we will quantify the width of the "corridor" used by all butterflies as (a) the number of patches that are visited by any butterflies divided by (b) the mean distance between starting and ending locations, over all butterflies (figure 5.1). This corridor width measure should be low (approaching 1.0) when all butterflies follow the same, straight path uphill, but should increase as the individual butterflies follow different, increasingly crooked, paths.

To analyze the model we are thus going to produce a plot of corridor width vs. q. This is important: when analyzing a model, we need to have a clear idea of what kind of plot we want to produce from what output, because this tells us what kind of simulation experiments we have to perform and what outputs we need the program to produce (see also chapter 22).

Now that we have determined what we need to change in the model to observe and analyze corridor width, let's implement the changes.

■ Because it is now obvious that we need to conduct experiments by varying q and seeing its effect, create a slider for it on the Interface tab. (The Interface Guide of NetLogo's User Manual will tell you how.) Set the slider so that q varies from 0.0 to 1.0 in increments of 0.01.

■ Change the `setup` procedure so that 50 individuals are created and start from the same position. Then vary q via its slider and observe how the area covered by all the butterfly paths changes.

■ Modify the `move` procedure so that butterflies no longer move if they are at a hilltop. One way to do this is to test whether the turtle is on a patch with elevation greater than that of all its neighbors and, if so, to skip the rest of the `move` procedure. The code for this test is a little more complex than other statements you have used so far, so we explain it. Put this statement at the start of the `move` procedure:

```
if elevation >=
    [elevation] of max-one-of neighbors [elevation]
  [stop]
```

This statement includes a typical NetLogo compound expression. Starting on the left, it uses the `if` primitive, which first checks a true-false condition and then executes the commands in a set of brackets. The statement ends with `[stop]`, so we know that (a) the true-false condition is all the text after `if` and before `[stop]`, and (b) if that condition is true, the `stop` primitive is executed.

You may be confused by this statement because it is clearly in turtle context—the `move` procedure is executed by turtles—but it uses the patch variable `elevation`. The reason it works is that in NetLogo turtles always have access to the variables of their current patch and can, as they do here, read and write them just as if they were turtle variables.

To understand the right side of the inequality condition (after ">="), you need to understand the reporter `max-one-of`, which takes an agentset—here, the set of patches returned by the reporter `neighbors`—and the name of a reporter. (Remember that a reporter is a procedure that returns or "reports" something—a number, an agent, an agentset, etc.—back to the procedure that called it.) The reporter here is just `[elevation]`, which reports the elevation of each neighbor patch. It is very important to understand that you can get the value of an agent's variables (e.g., the elevation of a neighbor patch) by treating them as reporters in this way.

You also need to understand the `of` reporter. Here, in combination with `max-one-of`, it reports the elevation of the neighbor patch that has highest elevation. Therefore, the `if` statement's condition is "if the elevation of my current patch is greater than or equal to the elevation of the highest surrounding patch."

Compound expressions like this may look somewhat confusing at the beginning, which is not surprising: this one-line statement uses five primitives. But you will quickly learn to understand and write such expressions and to appreciate how easy they make many programming tasks.

Now we need to calculate and record the corridor width of the butterfly population. How? Remembering our definition of corridor width (figure 5.1), we need (a) the number of patches visited and (b) the mean distance between the butterflies' starting and stopping patches. We can count the number of patches used by butterflies by giving each patch a variable that keeps track of whether a turtle has ever been there. The turtles, when they stop moving, can calculate the distance from their starting patch if they have a variable that records their starting patch. Thus, we are going to introduce one new state variable for both patches and turtles. These variables are not used by the patches and turtles themselves but by us to calculate corridor width and analyze the model.

▪ Add a new variable to the `patches-own` statement: `used?`. This is a *boolean* (true-false) variable, so we follow the NetLogo convention of ending the variable name with a question mark. This variable will be true if any butterfly has landed in the patch.

▪ Add a variable called `start-patch` to the `turtles-own` statement.

▪ In the `setup` procedure, add statements to initialize these new variables. At the end of the statement that initializes the patches (`ask patches [ . . . ]`), add `set used? false`. At the end of the turtles' initialization in the `crt` statement, set `start-patch` to the patch on which the turtle is located.

The primitive `patch-here` reports the patch that a turtle is current on. When we assign this patch to the turtle's variable `start-patch`, the turtle's initial patch becomes one of its state variables. (State variables can be not just numbers but also agents and lists and agentsets.)

Programming note: Initializing variables

In NetLogo, new variables have a value of zero until assigned another value by the program. So: *if* it is OK for a variable to start with a value of zero, it is not essential to initialize it—*except* that it is very important to form the habit of always assigning an initial value to all variables. The main reason is to avoid the errors that result from forgetting to initialize variables (like global parameters) that need nonzero values; such errors are easy to make but can be very hard to notice and find.

Now we need the patches to keep track of whether or not a butterfly has ever landed on them. We can let the butterflies do this:

▪ Add a statement to the `move` procedure that causes the butterfly, once it has moved to a new patch, to set the patch's variable `used?` to "true." (Remember that turtles can change their patch's variables just as if they were the turtle's variables.)

Finally, we can calculate the corridor width at the end of the simulation.

■ In the `go` procedure's statement that stops the program after 1000 ticks, insert a statement that executes once before execution stops. This statement uses the very important primitive `let` to create a local variable `final-corridor-width` and give it the value produced by a new reporter `corridor-width`.

■ Write the skeleton of a reporter procedure `corridor-width`, that reports the mean path width of the turtles. Look up the keyword `to-report` in the NetLogo Dictionary and read about reporters in the "Procedures" section of the Programming Guide.

In the new `corridor-width` procedure:

■ Create a new local variable and set it to the number of patches that have been visited at least once. (Hint: use the primitive `count`.)

■ Also create a new local variable that is the mean, over all turtles, of the distance from the turtle's current patch and its starting patch. (Look at the primitives `mean` and `distance`.)

■ From the above two new local variables, calculate corridor width and report its value as the result of the procedure.

■ In the `go` procedure, after setting the value of `final-corridor-width` by calling `corridor-width`, print its value in an output window that you add to the Interface. Do not only print the value of `final-corridor-width` itself, but also the text "Corridor width: ". (See the input/output and string primitives in the NetLogo Dictionary.)

If you did everything right, you should obtain a corridor width of about 25 when q is 0.4. The output window should now look like figure 5.2.

Figure 5.2
An output window displaying corridor width.

Programming note: Troubleshooting tips

Now you are writing code largely on your own so, as a beginner, you will get stuck sometimes. Getting stuck is OK because good strategies for troubleshooting are some of the most important things for you to learn now. Important approaches are:

▪ Proceeding slowly by adding only little pieces of code and testing before you proceed;
▪ Frequently consulting the User Manual;
▪ Looking up code in the Models Library and Code Examples;
▪ Using Agent Monitors and temporarily putting "show" statements (use the `show` primitive to output the current value of one or several variables) in your program to see what's going on step by step;
▪ Looking for discussions of your problem in the NetLogo User's Forum; and

- Using both logic ("It should work this way because . . .") and heuristics ("Let me try this and see what happens . . .").

And be aware that the next chapter is dedicated entirely to finding, fixing, and avoiding mistakes.

5.3 Analyzing the Model

Now we can analyze the Butterfly model to see how the corridor width output is affected by the parameter q. If you use the slider to vary q over a wide range, write down the resulting corridor widths, and then plot corridor width versus q, you should get a graph similar to figure 5.3. (NetLogo has a tool called BehaviorSpace that helps with this kind of experiment; you will learn to use it in chapter 8.)

These results are not very surprising because they mainly show that with less randomness in the butterflies' movement decisions, movement is straighter and therefore corridor width smaller. It seems likely that a less-artificial landscape would produce more interesting results, but our simplified landscape allowed us to test whether the observation output behaves as we expected.

There is one little surprise in the results: corridor width is well below 1.0 (about 0.76) when q is 1.0. How is it possible that the number of patches visited is 24% less than the distance between starting and ending patches? Do the butterflies escape the confines of Euclidean geometry and find a path that's shorter than the straight-line distance? Is there something wrong with the `distance` primitive? We leave finding an explanation to the exercises.

5.4 Time-Series Results: Adding Plots and File Output

Now let's try something we often need to do: examine results over time as the simulation proceeds instead of only at the end of the model run. Plots are extremely useful for observing results as a model executes. However, we still need results written down so we can analyze

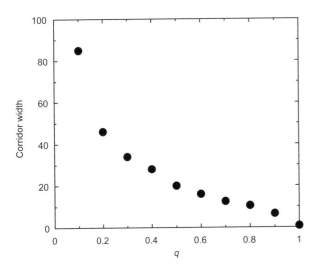

Figure 5.3
Corridor width of the Butterfly model, using the topography and start patch in figure 5.1, as a function of the parameter q. Points are the results of one simulation for each value of q.

them, and we certainly do not want to write down results for every time step from a plot or the output window. Instead, we need NetLogo to write results out in a file that we can analyze.

But before we can add any of these time-series outputs, we have to change the Butterfly model program so it produces results every time step. Currently we calculate corridor width only once, at the end of a simulation; now we need to change the model's schedule slightly so that butterflies calculate their path width each time step. The go procedure can look like this:

```
to go
  ask turtles [move]
  plot corridor-width
  tick
  if ticks >= 1000 [stop]
end
```

The new plot statement does two things: it calls the corridor-width reporter to get the current value of the corridor width, and then it sends that value to be the next point on a plot on the Interface tab. (Starting with version 5.0, NetLogo allows you to write the code to update plots directly in the plot object on the Interface. We prefer you keep the plotting statements in your code, where it is easier to update and less likely to be forgotten. In the dialog that opens when you add a plot to the Interface, there may be a window labeled "Update commands" that contains text such as plot count turtles. Just erase that text.)

- Change the go procedure as illustrated above.

- Read the Programming Guide section on "Plotting," and add a plot to the Interface. Name the plot "Corridor width." Erase any text in the "Update commands" part of the plot dialog.

- Test-run the model with several values of *q*.

Looking at this plot gives us an idea of how corridor width changes with *q* and over time, but to do any real analysis we need these results written out to a file. One way to get file output (we will look at other ways in chapter 9) is with the primitive export-plot, which writes out all the points in a plot at once to a file.

- Read the NetLogo Dictionary entries for export-plot and word, and then modify the statement that stops the program at 1000 ticks so it first writes the plot's contents to a file. The name of the file will include the value of q, so you will get different files each time you change q. Use this statement:

```
export-plot "Corridor width"
        word "Corridor-output-for-q-" q
```

Now, with several model runs and a little spreadsheet work, you should be able to do analyses such as figure 5.4, which shows how corridor width changes over time for several values of *q*.

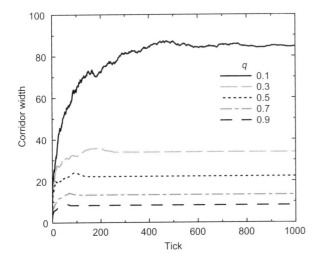

Figure 5.4
How corridor width changes during simulations, for five values of q. Each line is output from one model run.

Programming note: CSV files

When NetLogo "exports" a file via `export-plot` and other `export-`primitives, it writes them in a format called *comma-separated values* or *.csv*. This file format is commonly used to store and transfer data, and .csv files are readily opened by common data software such as spreadsheet applications. If you are not familiar with .csv format, see, for instance, the Wikipedia article on comma-separated values, or just open a .csv file in a word processor or text editor to see what it looks like. The format is very simple: each line in the file corresponds to a data record (e.g., a spreadsheet row); the values or fields are separated by commas, and values can, optionally, be enclosed in double quotes. In chapter 6 you will learn to write your own output files in .csv format.

The .csv files produced by `export-plot` and other primitives can cause confusion on computers that use commas as the decimal marker and semicolons as the field delimiter in .csv files (as many European ones do). The .csv files produced by NetLogo always use commas to delimit fields, and periods as the decimal marker. If you have this problem, you can solve it by editing the output to replace "," with ";" and then replace "." with ",". Or you can temporarily change your computer to use English number formats while you import your NetLogo output file to a spreadsheet or other analysis software.

5.5 A Real Landscape

Now we will see what happens to butterfly behavior and virtual corridors when we use a real, complex landscape. Importing topographies of real landscapes (or any other spatial data) into a NetLogo program is straightforward—once the data are prepared for NetLogo. On this book's web site, we provide a plain text file that you can import to NetLogo and use as the landscape. This file (called ElevationData.txt) is from one of the sites where Guy Pe'er carried out field studies and model analyses of butterfly movements (Pe'er 2003; Pe'er et al. 2004, 2005). The file contains the mean elevation of 25-meter patches.

Programming note: A simple way to import spatial data

Here is a way to import grid-based spatial data as patch variables. By "grid-based," we mean data records that include an x-coordinate, a y-coordinate, and a variable value (which could be elevation or anything else that varies over space) for points or cells evenly spaced in both x and y directions as NetLogo patches are.

The challenging part is transforming the data so they conform to the requirements of NetLogo's World. Real data are typically in coordinate systems (e.g., UTM) that cannot be used directly by NetLogo because NetLogo requires that patch coordinates be integers and that the World includes the location 0,0. The transformations could be made via NetLogo code as the data are read in, but it is safest to do it in software such as a spreadsheet that lets you see and test the data before they go into NetLogo. These steps will prepare an input file that provides the value of a spatial variable to each patch. (It is easily modified to read in several patch variables at once.)

The spatial input must include one data line for each grid point or cell in a rectangular space. The data line should include the x-coordinate, the y-coordinate, and the value of the spatial variable for that location.

NetLogo requires the point 0,0 to be in the World. It is easiest to put 0,0 at the lower left corner by identifying the minimum x- and y-coordinates in the original data, and then subtracting their values from the x- and y-coordinates of all data points.

NetLogo patches are one distance unit apart, so the data must be transformed so points are exactly 1 unit apart. Divide the x- and y-coordinates of each point by the spatial resolution (grid size, the distance between points), in the same units that the data are in. For example, if the points represent squares 5 meters on a side and the coordinates are in meters, then divide the x- and y-coordinates by 5. As a result, all coordinates should now be integers, and grid coordinates should start at 0 and increase by 1. Save the data as a plain text file.

In NetLogo's Model Settings window, set the dimensions to match those of your data set (or, in the `setup` procedure, use the primitive `resize-world`). Turn world wrapping off (at least temporarily) so that NetLogo will tell you if one of the points in the input file is outside the World's extent.

Place the spatial data file (here, ElevationData.txt) in the same directory as your NetLogo program. (See the primitive `user-file` for an alternative.)

Use code like this (which assumes you are reading in elevation data) in your `setup` procedure:

```
file-open "ElevationData.txt"
while [not file-at-end?]
[
  let next-X file-read
  let next-Y file-read
  let next-elevation file-read
  ask patch next-X next-Y [set elevation next-elevation]
]
file-close
```

The `ask patch` statement simply tells the patch corresponding to the location just read from the file (see primitive `patch`) to set its elevation to the elevation value just read from the file.

(If you expect to use spatial data extensively, you should eventually become familiar with NetLogo's GIS extension; see section 24.5.)

▪ Create a new version of your model and change its `setup` procedure so it reads in elevations from the file ElevationData.txt. (Why should you *not* place the new code to read the file and set patch elevations from it inside the old `ask patches []` statement?)

▪ Look at the elevation data file to determine what dimensions your World should have.

▪ Change the statement that shades the patches by their elevation so the color is scaled between the minimum and maximum elevations. (See the primitives `max` and `min`.)

   When we try the program now with the real landscape, we realize that the rule causing butterflies to stop moving when they reach a hilltop was perhaps good for quantifying corridor width in an artificial landscape, but in this real landscape it makes butterflies quickly get stuck on local hilltops. It is more interesting now to use the original model of Pe'er et al. (2005) and let the individuals keep moving for 1000 ticks even after they reach a hilltop.

▪ Comment out the statement causing butterflies to stop if they are at a local hilltop.

   We also might want to let our butterflies start not from a single patch, but a small region.

▪ In the `crt` block, assign `xcor` and `ycor` random values that are within a small square of 10 by 10 patches, located somewhere you choose.

   The model world should now look like figure 5.5, in which 50 individuals start in a small square and then move for 1000 time steps with *q* set to 0.4.

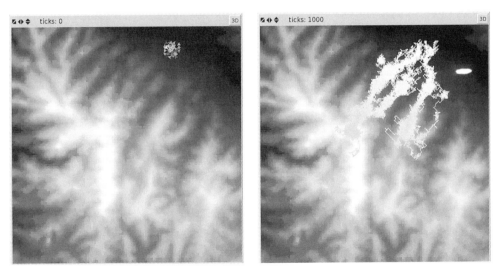

Figure 5.5
Topography of a real landscape showing (left) butterfly starting locations (initial x ranges 100–109; y is 130–139) and (right) virtual corridors that emerged from movement behavior and topography with *q* = 0.4.

The title of this chapter promises to take you from using NetLogo only for simple animations to doing agent-based science, but of course we made only a few first steps: modeling a system of multiple agents, adding quantitative observations that we can analyze, starting to use simulation experiments, and using real spatial data. We will, though, continue to build such skills for doing science with ABMs for the rest of the book.

The Butterfly model is extremely simple, but it already requires some difficult decisions—there are, for example, several different ways you could define path widths and corridor sizes, which likely would produce different results. It is thus important that you learn how to analyze simple models before you turn to more complex ones.

One reason the Butterfly model is good to start with is that people in all disciplines can understand it, even though it has been used in serious ecological research (Pe'er et al. 2004, 2005, 2006). But remember that ABMs and NetLogo are not restricted to organisms moving through landscapes. Almost exactly the same techniques and code can be used to model very different kinds of agents (people, organizations, ideas . . .) moving through many kinds of "space" (political or economic landscapes, social networks, etc.), and NetLogo can model agents that interact without moving at all.

5.7 Exercises

1. Implement the new versions of the Butterfly model that we introduce in this chapter.

2. The traces left by turtles in their "pen-down" mode are useful for seeing their path, but make it difficult to see the turtles' final location. Make a new button on the Interface of one of your butterfly models that erases these traces. (Hint: for moving and drawing, there is very often a primitive that does exactly what you want.)

3. Does your version of the model from section 5.2 produce the result that corridor width is less than 1.0 when q is 1.0? How can this be? (Hint: there are two reasons. Look at how the butterflies move, think about how space is represented, and make sure you understand how the primitive `distance` works.)

4. Try running the Butterfly model with the butterflies all starting in one corner of the space. Remember that we have world wrapping turned off, so the model would stop and report an error if a turtle tried to cross the edge of the World. Why do the butterflies not cross the edge and out of the World, causing an error, no matter how low a value of q you use?

5. Using the model version with a real landscape, vary the parameter q and see how it affects the interaction between landscape and butterfly movement. Start the butterflies in various places. From just looking at the View, what value of q seems best for finding mates on or near hilltops?

6. Analyze the question in exercise 5 in a more rigorous way. Have 50 butterflies start within an area of 10×10 patches near the edge of the landscape and move for 1000 ticks. Develop a measure of how closely clumped the butterflies are at the end of the run, to serve as an indicator of how likely they are to find a mate. One example is the average, over all butterflies, of the number of other butterflies within some radius. Another example is the number of patches containing butterflies (the lower this number, the more clumped the

butterflies). Conduct simulation experiments to see how this indicator of mating success varies with q.

7. Continuing from exercise 6, repeat the experiment with the butterflies starting in different parts of the landscape. Do the results depend on the local topography?

8. Try some alternative models of how butterflies move. For example, butterflies might not choose the highest neighbor patch, but instead randomly select among all the neighbor patches that are higher. What effect do the different traits have in the two landscapes (e.g., on the relation between q and corridor width)?

9. With your version of the Butterfly model, can you perform all the analyses that Pe'er et al. (2006) did with their version?

10. On this book's web site is another topographic input file for a rectangular area. Download it, make the necessary transformations, and use it in the Butterfly model.

11. Another ABM as simple as the hilltopping Butterfly model is that of Jovani and Grimm (2008). Obtain this publication (it is available through this book's web site), read its ODD model description (a version of which is also in section 23.4.1 of this book), and program the model in NetLogo. (The publication omits the value of parameter SD, which should be 1.0.) See if you can reproduce the general results reported in this paper.

Testing Your Program

6

6.1 Introduction and Objectives

In this chapter we discuss why and how you should search for the mistakes in your NetLogo programs and then document that they have been found and fixed. In your programming practice so far, you have no doubt already had some experience with *debugging*: figuring out and fixing the cause of obvious mistakes. Many of the techniques we present in this chapter are useful for debugging, but now the focus is more on rigorous and systematic efforts, *after* debugging is finished, to find the hidden errors that remain.

This practice is often called *software verification*: verifying that your software accurately implements your model formulation. To verify software, we need to do two things in addition to testing the program. We cannot verify that a program implements a model accurately unless we have described the model independently: one of the reasons we need a thorough written description of a model is so we have something to verify the software against. And, because to "verify" means to provide convincing evidence, we must document our tests in a way that convinces our clients (our thesis advisor, journal reviewers, decision makers who might use our results) that we made a sufficient effort to find and fix mistakes.

Why are we covering software testing here in the very first part of this book? The foremost reason is to emphasize the importance of software testing as part of the modeling cycle. Software testing is too often neglected by scientists who teach themselves programming. Models and conclusions from them are often published with little or no evidence that the models' software was free of serious mistakes. In contrast, computer scientists are taught that no software should be considered ready to use until it has been tested thoroughly: programming is not finished until testing is done. One disadvantage of NetLogo is that programs can be so easy to write that users too easily convince themselves that there could not possibly be any mistakes. However, as you develop modeling and programming experience, you will realize that software mistakes are very common, and many will go unnoticed if you do not look for them. Therefore, we introduce testing here so you can practice it throughout the rest of the course and make it a habit.

The second reason we introduce software testing early is that it makes modeling more efficient. Inexperienced programmers, if they do any software verification at all, often start testing only after writing a complete program. This practice makes mistakes even harder to identify and isolate (because they could be anywhere in the code), and is inefficient: you can spend many hours writing code that turns out to be unusable because of a mistake made at the beginning. Instead, software professionals treat testing as a *pervasive* part of programming: testing occurs continually as code is developed, and all parts of the software should be tested before the model is put to use. The sooner you find your mistakes, the more rapidly you will finish the entire modeling project.

Software testing is also important because it is impossible to separate from model analysis, especially for ABMs. As we build and start to use a model that produces complex results (as ABMs inevitably do), we look at the results and analyze them to see what we can learn from the model. When our model produces an interesting and unexpected result, we should immediately ask: Is this a novel and important result, or the consequence of a questionable model design decision, or just the result of a programming mistake? To answer this question we must do the detective work to figure out why the result arose, and the first step in this research is software verification: making sure the program is in fact doing what we want it to. Only after we eliminate the possibility of a programming mistake can we treat novel outcomes of a model as interesting results to analyze.

Software testing must of course be a trade-off: if we spend too little effort on it, then mistakes are likely to go undetected or to be found so late that they are very expensive—requiring us to repeat analyses, withdraw conclusions, apologize for producing misleading results, and so on. But it is virtually impossible to guarantee that there are no mistakes at all, and software testing takes time and effort. This chapter is about how to make a good trade-off that provides reasonable assurance that there are no important mistakes, with a reasonable effort. The most important thing we need is the attitude that programming mistakes are a common and inevitable part of modeling, nothing to be embarrassed about (as long as we find them), but not always obvious and easy to detect. The productive modeler simply assumes that software mistakes are inevitable and continually searches for them.

Learning objectives for chapter 6 are to understand:

- Seven common kinds of software errors;
- Ten important techniques for finding and fixing software errors, including writing intermediate model results to output files for analysis in other software;
- Why and how to document software tests.

6.2 Common Kinds of Errors

Here are some common kinds of mistakes you are likely to encounter with ABMs and NetLogo.

6.2.1 Typographical Errors

These errors result from simply typing the wrong text into your NetLogo code. Simple typographical mistakes (ask turtlse instead of ask turtles) will be found by NetLogo's syntax checker. However, it is easy to use the name of a wrong variable, especially when copying and pasting code, and this will not be caught by the syntax checker. For example, you might want to initialize turtle locations according to a normal distribution, and use this for the x-coordinate:

```
set xcor random-normal meanxcor stddev
```

Then you might copy and paste this to also specify the y-coordinate, and change xcor to ycor:

```
set ycor random-normal meanxcor stddev
```

What you forgot, however, is to also change `meanxcor` to `meanycor`, which can be important if these variables have different values. This error *might* be easy to detect from the View, but others will not, and NetLogo will not find these kinds of mistakes for you!

6.2.2 Syntax Errors

Syntax errors include forgetting to use brackets when required, not leaving spaces between numbers and operators in mathematical expressions (e.g., `set xcor max-pxcor/2` instead of `set xcor max-pxcor / 2`). NetLogo's syntax checker is quite good at finding these errors, but you cannot be sure that it will find all syntax errors.

6.2.3 Misunderstanding Primitives

It is very easy to introduce errors by using primitives that do not do exactly what you think they do. One example of such an error is not understanding that patch coordinates are integers while turtle coordinates are floating point numbers. (We previously referred to turtle coordinates as "continuous" variables, meaning they can have fractional, not just integer, values. In the computer, continuous variables are represented as "floating point" numbers.) Because of this difference, the following two statements can produce different results even though at first they look like they both create a new agentset `neighbor-turtles` containing the turtles within a distance of two:

```
let neighbor-turtles turtles in-radius 2
let neighbor-turtles turtles-on patches in-radius 2
```

Not only can these two statements produce different results, but each can produce different results depending on whether they are used in a turtle or a patch context (figure 6.1). The

Figure 6.1
The agentset produced by the statement `turtles-on patches in-radius 2` is quite different if the statement is executed by the center patch (left panel: all turtles in the white patches are in the agentset) instead of executed by a turtle (the large black one) in the same center patch (right panel: again, all turtles in white patches are in the agentset).

reporter `turtles in-radius 2` reports all the turtles in a circle with a radius of two from the agent executing this statement. If that agent is a patch, then the center of this circle is the center of the patch. If the executing object is a turtle, the center of the circle is that turtle's exact location, which may not be the center of its patch. On the other hand, `turtles-on patches in-radius 2` reports all the turtles that are anywhere on all the patches whose center is within two units of the agent executing the statement.

Another common mistake is not understanding a subtle difference between the `in-radius` and `neighbors` primitives. You might think that a patch would obtain the same agentset of nearby patches if it used the primitive `neighbors` or the statement `patches in-radius 1.5`, but there is a very important difference. (See it yourself by opening an Agent Monitor to a patch and entering this statement in its command line: `ask neighbors [set pcolor red]`. Then do it again using `ask patches in-radius 1.5 [set pcolor blue]`.) Another misunderstood primitive appears later in this chapter.

6.2.4 Wrong Display Settings

In NetLogo, some of the model's key characteristics—the size of the space, and whether the space "wraps" in either the horizontal or vertical dimension—are not written in the program statements but set via the World settings on the Interface tab. Forgetting to check or change these settings can induce errors that will not be found by focusing on the procedures. Starting with version 4.1, NetLogo lets you set the World size in your code via the primitive `resize-world`; using this primitive in your `setup` procedure could make it less likely that you forget to change display settings.

6.2.5 Run-time Errors

These are an important category of software errors that can be very difficult to deal with, but fortunately NetLogo is much more helpful than many programming languages. Run-time errors occur when there is no mistake in the program's syntax or logic except that sometimes when the program runs it does something that the computer cannot deal with. Examples include:

- Dividing by a variable when the variable has a value of zero;
- Trying to put an object at a location outside the World's dimensions (e.g., `setxy 50 20` when `max-pxcor` is 49 and world wrapping is turned off);
- Asking an agent to execute a procedure after that agent no longer exists (e.g., after it has executed the `die` primitive);
- Raising a number to a power so large that the result is larger than the computer can store; and
- Trying to open a file that was not closed during a previous model run (explained in section 9.4).

In some programming languages, run-time errors simply cause the code to stop running (with or without issuing an error statement, which may or may not be helpful) or, worse, to keep running after giving some variables completely wrong values. Luckily, NetLogo is quite good at stopping and issuing a helpful explanation when run-time errors occur.

But NetLogo is not completely perfect at catching run-time errors, so you still need to be aware of, and sometimes test for, run-time errors. Recent versions do not catch (for example) these run-time errors:

- Raising a negative number to a non-integer power, which produces a value of "NaN," which means "not a number." Watch carefully for "NaN" in Agent Monitors and output files. (Starting with version 5.0, NetLogo does catch this error.)
- Using a negative standard deviation for a normal distribution, which makes no sense (standard deviations are by definition positive). (Version 5.0 also catches this error.)
- Using a negative parameter for the Poisson distribution, which also makes no sense. The primitive `random-poisson` (explained in chapter 15) appears to always produces zero when its parameter is negative.

(Techniques for catching and dealing with such run-time errors are discussed in chapter 15.)

6.2.6 Logic Errors

These are typically the kinds of error most likely to cause real problems. They occur when the program is written in a way that executes and produces results, but the results are wrong because you made a mistake in the program's logic—you forgot to initialize a variable, programmed an equation wrong, used an incorrect condition in an `if` statement, and so on. Sometimes it is easy to tell that there is a logic mistake because it produces obviously wrong results, but often logic errors are discovered only by careful testing.

6.2.7 Formulation Errors

These are of course not programming errors but errors in the model's formulation that become apparent only after its software is written and executed. In building almost every model, some assumptions, algorithms, or parameter values will have unexpected consequences and require revision after the model is implemented and tested. When these errors are discovered and corrected, then of course the model's written documentation needs to be updated.

6.3 Techniques for Debugging and Testing NetLogo Programs

Software verification is a major issue in computer science, so there is a very extensive literature on it. Here, we discuss testing techniques that provide a reasonable compromise between effort and reliability for scientific models implemented in NetLogo.

When should these debugging and testing methods be used? Some of these are more useful in the debugging phase: checking for, diagnosing, and fixing mistakes as the program is written. In fact, all of the methods we present here are useful for diagnosing mistakes as a program is written. But some of the techniques, especially code reviews and independent implementation, should always be reapplied to models after they are completely coded and before they are used to support research and publication, decision-making, or any other application with consequences. These tests of the finished code are to find any remaining hidden mistakes and to provide evidence (as we discuss in section 6.4) that the software is ready to use.

6.3.1 Syntax Checking

NetLogo syntax is very easy to check: you only need to click on the Check button in the Procedures tab and NetLogo will tell you if it finds any mistakes. While this syntax checker is extremely useful, it does sometimes issue error statements that are unclear and point to places in the code far away from the actual mistake. For example, if you have a procedure that is in the observer context, but insert in it a statement that uses a turtle- or patch-context primitive, the syntax checker may issue a context error statement in a completely different part of the code. The key to efficiency is to click the Check button very often, so mistakes are found quickly and one at a time, and are therefore easy to isolate and fix. Make a habit of using the syntax checker after writing every line or two of program.

As you stepwise write your code, do it so the code remains free of syntax errors even when incomplete. For example, if you are writing a turtle procedure that must call a patch procedure called `update-resources`, the syntax checker will tell you there is an error if you have not yet written this `update-resources` procedure. However, you can simply add a skeleton procedure:

```
to update-resources
end
```

and the syntax checker will no longer consider it an error. Likewise, if you are writing a complicated `ifelse` statement, you can write and check the "if" part before completing the "then" and "else" parts. For example, the syntax checker can be used between each of the following four stages of writing this statement (which is from the model we use at the end of this chapter as a code testing exercise):

(1)

```
ifelse (xcor >= min-marriage-age)
  [ ]
  [ ]
```

(2)

```
ifelse (xcor >= min-marriage-age) and
    (random-float 1.0 < 0.1)
  [ ]
  [ ]
```

(3)

```
ifelse (xcor >= min-marriage-age) and
    (random-float 1.0 < 0.1)
  [set married? true]
  [ ]
```

(4)

```
ifelse (xcor >= min-marriage-age) and
    (random-float 1.0 < 0.1)
  [set married? true]
  [set married? false]
```

We used this technique already in chapter 2 and, although it seems trivial, it is extremely helpful in making programming efficient.

6.3.2 Visual Testing

Visual testing means taking a look at the model's results on NetLogo's Interface and seeing if anything is unexpected. Are there obvious problems (e.g., all the turtles are created in one patch when they are supposed to be initialized randomly over the space)? Do strange patterns appear after the model runs for a while? Visual testing is extremely powerful and important for ABMs: while building almost every model, there will be errors that are readily visible from the display while being unlikely to be detected soon, if ever, via other methods.

NetLogo's built-in World display can make visual testing quite easy—*if* you designed the software to use this display effectively. Characteristics of model agents and their environment can be displayed by setting their color, labels, and (for turtles) size and shape. Design your software so the display shows you the information that is most valuable for spotting problems and understanding the model's behavior. We will explore use of the display further in chapter 9, but here are some tricks for visual testing:

- You can start using the World display to test your software long before the software is complete. As with using the syntax checker, you should use visual testing early and often as the model procedures are written. For example, as soon as the `setup` procedure has been programmed to initialize patch variables (e.g., if patches are given initial values of some characteristic such as elevation or resource productivity) you can shade the patch colors by one of those variables, add the `setup` button to the Interface tab, set up the World, and check the display of patch characteristics. Then, as you continue writing the `setup` procedure to initialize turtles you can flip back to the Interface tab to look for mistakes in their location, size, etc.
- The color primitive `scale-color` makes it very easy to shade the color of patches or turtles by the value of one of their variables.
- Turtles and patches have built-in variables `label` and `plabel`. You can use these to attach a label to all, or some, agents. For example, if turtles have the state variable `energy-reserves`, you can follow how this variable changes with the simple statement (which would go after the statement updating the value of `energy-reserves`):

```
set label energy-reserves
```

(You can also limit the number of digits displayed in the label with the primitive `precision`, and set the label font size in the Model Settings window.)
- It often is useful to study a much smaller World and fewer turtles while testing. Just reduce the size of the World and increase the patch size in the Model Settings window or create fewer turtles.
- Often it is too hard to understand what is going on when a model runs at full speed. You can slow execution down via the speed control on the Interface tab, but commonly you will want to follow program execution tick by tick, or even procedure by procedure. A `step` button that executes the program's schedule only once is useful for every model. Simply add a new button to the Interface, tell it to execute the command `go`, but enter "step" in the "Display name" window and leave the "Forever" box unchecked. You can also add buttons that each execute just one of the procedures in your `go` procedure.

6.3.3 Print Statements

Perhaps the most tried-and-true method for isolating and diagnosing mistakes is a very simple one: inserting statements that write information out to the display or to a file so you can see what is going on. NetLogo has a variety of input/output primitives useful for this: look up `export-output`, `output`, `print`, `show`, `type`, and `write`.

One common use of print statements is to determine where in a program some problem occurs. For example, if you wrote a model that seemed to take forever to finish one step, you could isolate the procedure that executes so slowly by putting in statements like `show "Starting procedure X"` and `show "Ending procedure X"` at the start and end of each procedure, where "X" is the name of the procedure. Then when you run the model, you will see which procedure starts but never ends.

Another common use of print statements is to output the value of key variables at different times to help diagnose why a model behaves unexpectedly. Simply write out the name and value of any variables that you need to see to understand what is going on (make sure you understand the primitive `word`).

(Many other software platforms provide integrated debuggers: tools that let you see your program's variables while executing it one step at a time—which is much more efficient than print statements. Unfortunately, NetLogo does not.)

6.3.4 Spot Tests with Agent Monitors

You can right-click your mouse on the View and open an Agent Monitor to any patch, and to the turtles and links on the patch. (If you are not familiar with them, you should pause now and read the section on "Agent Monitors" in the User Manual's Interface Guide.) These Agent Monitor windows show the values of the variables of the selected objects, so they can be used to "spot test" calculations. This is typically done by manually recording the value of the variables, calculating by hand how they should change, and then stepping the model through one iteration of its schedule and seeing if the change reported by the Agent Monitor matches your expectation.

Agent Monitors can be very useful for quickly seeing agent state variable values and testing key calculations for obvious errors. However, these spot checks are not sufficient by themselves for testing a model because it is practical to look at only a very small number of calculations. The "independent reimplementation" method (section 6.3.10) is more thorough and often redundant with spot tests.

6.3.5 Stress Tests

"Stress" or "extremes" testing means running a program with parameters and input data outside the normal ranges. Errors that are hidden under normal conditions often expose themselves during stress tests. One reason that stress tests work is that we can make very clear predictions of how the model should behave under extreme conditions: if we set q in the butterfly hilltopping model to 1.0 and the turtles still wander about, then we know something is wrong with how they move. In a real example, a fish model was run under conditions that should have caused all small fish to die rapidly (high densities of large fish, which eat the small ones). But some small fish survived, indicating a problem. (Their length had been set to NaN by a runtime error mentioned in section 6.2.5.)

Stress tests can use extreme parameter values, which should make agents (for example) make immense profits or go out of business instantly, or reproduce rapidly or not at all; and use environmental conditions that are outside the normal range, or unrealistically variable, or unrealistically constant. Examine results at both the individual level (at a minimum, via Agent Monitors) and system level, and investigate anything that seems unexpected.

6.3.6 Test Procedures

A test procedure is a new procedure that you write just to produce intermediate output (either graphical output on the View or file output) used only for testing. For example, in the exercise in section 6.5 we provide code that has a test procedure called `tag-network`. In that model, agents interact with other agents within their "social network," a neighborhood in the two-dimensional world. The test procedure simply displays (a) the turtle that executed the procedure (by setting its label to the word "me") and (b) the other turtles within its social network (by setting their label to show whether they are male or female). The model user can open an Agent Monitor to a turtle, execute the test procedure via the Agent Monitor's command window, and see which turtles are in the network and their sex. Test procedures may require adding new state variables that are only for observing what is going on. You can create extra buttons to call these procedures, or call them simply by entering their name in the Command Center or in the command window of an Agent Monitor.

6.3.7 Test Programs

It can be hard to test a certain primitive, procedure, or algorithm within a large program. In such cases we recommend writing a separate short program that serves only to test the particular programming idea. For example, you might want a turtle to "remember" the sequence of patches it has visited (so it can, for example, move back to where it was in the past). You can model the turtles' memory as a NetLogo list: define a turtle state variable `path` and make it a list. Each time a turtle moves, it adds its current patch to this list. But the syntax for using lists can be confusing. Thus, it is a good idea to write a test program where one turtle randomly turns and moves through the landscape for, say, 100 ticks, and stores every visited patch in its `path` variable:

```
turtles-own [ path ]
```

To test if everything worked right, you can then ask the turtle to move backwards along its path.

A test program for this problem could include this as the `setup` procedure:

```
to setup
  ca
  crt 1
  [
   set color red ; So we can tell initial path from return
   set path (list patch-here) ; initialize the path
                              ; memory list
   pd            ; put pen down

   ; Move randomly and memorize the patches visited.
   ; Each new patch is put at the *start* of the list
   repeat 100
   [
     rt (random 91 - 45)
     fd 1
     set path fput patch-here path
   ]
```

```
      ]
    end
```

Now, a test procedure called `go-back` could include this:

```
ask turtles
[
  set color blue   ; make the turtle's return trace blue
  foreach path     ; see "foreach":
                   ; executes once for each patch on list
  [
    set heading towards ?
    fd 1
  ]
]
```

To test the memory algorithm, you need only type `setup` in the Command Center to create a turtle path, then type `go-back` to see whether the `go-back` procedure succeeds in making the turtle retrace its path.

6.3.8 Code Reviews

Professional programmers often have their code read and reviewed by peers, just as scientists have their research articles reviewed before publication. The reviewer's job is to compare the code to the model formulation carefully and look for such problems as typographical errors, logic errors, parts of the formulation left out of or misrepresented in the program, and important assumptions that are in the software but not documented in the written formulation. Reviews can be especially useful when the programmer suspects there is a mistake but cannot find it; a fresh set of eyes can often find a mistake that the programmer has been looking at for so long that she no longer notices it.

A second job of the reviewer is to make sure the code is well organized and easy to understand. Are the names of variables and procedures descriptive? Are the statements simple and clear? Are there accurate and helpful comments throughout? If not, it is unlikely that the programmer or anyone else will be able to find mistakes or make revisions painlessly.

You should always write your program as if you expect it to be reviewed by someone else. Then when you need help, or the comfort and credibility that come with having your work reviewed, it will be easy to get a useful review. (And if your model is really used for science, then other people very likely will want to look at and use your code. One of the surest signs of a good model is that other people want to use it.) Even if you are the only person who ever works with your code, you will be more efficient if the code is clearly organized, the procedures and variables are named in a way that tells you what they mean, and there are frequent comments.

6.3.9 Statistical Analysis of File Output

One of the more systematic ways to search for problems is to write file output from key parts of a model and then analyze the output to see if it meets your expectations from the model formulation. Do stochastic events happen with the expected frequency? Do variables stay within the range you expect them to? These tests are essentially hypothesis-testing experiments: the modeler predicts what the results should be from the model's formulation, then tests statistically whether the model produces the predicted results. If not, then the modeler must determine whether and how the code is wrong, or whether there was a mistake in how the prediction was made.

We can illustrate this kind of testing on the simple butterfly hilltopping model developed in chapter 4. In this model (see the code at section 4.4), turtles move to the highest neighbor patch with probability equal to the parameter q. We might therefore predict that the frequency with which turtles move to the highest neighbor patch is equal to q. However, this prediction is wrong: when turtles move randomly (with probability $1 - q$) they still have one chance in eight of landing in the highest patch. (We are neglecting the possibility that turtles are at the edge of the World, where they have fewer neighbors.) So, consulting our probability book, we revise our predicted frequency of moving to the highest patch to $q + \frac{1-q}{8}$. (It turns out that even this predicted frequency is wrong when more than one neighbor patch has the same, highest, elevation; at the peak of a conical hill, four neighbor patches all have the same elevation.)

We can test whether our code produces results meeting this prediction by having the turtles write out the elevation of each of their neighbor cells and then the elevation of the cell they moved to. First, using help from the programming note on output files below, we add statements to the `setup` and `go` procedures to open and close an output file (giving it a name something like "TestOutput.csv" because it will be in comma-separated value format). Next, we can add these output statements to the move procedure:

```
; The butterfly move procedure, which is in turtle context
to move
  ; First, write some test output:
  ; elevation of all neighbors
  ask neighbors [file-type (word elevation ",")]

  ; Decide whether to move uphill with probability q
  ifelse random-float 1 < q
    [ uphill elevation ]              ; Move uphill
    [ move-to one-of neighbors ]      ; Otherwise randomly

  ; Write the elevation turtle moved to,
  ; using "print" to get a carriage return.
  file-print elevation

end ; of move procedure
```

Programming note: A very brief introduction to output files

In section 9.4 we describe in depth how to create and use output files. Here is some quick example code for use in this chapter. First, in the `setup` procedure these lines will create an output file and erase any old copies of it, then "open" it so it is ready to write to:

```
if file-exists? "TestOutput.csv"
    [file-delete "TestOutput.csv"]
```

```
file-open "TestOutput.csv"
```

We also need to "close" the file when we are done writing to it. We can do this in code that is executed when the model's 1000 ticks are done: in the go procedure, change the stop statement to this:

```
if ticks >= 1000
   [
      file-close
      stop
   ]
```

Now, we can write to the file any time after setup is executed and before the model stops at 1000 ticks.

Then we change the setup procedure so 10 turtles are created (to generate a higher sample size) and run the model. The resulting file can easily be imported to a spreadsheet and analyzed as in figure 6.2.

6.3.10 Independent Reimplementation of Submodels

The only highly reliable way to verify software for any model is to program the model at least twice, independently, and then test whether these independent implementations produce exactly the same results. However, simply programming your model twice in NetLogo is probably not a good idea because it is too likely that the same logic mistakes would be made in both NetLogo programs, especially if the same person wrote both. But reprogramming a model in a platform other than NetLogo would be an extreme burden: why learn NetLogo if you still have to reprogram your model in some other language? A compromise approach that can work well for ABMs is to test—separately—each of an ABM's key submodels against a version coded in a platform that you already know how to use: a spreadsheet, or perhaps R or MatLab. While it

	A	B	C	D	E	F	G	H	I	J
1	Neighbor elevations								Elevation moved to	Did turtle move to highest neighbor?
2	19.3960	18.1648	18.5998	18.5994	19.3895	18.7842	18.7590	18.1096	18.7590	=IF(I2=MAX(A2:H2),1,0)
3	18.8552	20.1671	19.5205	17.9939	19.3895	18.5994	18.5998	18.1096	20.1671	1
4	20.2679	18.7590	19.5205	19.3895	21.5747	20.7938	20.0000	20.9311	19.5205	0
5	18.7590	18.8552	20.1671	18.1096	19.5862	20.9311	20.2679	19.3895	20.9311	1
6	20.7938	19.5205	21.5747	20.2679	22.3414	21.0000	20.1671	21.6804	21.6804	0
7	21.0000	21.5747	21.7157	22.4138	20.2679	22.3414	23.0928	20.9311	23.0928	1
8	22.3414	22.9815	22.4138	23.8275	23.7512	21.6804	23.1299	24.5049	24.5049	1
9	23.0928	25.9168	24.5442	25.2412	24.3875	25.1605	23.7512	23.8275	25.2412	0
10	25.9168	23.8275	24.5442	25.9584	25.2412	26.6548	24.5049	25.1605	26.6548	1

Figure 6.2
The first few lines of a spreadsheet used to analyze file output from the Butterfly model. Columns A–I were produced by the NetLogo program's file output statements: columns A–H are the elevations of the eight neighbor patches before the turtle moves, and column I is the elevation of the patch that the turtle then moved to. Column J contains a formula that produces a 1 if the turtle's new patch is the highest of its neighbors and a 0 if not (this formula is displayed in cell J2). Now the mean of column J is the fraction of moves that were to the highest neighbor patch, which can be compared to the predicted value of $q + (1 - q)/8$.

would be extremely hard to reprogram the whole model, it is usually quite easy to reprogram its submodels separately.

To test each submodel, add statements to the NetLogo code that write to a file the input to, and results from, the submodel. Then import that file to a spreadsheet or other platform, and reprogram the submodel in the spreadsheet. Compare the results produced by the spreadsheet code to the results produced by your NetLogo code, identify the differences, and figure out why they occurred. Make sure you test many, many cases over a wide variety of conditions: one of the most important advantages of this testing method is that you can easily test your model under the full range of conditions (parameter values, input data, etc.) that could possibly occur in the full model. It is very common for mistakes to occur only under limited conditions (as we will soon see).

To demonstrate independent reimplementation of submodels, we can again use the Butterfly model from the end of chapter 4 and test its movement submodel. This movement method is so simple that we certainly do not expect any mistakes, but it provides a useful example.

We need more output than we did in the previous section because our objective is to reimplement the movement behavior, not just test it statistically: we need to reproduce exactly and check every decision made by each turtle, each tick. These decisions are made by comparing a random number to the parameter q, so to test it we need to know what random number the turtle used. These output statements will work:

```
; The butterfly move procedure, which is in turtle context
to move

  ; First, write some test output:
  ; elevation of all neighbors
  ask neighbors [file-type (word elevation ",")]

  ; Decide whether to move uphill with probability q
  ; Now, save the random number in a local var
  ; so we can write it out
  let rand-num random-float 1

  ifelse rand-num < q
    [ uphill elevation ] ; Move deterministically uphill
    [ move-to one-of neighbors ] ; Otherwise randomly

  ; Now write out the random number and q,
  ; then the elevation moved to,
  ; using "print" to get a carriage return.
  file-type (word rand-num "," q ",")
  file-print elevation

end ; of move procedure
```

Now, the output can be tested in a spreadsheet that also contains the movement algorithm. The algorithm is that if the random number is less than q then the butterfly moves to the neighbor patch with highest elevation. We can program a spreadsheet, using the output file, to implement this decision and compare its results to those of the NetLogo code (figure 6.3).

	A	B	C	D	E	F	G	H	I	J	K	L	M	N
1	Neighbor elevations (cols. A–H)								Random number	q	Elevation moved to	Should move uphill?	Did move uphill?	Did not move uphill when should?
2	18.110	18.599	18.600	18.784	18.759	19.390	19.396	18.165	0.885	0.4	18.759	=IF(I2<J2,1,0)		
3	18.599	18.784	18.600	18.759	19.396	19.390	18.110	18.165	0.542	0.4	19.38954		=IF(K3=MAX(A3:H3),1,0)	
4	19.396	19.390	18.110	18.784	18.600	18.599	18.165	18.759	0.276	0.4	19.39603			=IF(AND(L4=1,M4=0),1,0)
5	18.784	19.390	18.110	18.759	18.600	19.396	18.599	18.165	0.866	0.4	19.38954	0	0	0
6	18.165	18.784	18.599	18.600	18.759	19.390	18.110	19.396	0.523	0.4	18.759	0	0	0
7	19.390	18.759	18.599	19.396	18.110	18.600	18.165	18.784	0.671	0.4	18.78424	0	0	0
8	19.396	19.390	18.110	18.165	18.784	18.759	18.599	18.600	0.584	0.4	19.39603	0	1	0
9	18.599	18.784	18.165	18.110	19.396	19.390	18.600	18.759	0.306	0.4	19.39603	1	1	0
10	18.784	19.390	19.396	18.600	18.759	19.390	18.110	18.165	0.300	0.4	19.39603	1	1	0

Figure 6.3

A spreadsheet reimplementation of the decision whether to move uphill in the Butterfly model. Columns A–K are output produced by the NetLogo code presented in the text. Column L reimplements the decision: it reports a 1 if the random number was less than q so the butterfly should move to the highest neighbor patch. Column M reports whether the butterfly actually did move to the highest neighbor. Column N reports a 1 if there is a discrepancy: if column L is 1 but the butterfly did not move to the highest neighbor patch. Keep in mind that (a) butterflies at the edge of the space would have fewer than eight neighbors, so the spreadsheet would need modification, and (b) column M will sometimes be 1 when column L is 0 because the highest neighbor can be chosen randomly.

	A	B	C	D	E	F	G	H	I	J	K	L	M	N	O
1	Neighbor elevations (cols. A–H)								Random number	q	Elevation moved to	Should move uphill?	Did move uphill?	Did not move uphill when should?	
9993	46.838	49.000	47.764	46.838	47.000	48.586	47.764	48.586	0.194	0.4	49	1	1	0	
9994	47.172	46.394	45.528	47.764	47.000	46.000	45.877	48.000	0.238	0.4	48	1	1	0	
9995	97.764	97.764	100.000	99.000	99.000	98.586	98.586	98.000	0.141	0.4	100	1	1	0	
9996	97.764	96.394	98.000	95.877	95.528	97.172	97.000	96.000	0.604	0.4	97	0	0	0	
9997	48.000	49.000	49.000	48.586	48.586	47.764	47.764	50.000	0.295	0.4	50	1	1	0	
9998	99.000	98.586	99.000	98.586	98.586	99.000	98.586	99.000	0.352	0.4	100	1	0	1	
9999	97.764	98.586	98.000	99.000	98.586	97.764	99.000	100.000	0.793	0.4	98.58579	0	0	0	
10000	99.000	98.586	98.586	98.586	99.000	99.000	98.586	99.000	0.483	0.4	98.58579	0	0	0	
10001	99.000	97.172	96.394	98.000	96.838	97.764	97.000	98.586	0.624	0.4	97.76393	0	0	0	
10002															
10003														Sum:	1021

Figure 6.4

Comprehensive comparison of NetLogo and spreadsheet results for the butterfly movement decision. Column N produces a 1 if the spreadsheet reimplementation conflicts with the NetLogo results. All the values in column N are summed at cell N10003. Because this sum is not zero, we know that there were discrepancies in 1021 butterfly moves.

After programming the spreadsheet to reimplement the decision of whether to move uphill (column L in figure 6.3), we can add columns M–N to identify any discrepancies between the NetLogo results and the spreadsheet results. The function in column N produces a 1 if there is clearly an error: the butterfly should have moved to the highest neighbor patch but did not.

Now we can check whether there were any errors in all the butterfly moves by going to the bottom of the spreadsheet and summing all the values in column N: if there are no errors, this sum should be zero (figure 6.4).

Much to our surprise, this simple test of our very simple program is telling us that there is a big problem! There are 1021 cases in which butterflies were supposed to move to their highest neighbor, yet ended up at a cell where the elevation is not equal to the elevation of their highest (pre-move) neighbor. An example is shown in figure 6.4, at spreadsheet row 9998. Examining this row, we see that the butterfly ended up *higher* than any of its eight neighbors, and in fact the butterfly was at the peak (100 units high) of our simple landscape. This clue leads us back to the NetLogo Dictionary to carefully reread the documentation for the primitive `uphill`. There we find this sentence: "*If no neighboring patch has a higher value than the current patch, the turtle stays put.*" This behavior of `uphill` conflicts with our model formulation (section 3.4), which says butterflies cannot stay put. This difference could cause substantial changes in model results as butterflies approach hilltops.

This test shows that we made one of the mistakes discussed in section 6.2: misunderstanding a NetLogo primitive. Instead of the statement

```
[uphill elevation]
```

our program should have used

```
[move-to max-one-of neighbors [elevation]]
```

(Yes, the authors did make this mistake and find it only via the spreadsheet reimplementation. That is one of many reasons why they insist that you test your programs early, often, and thoroughly.)

6.4 Documentation of Tests

Another habit we need to adopt from software professionals is documenting the tests we conduct. Normally there is no reason to document all but the most important tests and debugging done as a model is programmed, but once a code is ready for scientific use it is important to conduct and document a conclusive set of final tests.

If you think you have found all the mistakes in your program, why do you also need to document how you did it? There are two reasons. First, of course, is so you can give your work credibility by showing its "consumers" that you tested it adequately. Many research fields are still naive about simulation, and the people who review and use model-based research often do not expect documentation of software verification. That means your work might be accepted without such documentation, although it always makes you look experienced and professional when you offer to provide documentation that you tested your software. But this naiveté is actually a reason why documentation is important: simulation is often criticized as unscientific because so many models are untested. If we want our work accepted as valid science, we need to do our part by providing evidence that our models are in fact well tested. (The "TRACE" format of Schmolke et al. 2010 for documenting the modeling cycle prominently includes documentation of software testing.)

The second reason to document software tests is for our own use. Any important model is reused and revised many times, which means that some tests will need to be repeated or updated as the code evolves. A little bit of documentation can make it much easier to repeat software tests.

Luckily, it is quite easy to document software tests adequately. This documentation should describe (1) the kinds of tests that were used, and (2) the methods and results of the comprehensive tests of the "final" model (usually, statistical analyses and independent implementations as described in sections 6.3.9–10).

Describing the kinds of tests used can be as simple as listing:

- Who reviewed the code, at what stage(s) in its development;
- What patterns were observed visually and investigated;

- The parts of the model that were tested statistically and against an independent implementation; and
- If it may be helpful in the future, the kinds of errors found and how they were corrected.

The detailed methods and results of the final independent implementation tests can be documented by saving, and adding comments to, the spreadsheets (or other computer files) containing the tests.

Finally, software testing typically requires adding quite a bit of temporary code to your program: print statements, output to files, test procedures, etc. Part of the documentation process is to leave this temporary code in your program so readers know that you used it for testing, and so you can retest the code easily as changes are made. Simply comment the testing statements out when you are not using them.

6.5 An Example and Exercise: The Marriage Model

Now you have the opportunity to try out software testing on a real NetLogo program. We describe the model, then provide software that implements it but with several errors. (These are real errors that the authors of this book made—and found—while programming the model.) Then you can apply the methods described in this chapter to find as many of the errors as you can, and document your efforts to produce a mistake-free version of the program that is ready to use.

6.5.1 Model Description

The NetLogo program we use is a social science model loosely based on the Wedding Ring model of Billari et al. (2007). (This simplified model and its software were prepared by the authors of this book; Billari et al. have no responsibility for it!) You can obtain the NetLogo file via this book's web site. The model is fully described on the Information tab. Here we only include the ODD description of the model's state variables and scales, and the process overview (which include some details on the submodels).

Purpose

The objective of this model is to test a hypothesis for characteristic patterns in the ages at which people marry. In many modern cultures, few people marry at very young ages; as age increases the rate at which people marry increases sharply. Past a peak age, though, the fraction of single people who marry decreases gradually. This model simulates the hypothesis that this pattern results from individuals basing their decision about when to marry on a "social pressure" that increases nonlinearly with the fraction of social peers who are married. In the model, individuals feel very little pressure to marry when few of their peers (< 30%) are married, but great pressure when many (>70%) are married.

State Variables and Scales

The only entities in this model are agents representing people. People have two state variables (described in the following paragraph) for their location within a social network. People also have variables for their sex and their marriage status (married vs. unmarried).

The people in this model exist in a space, but it is not a geographic space. Instead, the space depicts social networks. The x-axis of the space represents age: a turtle's x-coordinate is the person's age in years. The y-axis represents social status, with people closer in the y dimension being more similar in social variables (interests, educational experience, income, etc.) that

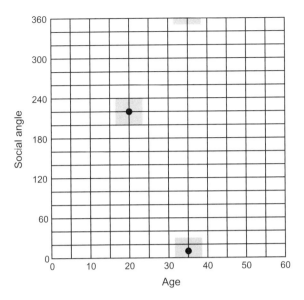

Figure 6.5
Social space in the marriage model. The two black dots represent people, one 20 years old with a social angle of 220°, the other 35 years old with social angle of 10°. The grey rectangles represent their social networks when social network is defined as having an age range of ±3 years and ±20° social angle.

make them more likely to know each other. The y-axis is wrapped and scaled from 0 to 359 so that social status is modeled as a "social angle" on a circle (figure 6.5; this is why the original model was called the Wedding Ring). Therefore, a person's social network—the other people who are close in both age and social status—is her neighborhood on this social space.

The model runs at a one-year time step, and simulations run for 200 years.

Process Overview and Scheduling

The model includes the following actions executed each one-year time step:

- *Aging and death*: The age of all individuals is incremented. Individuals exceeding age 60 die.
- *Childbirth*: Sixteen females produce one child each. These mothers are randomly chosen from among the married females with age less than 40. (If there are fewer than 16 such females, they all give birth.) New children are given an age (x-coordinate) of zero and a social angle (y-coordinate) chosen randomly between their parents' social angles.
- *Marriage*: Marriageable individuals—those still single and age 16 or higher—decide whether to try to marry, which depends on "social pressure." Social pressure is a nonlinear function of the married fraction of their social network. When a person decides to marry, it randomly selects a partner from all those within its social network who are unmarried and of the opposite sex. (The partner does not decide whether it wants to marry.) If there is such a partner, the two marry (their marriage status changes from `false` to `true`).
- *Output*: The distribution of marriage status is depicted by a histogram of the number of married people, by age.

This model is of moderate complexity, whereas so far we have only looked at very simple models. The Marriage model includes several programming concepts that should be familiar from the NetLogo tutorials, though we did not use them in the Mushroom Hunt and Butterfly models. With a bit of investigation in the NetLogo Dictionary and Users Guide, you should be able to understand these new concepts, which will all be explained in more detail in part II of this book. New things in this program include using:

- The primitive `stop` to jump out of a procedure.

- The primitive `myself`; see the NetLogo Dictionary to understand this reporter and how it differs from `self`.
- Mathematical expressions to evaluate equations.
- The logical expression `!=` to check whether two variables are not the same.
- Characters or strings as state variables (`M` for male, `F` for female).
- The primitive `and` to create a compound boolean expression, which is true only if several conditions are all true.
- Subsetting to identify agents with specific characteristics. For example, the program's variable `social-network` is the subset of turtles that are within an agent's social range (with social angle +/- 20) and age range (with age +/- 3 years); `potential-partners` then is a subset of `social-network` that contains only people who are of the opposite sex, at least 16 years old, and not married. (See also the agentset `moms`.)
- Variables that contain an agent, using a name (e.g., `my-spouse`) that makes the code easier to understand. Note that the statement

  ```
  let my-spouse one-of potential-partners
  ```

  creates a local variable `my-spouse`, which is a new name of one turtle that also belongs to the agentset `potential-partners`. Any changes to `my-spouse` affect just this one agent out of the whole population of turtles.
- A test procedure that can be executed from an Agent Monitor (see that procedure's comments).
- A histogram plot.

6.5.2 Testing the Software

The version we provide on this book's web site has several errors that you should find as you try the testing approaches on it. Some errors are obvious but others are not at all. In the exercises to this chapter, your job is to find the errors, fix them, and document the testing process. We suggest you follow these steps, fixing errors as you find them. At each step, document what you did and what you learned.

- Review the code. Read the Information tab carefully and compare the code to it. Look for any differences between the code and the model description.
- Run the model and watch it on the World display. Look for unusual patterns and think about why they might occur if the model is or is not implemented correctly. (In fact, there are strange patterns, some due to programming mistakes and some due to the model's formulation.)
- Use the file output to test the code for how an individual's social pressure is calculated. After running the model (for only a few ticks, to keep the output from being too large), import this file into a spreadsheet (or other software). Find ways to test whether the NetLogo code calculated social pressure correctly.

6.6 Summary and Conclusions

Not all programming errors have important effects on the ultimate use of a model; the mistake we found in the Butterfly model might have little effect on any general conclusions we draw from the model. Unfortunately, the possibility that a programming bug might not affect our conclusions does not make software verification unimportant: we cannot know whether a mistake is critical or not until we find and remove it. And we don't want to confirm the widespread belief that ABMs are unscientific because their software is not tested well enough to make results reproducible. Paying serious attention to software reliability and reproducibility is, therefore, just part of the job when we are doing scientific modeling instead of just playing around with NetLogo.

Your most important strategy for avoiding undetected errors and the delays, costs, and loss of credibility they can bring is having a productive attitude. Instead of hoping unrealistically that your program does not have mistakes, just assume from the start that mistakes are inevitable and that finding them is part of the fun. Test your ingenuity by coming up with new ways to test your code. Challenge yourself and your colleagues to find the most subtle and hidden mistake. Good programmers are known to run down the hall shouting about the devious bug they just found and squashed.

NetLogo provides tools (the syntax checker, the graphical interface, run-time error checking) that make it easy to find and fix many mistakes, but it is surprisingly rare for even the simplest NetLogo program to be free of errors before it is tested thoroughly. When your program runs and produces output, it is time for the second stage of programming: searching for and fixing the errors that almost certainly are there. This second stage actually requires as much creativity as the first stage of drafting the code, and can be even more fun. (There is, for example, a very simple and elegant way to test the Marriage model's code for calculating social pressure.) Software testing includes making predictions of how the model should behave, dreaming up ways to test those predictions, statistical analysis of output, and often some imaginative detective work to explain the discrepancies between prediction and output. A simple summary of what to do is: produce lots of output, look at it hard, and investigate any of it that seems even slightly unexpected.

Statistical and independent reimplementation tests (sections 6.3.9 and 6.3.10) are essential because they can test many thousands of executions—mistakes often happen only under some (often, rare) conditions. However, these approaches have their limitations. Statistical analysis of model output can produce unexpected results not only because of code errors but also because the probability of certain results occurring can be difficult to predict correctly. And sometimes it is difficult to completely reimplement whole procedures because the NetLogo primitives use algorithms and random numbers we do not have access to. (We can, however, write a test program to carefully explore what a primitive does, and even email the NetLogo developers for details that are not clear from the dictionary.)

Here are some other key strategies that should become habits:

- Find your mistakes early, not late. Make testing a pervasive part of software development.
- Plan for testing: make sure there is time for it, and time to repeat it as the model evolves.
- Write your code so it is clear and easy to understand, so you never have to hesitate before asking someone to review it for you. When your work is published and you are famous, you can then confidently share the model with all the people who want to use it. (This really happens.)
- Save and document your tests, assuming that you will need to repeat them as the code evolves and that you will need to "prove" to readers and clients that your software is reliable.

From now on, all your programming work for this course should be accompanied by documentation of how you tested the code to prove as well as you can that it implements the formulation faithfully.

1. Read the Wikipedia entry on "Software testing." This article contains a nice overview of software testing issues and methods, reinforcing and expanding on topics covered in this chapter. It also reinforces the concept that testing is an essential, important part of computer science and programming. Also, look at the books on software testing in your university's library (or at least look at the catalog to see how many there are).

2. Using one of NetLogo's library models, become familiar with techniques for getting low-level output from models: using Agent Monitors for patches and turtles, using labels, and using `show` (or `write`) to see variable values. Add a `step` button so you can execute one tick at a time.

3. Section 6.3.7 provides an example test program. Did the memory algorithm it tests work? How well? (How else could you make turtles return to where they have been?)

4. Conduct the statistical analysis of file output from the butterfly hilltopping model outlined in section 6.3.9 (but with the mistake identified in section 6.3.10 fixed). Predict how frequently butterflies should move to the highest neighbor patch for several values of q, then test the predictions by analyzing the file output. Is your prediction reproduced? How can you support your conclusion? (If you have never had a statistics class, consult someone who has.) Explain any discrepancies and how they could be dealt with.

5. Conduct the independent implementation test of the Butterfly model described in section 6.3.10.

6. Does the mistake found in section 6.3.10 affect the butterfly analysis results obtained in chapter 5?

7. Test the code provided on this book's web site for the Marriage model program, using the approach described in section 6.5.2. (Hint: it is not necessary to reprogram the calculation of social pressure to test the NetLogo program's calculation of it!) How many errors do you find in the code?

Model Design Concepts

Introduction to Part II

7

Part I of this course provided a crash course in modeling and NetLogo: we introduced many fundamental techniques very quickly, with little opportunity to practice and understand them thoroughly. Therefore, part II is designed to reinforce what you have already learned and to teach even more of NetLogo. But we also take time to explain in more detail how NetLogo is designed and why, to give you a deeper understanding of how to program in general. So if you feel a little lost and overwhelmed at this point, please be patient and persevere. Becoming proficient in NetLogo remains a major goal as we move ahead.

However, part II has a second major goal: learning about agent-based modeling as a specific, unique way to do science. We address this second goal by focusing on a conceptual framework for designing, describing, and thinking about agent-based modeling.

What is a "conceptual framework" for ABMs and why do we need it? Each general approach to modeling has a set of basic ideas, theory, notation, and conventions that provide a way to think about and design models. In many scientific fields, differential equations have traditionally been the most important conceptual framework for modeling. In statistical modeling, there are several conceptual frameworks (e.g., parametric, Bayesian), each with its own theory and methods. The conceptual framework we use for ABMs is simply a list of concepts that capture the most important characteristics of an ABM, especially the ways that it is different from other kinds of models. These concepts, as applied to ecology, are discussed extensively in chapter 5 of Grimm and Railsback (2005) and summarized there in a "conceptual design checklist." We introduced the concepts in section 3.3 as part of the ODD model description protocol, and table 3.1 provides a checklist of key questions for each concept.

Each time you build an ABM, you need to think about how each of these concepts is addressed, if it is, in the model. Why? The first reason is to make sure that you think about all the important model design issues and decide explicitly how to handle each; otherwise, important parts of the model's design could be determined accidentally or only by programming convenience. The second reason is to make it easier to describe your model to others. When modelers use the ODD protocol and these design concepts to describe ABMs, it is much easier for others to understand the model's most important characteristics. Not all of the concepts will be important to each ABM, but thinking about each of the concepts helps you convince yourself and others that all the important model design decisions were made explicitly and carefully. (For this reason, the design concepts are also very useful when you review someone else's ABM.)

Most chapters of part II addresses one of these design concepts, with the goals of teaching the concept and the programming skills associated with it. There are, however, some differences between the design concepts summary of table 3.1 and the organization of part II. Here, we skip the Basic Principles concept because it tends to be more specific to scientific domains and less directly tied to software. Chapter 11 addresses two closely related concepts, Adaptive Behavior and Objectives, and we have chosen not to address the rather complicated and less-often-used Learning concept. We also include chapter 14 on Scheduling, a very important modeling and programming topic that is not one of the ODD design concepts (it was, in earlier versions of ODD, but is now in the "Process Overview and Scheduling" element). The chapters of part II also do not follow exactly the same order that the design concepts appear in ODD. In table 3.1, the concepts appear in order from most important and overarching to most specific and detailed. In the following chapters, though, the concepts are addressed in an order chosen mainly to present the associated programming skills efficiently.

In each chapter, we summarize the concept, describe some typical methods for using it in models, and discuss how these methods can be programmed in NetLogo. Then we use example models to give you experience with the concept while also reinforcing and building your NetLogo skills. We insert notes and hints to help you learn new programming tricks, trying not to duplicate the User Manual but instead to point out parts of NetLogo that are particularly important for scientific users. And we introduce some software and modeling skills related not to NetLogo but to analyzing results after your model has run.

Mixing the two goals of understanding ABM concepts and learning NetLogo in each chapter may seem like inviting confusion and frustration, but these two goals are actually parallel and complementary, because many parts of NetLogo were built specifically to implement these same ABM concepts.

7.2 Overview

Table 7.1 provides an overview of part II. It briefly describes, for each chapter, the conceptual topics (corresponding to the "Key questions" of table 3.1), the general modeling topics, and the NetLogo and software topics addressed.

Table 7.1 Overview of Part II

Chapter	Conceptual topics	Modeling topics	NetLogo and software topics
8: Emergence	Which model results emerge in complex ways from agent behaviors, and which are instead relatively imposed and predictable?	Designing and analyzing simulation experiments	BehaviorSpace
9: Observation	What outputs and displays are needed to understand and test an ABM?	Types of output: animation, plots, file	Obtaining information from the View, using plots and other Interface elements, writing output files
10: Sensing	What information do agents use, and how do they obtain it?	How the extent over which agents sense information affects model results	The scope of NetLogo variables, obtaining and changing variables of other objects, writing reporters, using links
11: Adaptive Behavior and Objectives	What decisions do agents make in response to changes in their environment and themselves? What measure of their objectives (fitness, utility, etc.) do agents use to make decisions?	Optimization and satisficing methods for decision-making	Identifying and subsetting agentsets, identifying optimal members of agentsets
12: Prediction	What predictions do agents make in their adaptive behavior?	Independent testing and analysis of submodels, Bayesian updating as a model of prediction	Contour plotting in Excel or other platforms
13: Interaction	How do agents interact with each other, either directly or indirectly via mediating resources?	Modeling competition via a shared resource, effects of global vs. local interaction, modeling memory	Modeling persistent interactions using variables for another agent, using lists
14: Scheduling	How is time represented? How does the order in which actions are executed affect results?	Modeling time via discrete time steps vs. discrete events in continuous time, using sorted vs. unsorted execution to represent hierarchical or nonhierarchical systems, synchronous vs. asynchronous updating, stopping rules	The go procedure, controlling the order in which actions are executed and the order in which agents execute an action, loops, non-integer ticks
15: Stochasticity	When should processes be modeled as if they are random? What stochastic functions are appropriate for such processes?	Replication experiments to determine the effects of randomness, common discrete and continuous random number distributions	Random number primitives, writing reporters, using BehaviorSpace to replicate experiments, defensive programming
16: Collectives	Are there intermediate levels of organization: aggregations of individuals that affect the population and the individuals? How should such collectives be modeled?	Representing collectives as emergent characteristics of agents and as a separate type of agent, logistic functions to represent probabilities, stochastic models of behavior that use observed frequencies as probabilities	Breeds to represent multiple kinds of agent; representing collectives as patches, via links, or as a separate breed

Emergence

The most important and unique characteristic of ABMs is that complex, often unexpected, system dynamics emerge from how we model underlying processes. Emergence, therefore, is the most basic concept of agent-based modeling. The key question about emergence is this: what dynamics of the system and what model outcomes *emerge*—arise in relatively complex and unpredictable ways—from what behaviors of the agents and what characteristics of their environment? What other model dynamics and outcomes are instead *imposed*—forced to occur in direct and predictable ways—by the model's assumptions?

By "unpredictable" here we refer to outcomes that are difficult or impossible to predict just by thinking. The concept of emergence has sometimes been given mystic connotations such as being unexplainable in principle, but with ABMs we are focusing on just the opposite: on explaining things via simulation. Epstein and Axtell (1996) described this kind of explanation aptly with their famous question "Can you grow it?": can you make your model system look and behave like the real one?

There is no simple way to classify some particular outcome of an ABM as emergent vs. imposed, but we can think about these qualitative criteria for a model result being emergent:

- It is not simply the sum of the properties of the model's individuals;
- It is a different type of result than individual-level properties; and
- It cannot easily be predicted from the properties of the individuals.

As an example, we can address this emergence question for the Butterfly model. The first step is to identify the specific model outcomes we are considering. The specific numerical results of the Butterfly model that we have been focusing on are the widths of corridors that groups of hilltopping butterflies use as they migrate. These outcomes result, obviously, from the one decision that model butterflies make: whether to move uphill or randomly at each time step. Do the Butterfly model's corridor width results meet the three criteria for being emergent? Corridor width is not simply the sum of individual butterfly properties, and is a different type of property from the properties of individuals. The individual butterflies have only one property, their location, and the width of a migration corridor is a different property than the location of an individual. (And remember that we discovered in chapter 5 that butterfly corridors changed qualitatively when we switched from a simple cone landscape to a realistically complex landscape, so corridor width emerges from the environment as well as from butterfly behavior.)

But we must admit that the corridor width of a group of butterflies is, in a way, just the sum over time of where the individual butterflies were. Is corridor width easily predicted from the properties of the individuals? Not completely, but even without the model we would expect corridor width to decrease as the value of the parameter q (the probability of the butterfly moving directly uphill instead of randomly) increases, and to get very large as q approaches zero. So we could describe the butterfly migration model as having its key outcome emerge from the individual decision of whether to move uphill or randomly (and from the landscape), but because this behavior is so simple, there is at least one model result—how corridor width varies with q (figure 5.3)—that is quite predictable.

The extent to which ABM results are emergent can vary widely, and having more emergence does not necessarily make a model better. The best level of emergence is often intermediate. Models with highly imposed results are often not very interesting, and sometimes their problem can be solved with a simpler mathematical model. On the other hand, a model with too many of its results emerging in complex ways from too many complex individual behaviors and mechanisms can be very hard to understand and learn from. The Butterfly model is not a bad ABM just because some of its results are not surprising; it is a perfect model for the problem it was designed for, but that problem did not involve highly complex agent behavior.

Learning objectives for this chapter are to:

- Develop an understanding of what makes the results of ABMs more vs. less emergent, by looking at two of NetLogo's library models.
- Start using the most important technique in agent-based science: designing and analyzing simulation experiments to test hypotheses and develop understanding about how an ABM works and how well it reproduces the system it models.
- Master the use of BehaviorSpace, NetLogo's extremely useful tool for automatically running simulation experiments and producing output from them.
- For the first of many times, analyze output produced by NetLogo models by importing results to other software for graphing and statistical analysis.

8.2 A Model with Less-Emergent Dynamics

The Biology category of NetLogo's Models Library includes a model called Simple Birth Rates. It is indeed a very simple model, designed to simulate how the difference in birth rate between two co-occurring species (red and blue turtles) affects the number of individuals in each population. The two species differ only in the number of offspring each individual produces each time step; the probability of death is the same for both species and is adjusted so that the total number of individuals is approximately constant. (If you study business, think of these as two chains of fast-food restaurant that go out of business at the same rate but differ in how often they open new shops.) Open this model, read its Information tab, and play with it a bit.

Let us focus on one question that this model can address: how does the time until "extinction" of one species depend on the relative birth rates of the two species? If we hold the birth rate of the red species (global variable `red-fertility`) constant at 2.0 and vary `blue-fertility` from 2.1 to 5.0, how does the number of ticks until extinction of the red species change? In other words, what shape do we expect the graph at figure 8.1 to have?

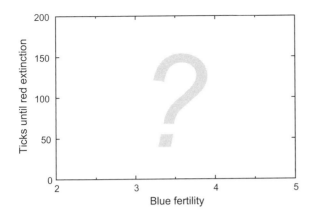

Figure 8.1
The problem we address with the Simple Birth Rates model: what shape will this graph have?

This model includes a button and procedure called `run-experiment` designed to help with this problem: it runs the model until one species is extinct, tells us how long it took, and then starts another run.

If we think about this problem for a minute, we can form a pretty strong expectation of the answer. When birth rates of the two species are close, they should coexist for a long time. As the difference in birth rate increases, the time to red extinction should decrease rapidly. But no matter how rapidly the blue reproduce, it seems unlikely that the reds will go extinct immediately (the model is initialized with 500 of them), so the time to red extinction will probably never get as low as one time step. We thus expect the curve on figure 8.1 to start high when `blue-fertility` is 2.1, drop rapidly, then flatten out at a value above 1 as `blue-fertility` increases.

Are these expectations correct? To find out, we need to do a simulation experiment.

8.3 Simulation Experiments and BehaviorSpace

To see how the time to extinction of red turtles in the Simple Birth Rates model varies with the fertility parameter of blue turtles, we need to run the model over a wide range of `blue-fertility` values and record the time (tick number) at which the number of red turtles reaches zero. But there is a complication: this model, like most ABMs, is stochastic, producing different results each time we run it because the turtles that die each tick are chosen randomly (see the procedure `grim-reaper`). (We will look more at stochasticity in chapter 15.)

Because the model produces different results each run, to really understand it we need to estimate the average results and the variability around those averages. Modelers often do this the same way that scientists study real systems that are highly variable: by executing *experiments* that include *replicates* of several different *scenarios* (statisticians often use the word "treatment" as we use "scenario"). In simulation modeling, a scenario is defined by a model and one set of parameters, inputs, and initial conditions; if the model had no stochastic elements, it would produce exactly the same results each time it executed the same scenario. Replicates are model runs in which only the stochastic elements of the model change. To be specific, *replicates* usually refer to model runs among which only the random numbers differ. (There are also other ways to replicate model scenarios, e.g., by randomizing input data or initial conditions.)

Now, to analyze how blue fertility affects the time to extinction of red turtles, we will use a simulation experiment that varies `blue-fertility` from 2.1 to 5.0 in increments of 0.1, producing a total of 30 scenarios. Further, we will run 10 replicates of each scenario. Each model run will continue until there are no more red turtles and then output the tick number at which this extinction occurred. These results need to be recorded in a file that we can then import to spreadsheet or statistical software and, finally, calculate the mean and standard deviation of time to red extinction and graph how they vary with `blue-fertility`.

This experiment is an example *sensitivity experiment*: a simulation experiment in which we vary one parameter over a wide range and see how the model responds to it (see also chapter 23). This experimental design is very useful for understanding how a model, and the system it represents, responds to one factor at a time.

Executing this experiment as we did in chapter 5—moving a slider to change the parameter value, hitting `setup` and `go` buttons, and writing down results—would take a long time, but now we will show you how to do such experiments very easily via BehaviorSpace. At this point you should read the BehaviorSpace Guide, which is under the "Features" heading of the NetLogo User Manual.

You can think of BehaviorSpace as a separate program, built into NetLogo, that runs simulation experiments on your model and saves the results in a file for you to analyze. By filling in a simple dialog, you can program it to do any or all of these functions:

- Create scenarios by changing the value of global variables;
- Generate replicates (called "repetitions" in NetLogo) of each scenario;
- Collect results from each model run and write them to a file;
- Determine when to stop each model run, using either a tick limit (e.g., stop after 1000 ticks) or a logical condition (e.g., stop if the number of red turtles is zero); and
- Run some NetLogo commands at the end of each model run.

The information that you enter in BehaviorSpace to run experiments is saved as part of the model's NetLogo file. Therefore, the first thing you need to do now, before using BehaviorSpace with the Simple Birth Rates model, is to save your own copy of this model with a unique name. From now on, use your own versions of the model instead of the one in the Models Library. Then:

■ Open up BehaviorSpace from NetLogo's Tools menu.

The BehaviorSpace dialog that opens lets you create new experiments, edit previously saved ones, or copy or delete experiments. By "experiment" (or "experiment setup"), this dialog refers to a set of instructions defining the scenarios, replicates, outputs, and other experiment characteristics. that you want. You can create and save a number of different experiment setups for the same model.

■ Create a new experiment setup by clicking on the New button.

This will open the Experiment dialog, which you will now fill out by modifying the default values. As you fill out the Experiment dialog, you can save your settings by clicking on the OK button at the bottom (to resume work on the dialog, reopen the experiment by clicking on Edit in the BehaviorSpace dialog).

■ The first thing you should do in the Experiment dialog is give the experiment a new name. The name should describe the experiment, so it should be something like "Blue_fertility_ effect_on_red_extinction."

■ Next, you can specify the scenarios to run by filling in the field labeled "Vary variables as follows." Note that NetLogo automatically inserts the global variables that are in sliders or other Interface elements. From the BehaviorSpace documentation (and the example provided right below the field), you know that you can create scenarios varying `blue-fertility` from 2.1 to 5.0 in increments of 0.1 by entering:

```
["blue-fertility" [2.1 0.1 5.0]]
```

■ In the "Vary variables as follows" field, it is a very good idea to also include the model variables you want to hold constant, fixing their values so they do not get changed accidentally by moving a slider:

```
["red-fertility" 2]
["carrying-capacity" 1000]
```

■ Set the "Repetitions" value to 10.

Now you can tell NetLogo what results you want by filling in the field labeled "Measure runs using these reporters." You put NetLogo statements in this field that report outputs; these statements will be executed and the results written to the output file. In this case, the result that you want is the tick number at which red turtles become extinct. To get this, you can tell BehaviorSpace to stop the model when the number of red turtles is zero, and output the tick number when this happens. The reporter that gives the tick number is just the NetLogo primitive `ticks`, so:

■ Put `ticks` in the "Measure runs . . ." field and uncheck the box for "Measure runs at every step." (See the programming note below to learn exactly when BehaviorSpace writes output.)

How do you tell BehaviorSpace to stop the model when reds go extinct? You need to put a condition in the "Stop condition" field that is true when the model should stop:

■ Fill in the statement `red-count = 0`. (From the program, you will see that `red-count` is a global variable containing the current number of red turtles.)

Nothing else in the Experiment dialog needs to be changed now, but make sure you understand what the boxes for "Setup commands," "Go commands," "Final commands," and "Time limit" do. The Experiment dialog should now look like figure 8.2.

■ Now click OK to close the dialog. Close the BehaviorSpace dialog and save your file.

You are now ready to run the experiment.

■ Open BehaviorSpace again from the Tools menu, select your experiment setup, and click "Run."

Try both the "spreadsheet" and "table" output formats; they each produce files in .csv format designed to be imported into spreadsheets or other software. You are also given a choice of how

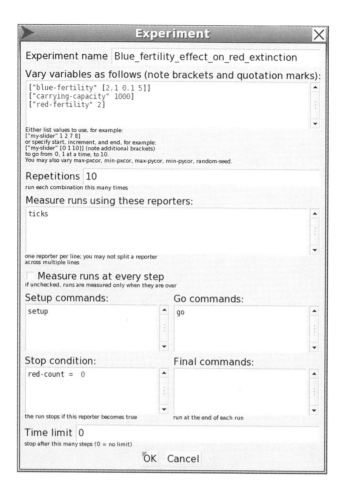

Figure 8.2
Experiment setup for Simple Birth Rates.

many of your computer's processors to use, if you have multiple processors (discussed in the second programming note below). After the experiment starts, you can deselect the updates of View and plots and monitors to speed up execution.

Programming note: How BehaviorSpace works

BehaviorSpace is an extremely important tool, but if you do not understand exactly how it works you will find it frustrating and you may misinterpret its results. You need to understand several details.

At the start of each run, BehaviorSpace changes the variable values as specified in its "Vary variables ..." box. BehaviorSpace sets these values *before* it executes the `setup` procedure. Therefore, if `setup` sets a variable that BehaviorSpace is supposed to be controlling, the value given by BehaviorSpace will be overwritten by the value set in `setup`. Consequently, any variable that you want BehaviorSpace to control must be defined and initialized only by an Interface element (typically, a slider or chooser).

When the "Measure runs at every step" box is *not* checked, results are calculated and written to the output file only after a model run has stopped. The run can be stopped by a `stop` statement in the `go` procedure (as we have in the Butterfly model), or by using the

"Stop condition" box in BehaviorSpace's experiment editing menu, or by using the "Time limit" box in the experiment menu.

When the "Measure runs at every step" box *is* checked, BehaviorSpace produces its first output *when the* `setup` *procedure completes*. This reports the state of the model after it is initialized and before simulations begin.

When the "Measure runs at every step" box is checked, BehaviorSpace then produces output *each time the end of the* `go` *procedure is reached*. If a `stop` statement terminates the run before the end of the `go` procedure (as in the Butterfly model), no output will be written for that last time step.

If you specify a "stop condition" or "time limit" in the BehaviorSpace experiment instead of in your procedures, BehaviorSpace checks this condition each time the `go` procedure has completed a time step and output for that step has been written. Hence, if you use a stop condition `ticks = 1000` NetLogo will completely finish the `go` procedure on which `ticks` reaches 1000, then write its BehaviorSpace output, then terminate the run.

There are therefore two ways to get output every time step as you expect it. One is to remove the `stop` statement from the `go` procedure and instead put it in the BehaviorSpace experiment (e.g., by setting the time limit to 1000). The second is to reorganize the `go` procedure so it starts with `tick`, followed by a statement that stops the model if the desired number of ticks is exceeded (not just met). This issue is considered further in section 14.2.6.

Programming note: BehaviorSpace, multiple processors, and output files

BehaviorSpace can execute multiple model runs at the same time if your computer has multiple processors. When an experiment starts, you tell BehaviorSpace how many of your processors you want it to use, and it will execute one model run at a time on each processor. But there are two consequences for file output.

First, files produced by BehaviorSpace will be affected because the lines of output for different model runs may be intermingled instead of in the expected order, for example:

[run number]	blue-fertility	[step]	ticks
1	2.1	0	0
2	2.2	0	0
1	2.1	1	1
1	2.1	2	2
2	2.2	1	1

This problem can be solved after you import the BehaviorSpace output to a spreadsheet, by simply sorting the output lines by run number and step.

Second, any output files that you code yourself are likely to cause unavoidable and fatal run-time errors because the multiple runs will each try to open the same file at the same time, which is not allowed. (The same problem could happen with input files.) The only solutions are to get rid of your output files, have each model run write to a different file, or run BehaviorSpace on only one processor.

When BehaviorSpace has finished running the experiment, you can import the results to software of your choice, calculate the mean and standard deviation over the 10 replicates of the time to red extinction for each value of `blue-fertility`, and draw a graph similar to figure 8.3.

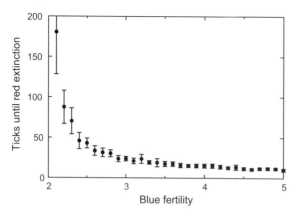

Figure 8.3
Example results of the Simple Birth Rates experiment. The dots are the mean time to red extinction, and error bars are +/– one standard deviation over the 10 replicates.

Does this graph look like you expected it to? While it seems reassuring that the model produced results that were quite as we expected, it also makes us wonder what the value of building an ABM was if it produces results we could easily anticipate. In fact, we chose this model to examine first because it is a clear example of an ABM whose dynamics are strongly imposed by its very simple, rigid agent behaviors: the agents make no decisions, they have no individual state variables, and their very few behaviors (reproducing, dying) are tightly specified, or imposed, by the global parameters. If the agents in this model adapted their fertility to (for example) local density of other turtles or food, or had to find mates within a certain time or distance before reproducing, the model would produce less predictable, more interesting results. While this model is fun to play with and provides a good BehaviorSpace exercise, it is not a good example of the kind of problem that requires an ABM to understand. So let's move on to a model that is, in contrast, a classic example of emergent dynamics.

8.4 A Model with Complex Emergent Dynamics

Another model in the Biology section of NetLogo Models Library is called Flocking. This is a NetLogo version of a well-known example (e.g., Reynolds 1987) of how complex and realistic dynamics can emerge from a few simple agent behaviors, in ways that could not be completely predicted. Schools of fish and flocks of birds can be thought of as emergent properties of how the individual animals move in response to each other. A school or flock can appear as a coherent entity, but it can also break up and re-form, and its shape changes continually. (There are fascinating videos of flocks of birds, especially starlings, on the web.)

In the Flocking model, the individuals have one behavior: adjusting their movement direction in response to the location and direction of other nearby individuals (their "flockmates," which are all other turtles within a radius equal to the `vision` parameter). They make this decision considering three objectives: moving in the same direction as their flockmates ("align"), moving toward the flockmates ("cohere"), and maintaining a minimum separation from all other individuals ("separate"). Parameters control the relative strength of these three objectives by limiting the maximum angle a turtle can turn to align, cohere, and separate. The `minimum-separation` parameter sets the distance at which turtles try to separate from each other.

You will quickly see that Flocking's results are in fact complex. The turtles form flocks that continually change characteristics, and these characteristics change as you vary the parameters. Further, parameters seem to interact (the effect of one parameter depends on the value of another); for example, see what happens when you vary `max-cohere-turn` when `minimum-separation` is low, then high. (To speed up the model, use the comments in the `go` procedure to make turtles move forward one unit instead of making five short moves.)

The Flocking model demonstrates two common (though not universal) characteristics of emergent dynamics. First, the most important and interesting results of the model seem qualitative and hard to describe with numbers. Whereas the state of the Simple Birth Rates model can be described by two numbers (how many red and blue turtles are alive), the characteristics of the emergent flocks clearly change as you vary parameters but in ways that are not easy to quantify. The second characteristic is that it takes some time before the flock characteristics emerge, both when the model starts and when you change parameter values in the middle of a run. ABMs often have a "warm-up" period in which their dynamics gradually emerge.

Let's think about how emergent Flocking's results are, using the criteria introduced in section 8.1. Are the results of this model different from just the sum of individual turtle properties, and of different types than the properties of the individuals, and not easily predicted from the rules given to the turtles? It is tempting to immediately answer "yes" to these questions, but there is one problem: we have not yet defined exactly *what* results of this model we are interested in. When we used the Simple Birth Rates model, we could clearly state what output we wanted to analyze: the time until red turtles went extinct. But we have not yet stated which results of Flocking we are analyzing.

This problem is a reminder that this course is about using ABMs for science. Flocking is a fascinating simulator: we specify some simple agent rules and observe the complex flocking dynamics that emerge. But when we do science we are instead trying to do the opposite: we identify some complex dynamics observed in the real world and then try to figure out what behaviors of the system's agents explain those emergent dynamics. Models similar to Flocking are in fact used for important science, as profiled in chapter 11 of Camazine et al. (2001) and section 6.2.2 of Grimm and Railsback (2005). In particular, Huth and Wissel (1992) combined ABMs with studies of real fish to address the question, "What assumptions about movement of individual fish explain the emergent characteristics of real schools of fish?"

We can use NetLogo's Flocking model to do simplified versions of Huth and Wissel's simulation experiments. Huth and Wissel (1992) started by specifying some specific model outputs that they could compare to real fish schools. These were statistics on properties such as how close the individuals are to the nearest other fish and how much variation there is in the direction they move. We can easily obtain similar statistical results from Flocking by using BehaviorSpace. (The model of Huth and Wissel is not directly comparable to Flocking; it uses slightly different behaviors and simulated only one small school of fish, whereas in Flocking there are typically several flocks.) We can, for example, calculate:

- The number of turtles who have flockmates;
- The mean number of flockmates per turtle;
- The mean distance between a turtle and the nearest other turtle; and
- The standard deviation in heading, over all turtles—an imperfect but simple measure of variability in direction.

Let's set up BehaviorSpace so it produces these results. For the Simple Birth Rates exercise we needed only one output at the end of each model run; now, let's get output from every tick of model runs that last 500 ticks. For now we do not want to vary any of the parameters, so we will set up no scenarios other than the "baseline" scenario with default parameter values. But we will run 10 replicates of this scenario. The Experiment dialog should look like figure 8.4. (Don't worry if you do not understand the reporters in the "Measure runs using . . ." field; you will learn about them in chapter 10.)

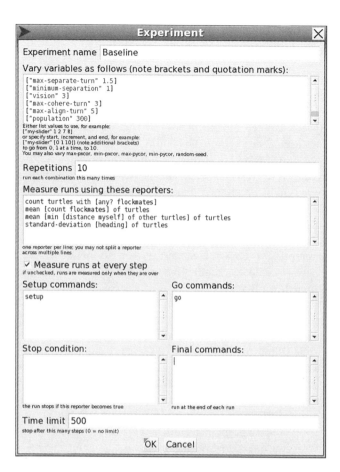

Figure 8.4
Experiment setup for Flocking.

When you try to run this experiment, you may get a run-time error because the experiment uses the turtle variable `flockmates`, the agentset of other nearby turtles. Remember that BehaviorSpace calculates its first output when the `setup` procedure has finished, before the run actually starts. At this point, `flockmates` has not yet been given a value, so NetLogo does not know that it is an agentset. You can solve this problem by initializing `flockmates`: giving it a value that is an agentset, at the time each turtle is created. Simply add the statement `set flockmates no-turtles` to the `create-turtles` statement in `setup`:

```
crt population
[
  set color yellow - 2 + random 7  ; random shades look nice
  set size 1.5                      ; easier to see
  set flockmates no-turtles         ; New for BehaviorSpace
  setxy random-xcor random-ycor
]
```

(Look up `no-turtles` to understand this.) Just remember to ignore the first line of output, which will say that all turtles have zero flockmates, because it reports the model before the first tick is executed.

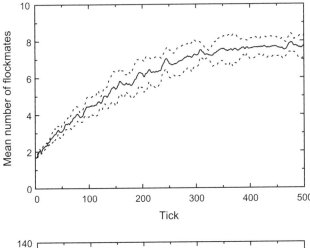

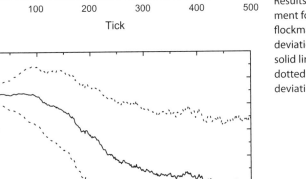

Figure 8.5
Results from Flocking experiment for (top) mean number of flockmates and (bottom) standard deviation of turtle headings. The solid lines are the mean, and dotted lines +/− one standard deviation, over 10 replicates.

Now you can graph and look at the results. For example, we found the mean number of flockmates to increase from less than two to over seven as the simulation proceeded (figure 8.5, top panel), while the standard deviation in heading decreased from over 100 to less than 40 degrees (bottom panel). Note that these results reflect the model's "warm-up" period when flocking patterns emerge. It appears that the number of flockmates and turtle heading outputs start to stabilize at about 400 ticks.

Now, let's conduct an experiment similar to one performed by Huth and Wissel (1992). What happens to the emergent schooling patterns if turtles adapt their direction considering only the one closest neighbor turtle? We can easily simulate this "one-flockmate" scenario with a small code change. Save a new version of Flocking and then change its `find-flockmates` procedure from this:

```
to find-flockmates  ;; turtle procedure
  set flockmates other turtles in-radius vision
end
```

to this:

```
to find-flockmates  ;; turtle procedure
  set flockmates other turtles in-radius vision
  set flockmates flockmates with-min [distance myself]
end
```

The new statement modifies the flockmates agentset so it contains only the nearest turtle. (In the unlikely case that the two nearest others are exactly the same distance from a turtle, flockmates would include both.)

NetLogo brainteaser

In the above code change that sets flockmates to the single nearest other turtle:

1. Why not simply use set flockmates min-one-of other turtles [distance myself] instead of two separate "set" statements? min-one-of selects the turtle with smallest distance from the calling turtle. (Try it!)
2. Why not simply use set flockmates other turtles with-min [distance myself]? (What seems strange if you try it?)

Answers:

1. min-one-of returns a single agent, not an agentset. This would turn the variable flock-mates into a turtle instead of an agentset, but other code in this model requires flock-mates to be an agentset. (The primitive turtle-set could be used to turn the result of min-one-of into an agentset.)
2. This statement always sets flockmates to an empty agentset. It first evaluates other turtles with-min [distance myself], which is simply the calling turtle. Then, other by definition removes the calling turtle from the agentset, leaving it empty. However, this statement *does* work if you put parentheses around other turtles to make NetLogo identify the other turtles before selecting from them the nearest one. We will study these agentset primitives in chapter 11.

With this change in turtle behavior, see how Flocking now looks. To see if these differences show up in the statistical output, repeat the BehaviorSpace experiment. You should see, in fact, that results are quite different. For example, the variation in turtle headings now changes little over time (figure 8.6) instead of decreasing sharply as it did in the baseline version (figure 8.5). Huth and Wissel (1992) used an experiment like this to conclude that real fish adjust their swimming direction considering several neighbors, not just one.

This experiment is an example of a second common kind of simulation experiment: contrasting alternative scenarios. The sensitivity experiment design we discussed in section 8.3 varies a parameter across a wide range, but scenario contrast experiments look at the differences between two (or several) distinctly different scenarios. (Scenario contrasts are more similar to the experimental designs typically used on real systems: specifying two or more treatments, then using replicates of each to determine how different the treatments' effects are.) Often, the different scenarios are different agent behaviors: in this example, the difference is defining flockmates as all other turtles within a range vs. only the closest other turtle. When we analyze scenario contrast experiments, we typically look at how different the results are between the two scenarios.

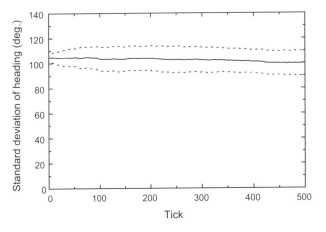

Figure 8.6
Flocking results for standard deviation in heading, with turtles using only one flockmate.

8.5 Summary and Conclusions

Simulation models, and especially ABMs, are useful because they can reproduce how complex dynamics of real systems emerge from the characteristics, behaviors, and interactions of the systems' members and their environment. ABMs can range from having relatively simple results that are closely imposed by simple, rigid rules for agent behavior, to producing many kinds of complex results that are very difficult to predict from agent behaviors. When we want to describe the emergence characteristics of an ABM, we can start by defining what the key outcomes of the model are and the agent behaviors that those outcomes emerge from. And we need to describe how model outcomes emerge from the agents' environment, if they do. In the Flocking and Simple Birth Rates models, results emerge only from agent behaviors; but the results of the hilltopping butterfly model depend qualitatively and quantitatively on behavior *and* on topography, an environment characteristic.

Making model results more emergent is not always better. Very simple ABMs with highly imposed dynamics may not be very different from an equation-based model of the same problem and hence not extremely exciting. But such simple ABMs can still be useful and appropriate for many problems, and can have advantages such as being easier and more intuitive to build and understand than equation-based models. At the other extreme, an ABM with such extremely emergent outcomes that they are completely unpredictable would be very hard to use for science because our goal is to figure out what underlying processes give rise to the emergent outcomes. Models such as Flocking that merely demonstrate emergence are not inherently scientific. However, they can be put to scientific use when we use them, with simulation experiments, to understand some specific emergent properties of a real system.

We observed two common characteristics of emergent dynamics in the Flocking model. First, important emergent outcomes may seem more qualitative than quantitative, but we can find numerical outputs to describe them. Second, it often takes a considerable "warm-up" period before emergent outcomes become apparent and stabilize.

In this chapter we demonstrated two very common and important kinds of simulation experiment. With the Simple Birth Rates model, we varied a parameter over a wide range and analyzed how model results changed in response; such *sensitivity* experiments analyze how sensitive the model is to a particular parameter or input. With the Flocking model, we compared two different versions of the model, a *contrast* of alternative model *scenarios* or versions. This kind of experiment is especially important for testing alternative model rules and determining which cause the ABM to be better at reproducing system behaviors observed in the real world.

In part III of this book we will delve more deeply into these simulation experiment issues. We will learn how to use qualitative patterns of emergent dynamics and contrasts of alternative model versions to develop theory for how agent-based systems work. In parts III and IV we will also learn how to use sensitivity experiments and other approaches for calibrating models to match real-world observations and to understand our model, and the system it represents, once it is built.

8.6 Exercises

1. Repeat the experiment on the Simple Birth Rates model in section 8.3, but using different values for the parameter `carrying-capacity`. How does the relationship between blue fertility and time to red extinction in figure 8.3 change when the carrying capacity is, for example, 500 and 1500 turtles?

2. "Risk analysis" is used in management of all kinds of systems. Conduct a new experiment on the Simple Birth Rates model using BehaviorSpace to determine how frequently the red turtles go extinct within 100 ticks. This output is an estimate of the probability ("risk") of red extinction within 100 ticks; how does it vary with blue fertility? You will clearly need different reporters to produce output and stop each run, but what else about the experiment could change?

3. Conduct a sensitivity experiment on the Flocking model. How do results such as those in figure 8.5 change if you use, for example, five values of `vision`, `max-align-turn`, or one of the other parameters?

4. Recent research on real starling flocks indicates that these birds do not respond to the flockmates within a limited distance but instead respond to the 6–7 nearest other birds, no matter how close or far they are (Ballerini et al. 2008). Conduct another scenario contrast experiment on the Flocking model. How do results of the experiment we did in section 8.4 change when you modify the `find-flockmates` procedure so flockmates are the nearest six turtles? (Hint: see the primitive `min-n-of`.) Include tests to prove that turtles always have six flockmates, and that a turtle's flockmates do not include itself.

5. Many models similar to Flocking assume that individuals adjust their direction not in response to all nearby others, but only in response to nearby others who are also within a particular vision angle (indicated by the arcs in figure 8.7, which has a vision angle of 60°). For example, if we assume turtles only see and respond to other turtles who are ahead, not behind, them, then flockmates could be set to all turtles that are within the vision distance *and* are at a vision angle less than 90° from the turtle's heading. Make this change to Flocking (which is extremely easy with the right primitive) and see how its results change as you vary the vision angle. Are the effects only qualitative or are they reflected in the model's statistical results?

Figure 8.7

Observation

The "observation" design concept addresses the fact that ABMs (like the systems they represent) produce many kinds of dynamics, so what we learn from them depends on how we observe them. A key part of designing a model and its software is deciding what results we need to observe and how.

We use observations for different purposes as we go through the cycle of building, testing, and using an ABM. As we build and start testing a model, we always need to see what the individuals are doing to understand the model's behavior and find obvious mistakes. Therefore, graphical displays with the power to observe any part of a model are essential in the early part of the modeling cycle. Graphical displays are also extremely important at the end of the modeling cycle, when it is time to show the model to other people, teach them how it works, and convince them that the model is understandable and testable. But as soon as we are ready to really test and analyze a model, we need to write results out to files so we can graph and analyze them in other software.

In this chapter we focus on ways to produce and use visual and file output from NetLogo. In part I you already learned basics such as using the View and its Agent Monitors, creating line plots and output windows, and writing simple file output. Now you will learn how to display more information on the View, produce other kinds of visual output, and produce output files with statistical summary data. In chapter 6 we discussed ways to get output for testing software, and here we focus on observation of the model itself. However, there is overlap between these objectives, and in this chapter we will review and use the observation methods presented in chapter 6.

In chapter 9 your learning objectives are to:

- Learn to use a variety of tools—colors; turtle sizes, shapes, headings, etc.; labels; and Agent Monitors—to observe models via the View.
- Master the use of NetLogo's time-series plots and histograms.
- Thoroughly understand how output files are created and used, and how to write custom output in .csv format.
- Learn use of BehaviorSpace and the `export-` primitives to write output files.

- Understand the benefits and ethics of copying code written by others.
- Practice teaching yourself NetLogo primitives and built-in variables; starting here, we commonly use many NetLogo tools without explaining them in detail, assuming you can understand them from the User Manual.

9.2 Observing the Model via NetLogo's View

The View—NetLogo's graphical display of the world of patches, turtles, and links—is a powerful observation tool. It provides many different ways to get information about the model: we can display information using the color of patches; the color, shape, size, and heading of turtles, and traces of their movement; the color, shape, and thickness of links; labels (and even label colors) for turtles, links, and patches; and "monitors" that can be opened for individual agents. (We have not yet used NetLogo's "links," but will soon.)

The built-in variables for color, shape, size, heading, label, and label color are changed via the set primitive. They can be used to display agent state or behavior by setting them to values that depend on the agents' variables. In chapter 4 we did this when we used the scale-color primitive to set patch colors to represent patch elevation. Another example is setting the color of a turtle to display how it made some decision; we can make butterflies tell us whether they moved uphill or randomly by changing this code:

```
; Decide whether to move uphill with probability q
ifelse random-float 1 < q
  [ move-to max-one-of neighbors [elevation]] ; Uphill
  [ move-to one-of neighbors ]  ; Otherwise move randomly
```

to this:

```
; Decide whether to move uphill with probability q
ifelse random-float 1 < q
  [
    move-to max-one-of neighbors [elevation] ; Uphill
    set color red ; Set color red to indicate uphill
  ]
  [
    move-to one-of neighbors ; Otherwise move randomly
    set color blue ; Set color blue to indicate random
  ]
```

For another example, we can have butterflies tell us what elevation they are at by setting their label to the elevation. Simply insert this statement at the end of the move procedure:

```
set label elevation
```

You can limit the number of decimal places in the label using the "precision" primitive:

```
set label precision elevation 0
```

It is important to understand that the View will not automatically display anything that happens between its normal updates. For example, if you program a turtle to take ten steps forward and turn left, as part of a procedure executed once per tick, you will not see these ten steps forward because the display is not updated until after the procedure is over.

One limitation of the View is that we often want to observe several patch variables, but we can only display one variable at a time using patch color. What if, in the Butterfly model, we want to see not only patch elevation but also the number of butterflies in each patch? There are several potential solutions to this problem.

First, we can use patch labels to display an additional variable. To see how many butterflies are in each patch at the end of each time step, we can add this line to the `go` procedure after the turtles have moved:

```
ask patches [set plabel count turtles-here]
```

If you try this solution, you may see that it is not satisfactory for this model: the patches are small, so their patch labels cover up the entire view.

Instead we can label only the patches that contain turtles:

```
ask patches with [any? turtles-here]
    [set plabel count turtles-here]
```

But this approach is often still not very satisfactory, because it requires us to patiently read all the numbers before we can detect any patterns, whereas colors let us see patterns very readily.

Another alternative is to have several simple procedures that each reset the patch color to display a different variable:

```
to display-patch-elevation
  ask patches
    [set pcolor scale-color green elevation 0 100]
end

to display-turtle-count
  ask patches
```

```
    [set pcolor scale-color red count turtles-here 0 10]
  end
```

Buttons on the Interface tab can then be attached to these procedures, so that clicking on each button redraws the View to display a different patch variable. (Or: these `ask patches` statements can be written right into the Commands field of the dialog that defines a button and never appear at all in the Procedures tab!)

Mistakes in how you program graphical displays can be especially troublesome because our instinct is to trust our eyes and assume that the problem is with the model itself. It is easy to waste time trying to fix a model that only looks wrong because the display is not correct. Therefore, it is always a good idea to test your outputs. To spot-check values displayed on the View, right-click on displayed agents and use the Agent Monitor. To test plots, create monitors (described below in section 9.3) to display the same information and verify that the two outputs match.

Programming note: Updating display variables to represent agent states

Not understanding how agents are updated on the View is a common source of confusion for new NetLogo users. You must remember that agent colors, labels, etc. only change when your program changes them; they are *not updated automatically* by NetLogo when the agent's state changes. For example, in the `setup` procedure you can add the statement `set label elevation` to the block of statements executed by turtles when created. (Try this, creating only 1 or 2 turtles.) The turtles now show their elevation when created. But this statement does *not* cause the labels to always show the turtles' *current* elevation; the turtle labels continue to show only their starting elevation. To continuously show the current elevation, you need another `set label elevation` statement that is executed each time a turtle moves.

Similarly, if you use patch colors to represent variables that change, you must reset these colors each time the variable changes. In the Butterfly model, we use the `scale-color` primitive to represent elevation. Patch colors are set only in the `setup` procedure because patch elevations do not change in this model. However, if we use the same approach to scale patch color by a variable that does change (e.g., the number of turtles in the patch), then the `scale-color` statement must be repeated each time step. Forgetting to update patch colors in this way often causes confusion. For example, assume you program turtles to consume some resource contained in the patches, and scale `pcolor` to display how much resource remains in the patches. Then, when the View shows that patch colors never change as the model runs, you assume there is something wrong with how the turtles or patches are programmed. But the problem often is that you simply forgot to recalculate `pcolor` every time step. So: when a graphical display makes you think you have a bug in your program but you cannot find it, first use Agent Monitors to check whether the bug is instead in how you programmed the display.

It is best to separate all these display updates from the model code by creating a special procedure (called something like `update-outputs`) that is called from the `go` procedure after all the agent behaviors are done. This procedure asks the patches, turtles, and links to update their colors and other display attributes so you do not have to mix these statements into the model code.

As you know by now, NetLogo provides several other kinds of displays ("interface elements") that can be added by selecting them from the drop-down menu on the Interface. *Monitor* interface elements provide an easy way to see how system-level properties of the model change as the model runs. When you add a monitor to the Interface, a dialog opens where you enter a reporter expression; the monitor then continuously displays the updated value of this expression. For example, you can see the mean elevation of all the butterflies by adding a monitor with this reporter (figure 9.1). Note the option that lets you reduce the number of decimal places displayed.

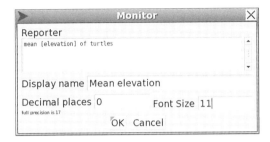

Figure 9.1
An Interface monitor set up to display mean butterfly elevation.

An *output* is simply a window where output can be written by your program (and perhaps saved to a file; see section 9.6). Use the primitives that start with output-.

Plots are often even more useful because they display more information than a monitor can. In section 5.4 we added a plot to the Butterfly model. The plot primitive simply adds a point to a plot, with its y-value equal to the value of a variable or reporter (e.g., mean [elevation] of turtles). If plot is called once per tick, then the number of ticks is on the x-axis. The plotxy primitive lets you write your own custom kinds of plots.

Histograms are a particularly useful kind of plot because they can display information about the whole population of agents instead of just one average value. With histograms you can see whether the agents are similar with respect to the variable being plotted or whether different agents have very different values.

NetLogo's histogram primitive and the Programming Guide make it easy to add histograms, but be aware that when you edit the plot settings you need to manually set the value of "Xmax" to the highest possible value of the variable being histogrammed. Even with the "autoplot" option turned on, NetLogo does not show histogram bars above the value of Xmax. For example, if you histogram the elevations of the butterflies (using histogram [elevation] of turtles), you need to set the plot's Xmax value to the highest value of elevation that you want to see in the plot. (It often takes some experimenting to find a good value of Xmax.) Failing to set Xmax to a high enough value can provide a misleading picture of results, or make it look like the histogram is not working, because all the bars are off the plot to the right. You also need to set the plot's "Mode" to "bar," and its "Interval" to the number of X units to be represented by each bar.

Here are a few other tips for plotting.

- You can put the plot or histogram statements directly in the go procedure. However, if your model has more than one plot and some file output (as it usually will), it is better to use a separate update-outputs procedure that updates all the plots and displays after all the agents have executed their behaviors. (Starting with version 5.0, NetLogo also lets you write the update code directly in the plot dialog on the Interface; we find it preferable to keep it with the rest of the code where it is easier to control and remember exactly when and how plots are updated.)

■ It is usually good to also add a statement at the end of the `setup` procedure that executes this output procedure. Then, when you set up the model, the plots will show its initial state before the `go` button is pushed.

■ If you use more than one plot, make sure you use the `set-current-plot` primitive before each `plot` or `histogram` statement. If you fail to do so, NetLogo can plot nonsense without producing an error statement.

9.4 File Output

File output is essential for ABMs: sooner or later—usually sooner—we need more information for testing and analyzing a model than we can get from NetLogo's View, and we need it in a numerical, not visual, format. NetLogo's designers know this and provide many tools for writing output files.

We often write output files to look at either (a) the variables and behaviors of the individual agents, most often for testing and understanding behavior of individuals, or (b) summary statistics on the population of agents.

Writing output involves four phases:

1. *Open an output file.* The computer identifies the file to be written to and prepares to write to it. Once the file is opened by NetLogo, no other program may write to it until the file is closed again. (And—NetLogo will have a run-time error if it tries to open a file that some other program already has open.) The file name must be specified either by your NetLogo program or by an interactive window in which the user selects or creates a file name. If the file already exists, the program or user must also decide whether to erase (overwrite) it or to append the new output to the end of it. Normally we program NetLogo to erase and overwrite existing files, because results from multiple model runs in one file can be confusing. To save output files so they are not overwritten the next time you run the model, simply rename them.

2. *Write "header" information to the file.* The program writes information at the top of the file to identify the file's contents. Usually this information includes at least one word or phrase to describe each column of output. Because the header information should be written only once each time the model runs, this is usually done just after the file is opened. Savvy programmers also write the current date and time in the file header, to avoid confusion about which model run the output is from. NetLogo's `date-and-time` primitive is useful for this.

3. *Write output to the file.* As the model executes, program statements write the selected variable values or statistics to the file. Typically, we program models to write one line of file output for each agent on each tick (to get output describing the individual agents) or one line of summary results per tick.

4. *Close the file.* Once the program has finished a model run, it closes the file so it can be opened in other programs. (If your NetLogo program opens a file and fails to close it, you will get a run-time error if you try to reopen the file.) Files are automatically closed when you quit NetLogo, but it is much better practice to have your program close them. You can open and close a file repeatedly within a program, which is sometimes convenient (e.g., to make sure the file gets closed if you stop the program at any time). However, keep

in mind that opening and closing files extremely often (e.g., by each turtle, each tick) can make your model run extremely slowly.

When you use BehaviorSpace or the `export` primitives, they do all these steps for you. But often you will need to write your own file output code.

9.4.1 Example Code for Individual-Level Output

You can modify the following example code for writing individual-level output on each turtle (or other type of agent) in a model. You will use such output on each individual agent often to test the model and its software, and sometimes to analyze results of models with few individuals. This code writes output in the .csv format that is easily imported to spreadsheets. One line is written for each turtle on each tick. The example assumes that turtles own state variables `bank-balance` and `profit`.

Programming note: Copying code

Here we provide example code for you to use, and at the very beginning of the book we pointed you to NetLogo's Code Examples library as a place to find code to reuse. Copying published, public pieces of code is acceptable and even encouraged in programming. Using reliable existing code saves time and avoids mistakes. However, it is unethical (and sometimes illegal) to copy extensive pieces of code and imply that you wrote them. If you copy a large or unique or clever piece of code from a public source, you need to include a comment saying that it is copied and from where. Giving credit in this way is not only ethical, it shows that you were smart enough to find code that already works instead of wasting time writing it yourself. So: please copy code pieces as needed from NetLogo's Code Examples or from this book, but give credit when the code is not trivial.

It is never OK to copy whole procedures or programs—such as your homework assignment—without saying that you did so. And be aware that you are still responsible for making sure that any code you use works correctly, even if you "borrowed" it.

To open an output file and write column headers to it, in the `setup` procedure:

```
; Open an output file for testing and analysis
; First, delete it instead of appending to it

if (file-exists? "IndividualBankBalanceOutput.csv")
  [
    carefully
        [file-delete "IndividualBankBalanceOutput.csv"]
        [print error-message]
  ]

file-open "IndividualBankBalanceOutput.csv"
file-type "id,"
file-type "tick,"
file-type "Size,"
file-type "BankBalance,"
```

```
file-print "CurrentProfit"
file-close
```

Some notes on this example:

- The code deletes any previous copies of the output file, with the `carefully` primitive making execution continue (while notifying the user) if a run-time error occurs.
- The code also temporarily closes the file after opening it and writing column headers to it. See the code for writing output to the file, below.
- Note the difference between the primitive `file-type` and `file-print`, which puts an end-of-line character after the last header. Hence, the five column labels are all written on the first line and separated by commas.
- If you use a European computer that uses "," as a decimal point, you will need to use ";" instead of a comma in .csv files. Alternatively, you can separate values with the tab character, which in NetLogo is written as "\t"; or temporarily set your computer to U.S. English settings.
- This example assumes that the first output will be the identification (`who`) number of an agent. If your output file starts with the *capital* letters "ID" instead of "id," Microsoft Excel will assume it is a different file type (SYLK) and refuse to open it.

Now you can use a block of code like this to write the state of each agent at each tick. This code could be in the `go` procedure or (preferably) in a separate observer-context procedure such as `update-outputs`.

```
; Write turtle states to output file
file-open "IndividualBankBalanceOutput.csv"
ask turtles
[
  file-type (word who ",")
  file-type (word ticks ",")
  file-type (word size ",")
  file-type (word bank-balance ",")
  file-print profit
]
file-close
```

Notice that the file is reopened and re-closed each tick. (`file-open` opens a file without erasing it, and subsequent output is written at the *end* of the file.) This keeps the file neatly closed if the model is stopped at any point. Also note use of the `word` primitive to add commas between the values on a line.

9.4.2 Example Code for Output of Summary Statistics

This example code writes summary statistics on all the turtles. The code to open the file is similar to the previous example except we no longer output a turtle identification number, and we change the column headers to reflect that the output is now statistics on turtle variables, not the values of the variables for each turtle. We now output three statistical measures of the turtle variable `bank-balance`: the minimum, mean, and maximum over the population. As previously, this code to write header information should be in `setup`.

```
; Open an output file for statistical results
; First, delete it instead of appending to it

if (file-exists? "BankBalanceStatsOutput.csv")
[
  carefully
      [file-delete "BankBalanceStatsOutput.csv"]
      [print error-message]
]

file-open "BankBalanceStatsOutput.csv"
file-type "tick,"
file-type "MeanSize,"
file-type "MinBankBalance,"
file-type "MeanBankBalance,"
file-print "MaxBankBalance"
file-close
```

Then the following code can be executed once per tick to write a line to this file. This code should also be in an observer-context procedure such as go or update-outputs.

```
; Write summary data on turtle states to output file
file-open "BankBalanceStatsOutput.csv"
file-type (word ticks ",")
file-type (word mean [size] of turtles ",")
file-type (word min [bank-balance] of turtles ",")
file-type (word mean [bank-balance] of turtles ",")
file-print max [bank-balance] of turtles
file-close
```

9.5 BehaviorSpace as an Output Writer

In chapter 8 we learned to use NetLogo's BehaviorSpace tool for automating simulation experiments. One very nice characteristic of BehaviorSpace is that it writes output files automatically, without adding any code to your model. You could, for example, produce BehaviorSpace output with the same information that the example output file code in section 9.4.2 does, by putting these statements in the "Measure runs using these reporters:" field of BehaviorSpace's experiment setup dialog:

```
mean [size] of turtles
min [bank-balance] of turtles
mean [bank-balance] of turtles
max [bank-balance] of turtles
```

It is important to remember exactly when BehaviorSpace writes its output, which we explained in chapter 8.

The one major limitation of BehaviorSpace for producing output is that it is not good at writing data on the individual agents. You can produce output for each agent (via reporters such as [size] of turtles), but its format—a string of values separated by spaces—is inconvenient.

9.6 Export Primitives and Menu Commands

NetLogo has several primitives (export-*) that save different parts of a model to .csv files. A particularly useful one is export-plot (also see export-all-plots), which writes all the points on a plot to a file. (Try using this primitive from the Command Center at the end of a model run.) The primitive export-world writes the values of all agent variables and all displays to one file, so you can investigate it thoroughly or even start new runs (via import-world) with the model in exactly the state it was in when saved.

The export primitives produce one output file each time they are executed, so they would be clumsy for creating output during a model run instead of just at the end of it. And, instead of using these primitives, you can export the same information from the File/Export option of NetLogo's main menu.

9.7 Summary and Conclusions

To use any model for science, you must be able to collect and analyze observations of the model's dynamics. Models as complex as ABMs produce many kinds of results, so we need a variety of observations tools, which NetLogo provides. By now you should be thoroughly familiar with:

- The "Interface tab" section of the Interface Guide part of the User Manual;
- The "Variables" special category of the NetLogo Dictionary; and
- The "View Updates," "Plotting," "Output," and "File I/O" sections of the Programming Guide.

Graphical displays are especially important as we build and test models, and when we need to demonstrate a working model to others. NetLogo's View and plots are extremely easy to use and can display an impressive amount of information. However, you must be careful in programming when and how these displays are updated. It is easy to make mistakes (especially, forgetting to update display variables such as pcolor after the agent variables they illustrate have changed) that can cause you to completely misunderstand what the actual model is doing. You should always test graphical displays by using tools like monitors and Agent Monitors.

To test and analyze behavior of the individual agents in your models, you will have to write your own code to produce output files. The NetLogo primitives for file output make it easy to produce custom output files that are easily read into spreadsheets or other analysis software.

When it is time to analyze system-level results from your model, you can also easily produce custom output files with summary statistics on your agents. However, NetLogo's BehaviorSpace tool and the export primitives provide very simple ways to create summary output files in .csv format.

We recommend that you put display updates and file outputs in a separate procedure that is executed at the end of each time step. Create a procedure called something like update-outputs that includes the statements to update the colors or any other display attributes of turtles, patches, and links; and to write output files. Call this procedure from the end of the go procedure (and at the end of setup). Then, using the controls on the Interface tab, set the View so it is updated once per tick. Now, you will know exactly when all observations were made: at the end of the time step, after all agent behaviors were executed. And putting all the observation code in one place keeps it from being mixed in with model code, where it is more likely to be neglected and messy.

9.8 Exercises

Here "the Butterfly model" refers to the version produced at the end of chapter 5.

1. In the Butterfly model, make the turtles display the direction they just moved by pointing (setting their heading) from the patch they moved from to the patch they moved to. You may have to change the size or shape of turtles to see the result. (You can do this very simply without knowing any geometry.)

2. In the Butterfly model, add these statements to the end of the setup procedure to cause one of the turtles to "take a walk" before the model starts:

```
ask one-of turtles
[
  repeat 20
  [
    set heading (heading - 30 + random 60)
    fd 1
  ]
]
```

What do you see when this executes? Why? What do you need to do to see this walk?

3. Add a plot of mean butterfly elevation to the Butterfly model.

4. Add another plot to the Butterfly model, this time a histogram showing the distribution of elevations of all the individuals. Make sure it is updated when the model is set up, and after each tick.

5. If you add a monitor to a model's interface (e.g., that displays mean [count flockmates] of turtles of the Flocking model), when is its value updated? Answer this question two ways: first by creating and experimenting with a monitor, and then by looking it up in the NetLogo documentation.

6. In chapter 8 we used the NetLogo library model Flocking, and saw that it took about 500 ticks for model-level variables such as mean number of flockmates and variability in agent headings to stabilize. Imagine that you want to run an experiment (e.g., the effect of the parameter minimum-separation on mean number of flockmates and standard deviation of headings) that starts only after this 500-tick "warm-up" period. How can you set up BehaviorSpace so that its model runs all start after exactly the same 500-tick warm-up?

7. Get a copy of Kornhauser et al. (2009) from the web and read it. This article contains many creative uses of NetLogo's View to visualize the states of your ABMs patches and turtles.

8. This is an excellent time to review the basic programming lessons from part I of the course, because the remaining chapters of part II require much more programming. There is a NetLogo refresher exercise at the chapter 9 section of this book's web site for you to download and complete.

Sensing

10.1 Introduction and Objectives

The sensing concept addresses the question, "What information do model agents have, and how do they obtain that information?" The ability of agents to respond and adapt to their environment and other agents depends on what information they have, so our assumptions about what agents "know" can have very strong effects. In many models, agents are simply assumed to have access to the variables of certain other agents or entities; for example, NetLogo models typically let turtles use any information held as variables of their patch. However, in some models it may be more realistic and useful to assume that sensing is imperfect, for example by assuming a certain level of error in the information agents obtain. And in some models it is important to represent *how* agents obtain information: the sensing process itself can be important to the question addressed by the model. Key decisions about modeling sensing are: (1) What variables of what kinds of model entities does an agent use? (2) Which particular agents or entities can the agent sense information from—e.g., only the closest neighbors, or other agents linked in some way? (3) How is the information sensed—does the agent simply know the information with no error, or with random uncertainty or a consistent bias, or does the model represent the mechanism by which information is obtained?

In this chapter we focus on the most common approach to sensing: assuming that agents can simply read information from other model entities. Once the modeler has decided what information the agents should "know," the programming problem is how to give the agents access to the variables that represent that information. This problem requires (1) identifying which of the model's entities contain the sensed information, in which of their variables; and (2) writing code to get the variable's value. We postpone detailed discussion of the first of these elements until the next chapter. Here, we examine how agents get information from other model entities: how does a turtle get the value of a variable belonging to its patch, or to neighboring patches, or other turtles? And how does a patch get values belonging to turtles or other patches?

We also introduce an example model that is a little more complex than the Butterfly model used in previous chapters. In this model, turtles represent people making business investment decisions, patches represent investment opportunities, and "sensing" has to do with what information investors have about which investments. We will continue to use this Business Investor model in chapters 11 and 12.

A final objective of this chapter is to introduce NetLogo's *links*, which are a type of agent designed to represent relationships among turtles. We explore links as a way to simulate sensing of information via a network, in one version of the investor model.

Chapter 10 learning objectives are to:

- Thoroughly understand what the "scope" of a variable is, and what entities in a NetLogo model can access global, patch, turtle and link, and local variables.
- Know how one agent can read the value of another agent's variables via the primitive `of`; and how an agent can change another agent's variables using `ask`.
- Know when and how to use local variables created via the primitive `let`.
- Reinforce the ability to write and use reporter procedures.
- Learn what NetLogo's links are, how to create them, and how to identify and get information from "link-neighbors."
- Be able to merge two or more agentsets into one.

10.2 Who Knows What: The Scope of Variables

When computer scientists talk about the "scope" of a variable, they refer to which entities in the program have access to directly read or change the variable's value. One of the basic ideas of the object-oriented programming style used by NetLogo is that most variables should have a limited scope: a turtle or patch should be the only part of the program that can directly read or change its own state variables. If other entities want access to one of those variables, they must "request" it by asking for the variable's value or telling the turtle or patch to change its variable. In general, it is safest for variables to have as narrow a scope as possible: the fewer entities with access to a variable, the less likely it is that the variable is ever mistakenly given a wrong value and the less likely it is that such an error would have widespread consequences.

In NetLogo there are several kinds of variables that differ in their scope, and it is very important to understand how these kinds of variables differ. It takes special primitives to use variables outside their scope, the topic we address in the next section. These sections should help clear up some of the confusion that you might have by now.

10.2.1 Global Variables

The global variables defined via the `globals` statement at the top of a NetLogo program have the broadest scope. They are called global because any object in the program can read and even change their value. If you define a global variable called `max-num-turtles` then this variable can be used anywhere in the code. Typically global variables are given a value in the `setup` procedure:

```
set max-num-turtles 100
```

Then any patch or turtle or link (or the observer) can use such code as

```
if count turtles > max-num-turtles
        [ask one-of turtles [die]]
```

And any object in the program can change the value of a global variable at any time; a patch procedure could have code like this:

```
if count turtles-here > 10
    [set max-num-turtles (max-num-turtles - 10)]
```

Global variables are particularly risky because anything in the model can change their value; and letting several different objects change the same global variable can make it difficult to keep track of its value. To avoid these risks, we typically use global variables for only a few purposes:

- Model-level parameters that control things like the initial number of agents or the number of ticks to be executed.
- Agent parameters: variables such as equation coefficients that are used by all the turtles or patches or links. Global variables are useful for parameters because their value can be set once (in `setup`, on the Interface, or by BehaviorSpace) then used by any objects in the model. But you must be extremely careful to make sure that no object in the model changes these parameters unless you really want it to.
- Variables representing conditions that do not vary among turtles or patches. Global "environmental" conditions (weather, external market prices, etc.) that vary over time but not among patches are typically represented using global variables.
- Any variable that we want to control from the Interface with a slider, chooser, switch, or input box. These Interface elements can only control global variables.
- Any variable that we want to change during a BehaviorSpace experiment. BehaviorSpace can only alter global variables.

Keep in mind that it is very easy to cause errors by forgetting that you changed a global variable by playing with its slider (or chooser, switch, or input box) on the Interface. When it is time to make serious model runs it is best to define and set the value of global variables on the Procedures tab (or in BehaviorSpace; section 8.3) instead of on the Interface.

10.2.2 Patch Variables

Patch variables include those defined by the programmer in the `patches-own` statement at the top of the program (if any), plus all the built-in patch variables such as `pcolor` and `pxcor` and `pycor`. The scope of these variables is an individual patch and any turtles on the patch. Each patch has its own value of each patch variable and can use or change it directly with statements such as

```
if pxcor < 0 [set pcolor red]
```

Turtles can directly access the variables of the patch they are on: they can read and set their patch's variables exactly the same way the patch can. In fact, the above statement reading `pxcor` and changing `pcolor` could be executed in either patch or turtle context. This capability is convenient but also a little dangerous: you must be careful that turtles do not change patch variables when they should not, or accidentally use patch variables instead of their own variables for location or color. (NetLogo of course never lets you change a patch's values of `pxcor` and `pycor` because patches cannot change their location.)

10.2.3 Turtle and Link Variables

Turtle and link variables are those defined in the `turtles-own` and `links-own` statements (if any) at the top of a program, plus the built-in variables for these two kinds of object. (In chapter 16 we will learn about *breeds*, which can also have their own variables.) The scope of these variables is limited to the individual turtle or link: each individual turtle has its own copy of each turtle variable with its own value, and only that turtle can directly read or set its variables. For example, we can define turtle variables for age and sex:

```
turtles-own [ age sex ]
```

and initialize the turtles with a sex value of either "M" or "F," and an age value from 0 to 60. Then, turtles could execute this procedure:

```
to age-and-die
  ifelse (sex = "M" and age > 60)
    or (sex = "F" and age > 65)
    [die]
    [set age age + 1]
end
```

This procedure directly accesses a turtle's age and sex variables, which can be altered only by the turtle. Therefore, this is a turtle-context procedure. The procedure could be executed via ask turtles [age-and-die]. But if you tried to get a patch to run this procedure on its turtles via ask patch 0 0 [age-and-die], NetLogo would not let you, for two reasons. First, patches cannot change turtle variables. Second, while NetLogo assumes turtles can access their patch's variables, it does not assume a patch even knows what turtles are on it (unless it finds out via the primitive turtles-here).

10.2.4 Local Variables

Anywhere in the code, local variables can be defined and used temporarily, and their scope is very limited. Local variables are created using the let primitive. They can be used by the object creating them only at the place where they are created, and they cannot be accessed by any other object.

A local variable disappears at the end of the procedure, or bracketed block of statements, in which it was created. This rule explains why these statements produce a syntax error saying that the variable number-of-offspring is undefined when the hatch statement tries to use it:

```
ifelse random-float 1.0 < probability-of-twins
  [let number-of-offspring 2]
  [let number-of-offspring 1]
hatch number-of-offspring
```

while these statements are legal:

```
let number-of-offspring 1
if random-float 1.0 < probability-of-twins
  [set number-of-offspring 2]
hatch number-of-offspring
```

In the first set of statements, let is inside brackets, so the local variable it creates—number-of-offspring—no longer exists after the code inside those brackets is done executing. In the second set, the let statement is outside the brackets, so number-of-offspring persists until the end of the procedure.

Note that after creating a local variable using `let`, you must use `set` to change its value, just as you do for any other variable.

Because local variables are so narrow in scope, they are especially safe to use. If a variable needs to be used only within a single procedure, and its value does not need to be saved from one time step to the next, then it should be a local variable. At the end of section 10.3 we mention one other unique characteristic of local variables.

10.3 Using Variables of Other Objects

The previous section discussed when NetLogo programs can "directly" access variables: using or setting variable values within their scope. But we often need to access variables outside their scope: one turtle needs to sense or change the value of some variable of some other turtle, a patch needs to sense and set the value of variables of the turtles on it or of neighbor patches, and so on. NetLogo puts no limits on one entity's use of another entity's variables, but this can be done only by stating explicitly which variable of which entity we want to access. The `of` and `ask` primitives are how we do this.

The `of` primitive is how an entity reads the value of another entity's variable. For example, if a patch needs the patch variable `elevation` of its neighbor patches, we use `[eleva-tion] of neighbors`. If a turtle needs the value of the variable `bank-balance` of another turtle it is linked to, we use `[bank-balance] of one-of link-neighbors`. (We explain `link-neighbors` later in this chapter.)

Note that `of` can be applied to one agent (e.g., `[bank-balance] of one-of link-neighbors`), in which case it produces a single number, or `of` can be applied to an agentset (e.g., `[elevation] of neighbors`), in which case it produces a NetLogo list of values. Because `of` produces a list when applied to an agentset, we can combine it with the primitives like `max` and `mean` that produce a statistic on a list:

```
let max-neighbor-elevation max [elevation] of neighbors
let mean-neighbor-elevation mean [elevation] of neighbors
```

(At this point it might be helpful to reread the Programming Guide section on lists.)

Now, what if an agent needs not just to read the value of another entity's variable, but to change that variable? If a turtle gets the elevations of its neighboring patches via `[eleva-tion] of neighbors`, you might think it could set the elevation of its neighbors to 23 via the statement:

```
set [elevation] of neighbors 23
```

But if you try this statement you will quickly learn that NetLogo does not allow it: you cannot use `of` to change a variable of some other entity. Instead, you must `ask` the other entities to change their variables themselves:

```
ask neighbors [set elevation 23]
```

The `ask` primitive identifies an agent or an agentset (here, `neighbors`) and tells it to execute the statement or statements within the brackets. The statements within the brackets are then in the context of whatever type of agent (turtle, patch, link, breed) is asked to execute them.

What if we want one entity to set the value of other entities' variables to its own value? For example, what if a patch should set the elevation of its surrounding patches to its own elevation? Doing this requires the very important primitive `myself`, which represents the entity that told the current entity to execute the code. (Look `myself` up in the NetLogo Dictionary.) The patch would use this statement (the parentheses are optional but make the statement clearer):

```
ask neighbors [set elevation ([elevation] of myself) ]
```

Now, if the patch is feeling superior and wants its neighbors to have only half of its elevation, it could use:

```
ask neighbors [set elevation ([elevation] of myself / 2)]
```

However, there is another way to do this:

```
let half-my-elevation (elevation / 2)
ask neighbors [set elevation half-my-elevation]
```

Why did we not have to use `of myself` to set the neighbor elevations to `half-my-elevation`? In NetLogo, you can use *local* variables directly, without `[] of myself`, in an `ask` statement. Because `half-my-elevation` is a temporary variable, it exists only for the patch creating it (the asking patch), so there is no ambiguity about which patch's value of `half-my-elevation` to use.

10.4 Putting Sensing to Work: The Business Investor Model

We demonstrate the sensing concept via a simple business model. In fields dealing with people and organizations, other terms such as "information acquisition" might be more intuitive than "sensing," but the concept is the same: when building an ABM we need to decide, and then program, exactly what information our agents have and where they get it.

This model simulates how people decide which business to invest in, when the alternative businesses they choose from differ in their annual profit and in their risk of failing such that the investors lose all their wealth. These business alternatives are modeled as NetLogo patches, a use of patches to represent something other than geographic space. In the first version, investors (modeled as turtles) are assumed to sense financial information only from neighbor patches; in a second version we will add links among turtles and simulate sensing through a network. As exercises, we will explore how different assumptions about sensing affect the results of this model. (In later chapters we will explore the effects of other assumptions, especially the way investors make trade-offs between profit and risk.)

10.4.1 The ODD Description

First, we describe the model in the familiar ODD format.

Purpose

The primary purpose of this model is to explore effects of "sensing"—what information agents have and how they obtain it—on emergent outcomes of a model in which agents make adaptive decisions using sensed information. The model uses investment decisions as an example, but is not intended to represent any real investment approach or business sector. (In fact, you will see by the end of chapter 12 that models like this one that are designed mainly to explore the system-level effects of sensing and other concepts can produce very different results depending on exactly how those concepts are implemented. What can we learn from such models if their results depend on what seem like details? In part III of this course we will learn to solve this problem by tying models more closely to real systems.)

This model could be thought of as approximately representing people who buy and operate local businesses: it assumes investors are familiar with investment opportunities within a limited range of their own experience, and that there is no cost of entering or switching investments (e.g., as if capital to buy a business is borrowed and the repayment is included in the annual profit calculation).

Entities, State Variables, and Scales

The entities in this model are investor agents (turtles) and business alternatives (patches) that vary in profit and risk. The investors have state variables for their location in the space and for their current wealth (W, in money units).

The landscape is a grid of business patches, which each have two static variables: the annual net profit the business there provides (P, in money units such as dollars per year), and the annual risk of the business there failing and its investor losing all its wealth (F, as probability per year). This landscape is 19×19 patches in size with no wrapping at its edges.

The model time step is one year, and simulations run for 25 years.

Process Overview and Scheduling

The model includes the following actions that are executed in this order each time step.

Investor repositioning. The investors decide whether any similar business (adjacent patch) offers a better trade-off of profit and risk; if so, they "reposition" and transfer their investment to that patch, by moving there. Only one investor can occupy a patch at a time. The agents execute this repositioning action in randomized order.

Accounting. The investors update their wealth state variable. W is set equal to the previous wealth plus the profit of the agent's current patch. However, unexpected failure of the business is also included in the accounting action. This event is a stochastic function of F at the investor's patch. If a uniform random number between zero and one is less than F, then the business fails: the investor's W is set to zero, but the investor stays in the model and nothing else about it changes.

Output. The World display, plots, and an output file are updated.

Design Concepts

Basic principles. The basic topic of this model is how agents make decisions involving trade-offs between several objectives—here, increasing profit and decreasing risk.

Emergence. The model's primary output is mean investor wealth, over time. Important secondary outputs are the mean profit and risk chosen by investors over time, and the number of investors who have suffered a failure. These outputs emerge from how individual investors make their trade-offs between profit and risk, but also from the "business climate": the ranges of P and F values among patches and the number of investors competing for locations on the landscape.

Adaptive behavior. The adaptive behavior of investor agents is their decision of which neighboring business to move to (or whether to stay put), considering the profit and risk of these alternatives. Each time step, investors can reposition to any unoccupied one of their adjacent patches, or retain their current position. In this version of the model, investors use a simplified microeconomic analysis to make their decision, moving to the patch providing highest value of an objective function.

Objective. (In economics, the term "utility" is used for the objective that agents seek.) Investors rate business alternatives by a utility measure that represents their expected future wealth at the end of a time horizon (T, a number of future years; we use 5). This expected future wealth is a function of their current wealth and the profit and failure risk offered by the patch:

$$U = (W + TP)(1 - F)^T$$

where U is expected utility for the patch, W is the investor's current wealth, and P and F are defined above. The term $(W + TP)$ estimates investor wealth at the end of the time horizon if no failures occur. The term $(1 - F)^T$ is the probability of not having a failure over the time horizon; it reduces utility more as failure risk increases. (Economists might expect to use a utility measure such as present value that includes a discount rate to reduce the value of future profit. We ignore discounting to keep this model simple.)

Prediction. The utility measure estimates utility over a time horizon by using the explicit prediction that P and F will remain constant over the time horizon. This assumption is accurate here because the patches' P and F values are static.

Sensing. The investor agents are assumed to know the profit and risk at their own patch and the adjacent neighbor patches, without error.

Interaction. The investors interact with each other only indirectly via competition for patches; an investor cannot take over a business (move into a patch) that is already occupied by another investor. Investors execute their repositioning action in randomized order, so there is no hierarchy in this competition: investors with higher wealth have no advantage over others in competing for locations.

Stochasticity. The initial state of the model is stochastic: the values of P and F of each patch, and initial investor locations, are set randomly. Stochasticity is thus used to simulate an investment environment where alternatives are highly variable and risk is not correlated with profit. Whether each investor fails each year is also stochastic, a simple way to represent risk. The investor reposition action uses stochasticity only in the very unlikely event that more than one potential destination patch offers the same highest utility; when there is such a tie the agent randomly chooses one of the tied patches to move to.

Observation. The World display shows the location of each agent on the investment landscape. Graphs show the mean profit and risk experienced by investors, and mean investor wealth over time. An output file reports the state of each investor at each time step.

Learning and collectives are not represented.

Initialization

Four model parameters are used to initialize the investment landscape. These define the minimum and maximum values of P (1000 and 10,000) and F (0.01 and 0.1). The values of P and F for each patch are drawn randomly from uniform real number distributions with these minimum and maximum values.

One hundred investor agents are initialized and put in random patches, but investors cannot be placed in a patch already occupied by another investor. Their wealth state variable W is initialized to zero.

Input Data

No time-series inputs are used.

Submodels

Investor repositioning. An agent identifies all the businesses that it could invest in; the neighboring (eight, or fewer if on the edge of the space) patches that are unoccupied, plus its current patch. The agent then determines which of these alternatives provides the highest value of the utility function, and moves (or stays) there.

Accounting. This action is fully described above ("Process overview and scheduling").

Programming note: Sensing the members of an agentset

In this model the investors must calculate the utility they would get at each of their neighboring patches. It seems natural to program this by having the investors loop through the neighbor patches and "sense" the profit and risk at each, while remembering which patch had the best utility. This approach is in fact what we would do in most programming languages, and we could do it in NetLogo by using the `foreach` primitive to loop through a list of neighbor patches. However, this approach is not natural or simple in NetLogo.

The natural way in NetLogo to sense which member of an agentset has the highest value of some function is to use the `max-one-of` primitive. In the Business Investor model the turtles can identify the patch offering highest utility using `max-one-of` in combination with a specific reporter that the *patches* use to calculate the utility that the turtle would obtain if in the patch. To identify and then move to the patch with highest utility, a turtle can execute this code:

```
; Identify the best one of the destinations
let best-patch max-one-of potential-destinations
                [utility-for myself]

; Now move there
move-to best-patch
```

In this code `potential-destinations` is an agentset containing any neighbor patches not already containing an investor, plus the current patch. (The following programming note, "Using -set to merge agentsets," will help you create this agentset.) `utility-for` is a reporter we still need to write (see the programming note "Writing reporters," below). It makes a patch calculate the utility an investor would get in it.

Programming note: Using `-set` to merge agentsets

It is not unusual to need to create an agentset by merging several other agentsets together: we want a new agentset that contains any agent that is in any of several other agentsets. In the Business Investor model, a turtle can choose its patch by creating an agentset of potential destination patches, then finding the best one. The set of potential patches must merge an agentset of all the empty neighbor patches with the turtle's current patch.

The agentset of empty neighbor patches can be created with

```
let potential-destinations neighbors with
    [not any? turtles-here]
```

Now, how do we add the turtle's current patch to the agentset? The primitive `patch-set` can do this because it generates a new patch set from whatever inputs you give it. (Similarly, the primitives `turtle-set` and `link-set` merge sets of turtles and links.) Read the documentation for `patch-set` carefully. Note that parentheses are needed around the word `patch-set` and its inputs whenever it is used to combine more than one input into an agentset. To create the agentset we want, follow the above statement with this one:

```
set potential-destinations
    (patch-set potential-destinations patch-here)
```

Or combine the two statements:

```
let potential-destinations (patch-set patch-here neighbors
    with [not any? turtles-here])
```

In section 10.4.3, we will need to use `patch-set` to combine several agentsets of potential destination patches.

Programming note: Writing reporters

Until now, we have used some of NetLogo's built-in reporters and we have written our own procedures, but we have not written our own reporters. Reporters are different from other procedures because they calculate a value and send it back to the code that called the reporter. Other programming languages use the term "function" for such blocks of code that calculate a number and return the result.

Why use reporters instead of just putting all the code in one procedure? Putting a calculation such as the determination of an investor's utility for a business into its separate reporter is what computer scientists call "modularity": writing isolated modules of code. Modular code has several benefits: it keeps the code that calls the reporter shorter, simpler, and easier to check; it lets us revise the reporter with less risk of messing up other code; and because we can call the same reporter from any place in our code, it often keeps us from having to code one behavior several times in different places. In NetLogo, the very handy `max-one-of` and `min-one-of` primitives can only be used with a reporter, making reporters especially useful for the methods that agents use to make decisions.

Reporters commonly take an input from the statement calling them. Here is an example reporter that calculates the square root of an input. You could call the reporter like this:

```
let side-length square-root-of area
```

where `area` is the number we want the square root of. The reporter looks like this:

```
to-report square-root-of [a-number]
  if a-number < 0.0 [user-message
    "Tried to take square root of a negative number!"]
  let root a-number ^ 0.5
  report root
end
```

The procedure `square-root-of` starts with the keyword `to-report`, so NetLogo knows that it is a reporter. The variable inside the brackets in the `to-report` statement (here, a-number) is a local variable that obtains the value of the input when the reporter is called (here, area). When the reporter is finished, the `report` statement sends the value of `root` back to the statement that called the reporter, which gives that value to the variable `side-length`.

In the previous programming note is this example statement:

```
let best-patch max-one-of potential-destinations
    [utility-for myself]
```

The reporter `utility-for` is passed an input `myself`, which is the turtle executing the statement. Because `potential-destinations` is an agentset of patches, the `utility-for` reporter must be in patch context. How would this patch reporter calculate the utility that the turtle would obtain in the patch? By sensing the information it needs from the turtle via the `of` primitive:

```
to-report utility-for [a-turtle]
  ; A patch-context reporter that calculates utility
  ; for turtle "a-turtle" in this patch
  ; First get the turtle's current wealth
  let turtles-wealth [wealth] of a-turtle

  ; Then calculate turtle's utility given its wealth and
  ; relevant patch variables
  let utility . . . (code that calculates utility)

  report utility
end
```

Now we will do some experiments to explore how different assumptions about sensing affect results of the investor model. First you should program the model using the hints and examples from the above programming notes. Test how you programmed the calculation of investor utility by producing an output file that contains the variables that go into the calculation and the utility determined each time step by each investor, and by reimplementing the calculation in a spreadsheet.

10.4.2 Effect of Sensing Range

Are investors more successful if they can sense more opportunities over a wider range? We can address this question with a simulation experiment that varies the distance over which investors can sense and consider patches. For example, we can look at what happens when we vary this distance from zero (investors cannot move from their initial patch) to 10. To do this, we will need a global variable for the sensing radius; then we can use BehaviorSpace to run a simulation experiment that increases this sensing radius parameter.

We can use the `in-radius` primitive to create an agentset of patches within the specified distance. But we also need to be careful to exclude patches that already contain a turtle while still including the turtle's current patch. We can use:

```
; First identify potential destination patches - within
; the sensing radius. The current patch is in the sensing
; radius but excluded because it contains a turtle.
let potential-destinations patches in-radius
    sensing-radius with [not any? turtles-here]

; Now add our current patch to the potential destinations
set potential-destinations
    (patch-set potential-destinations patch-here)
```

(Note that `patches in-radius sensing-radius`, unlike `neighbors`, includes the patch that the turtle is on. But the turtle's current patch is excluded from `patches in-radius sensing-radius with [not any? turtles-here]` because it always contains a turtle—the one executing the code.)

Now, if we vary the parameter `sensing-radius`, we get results like those shown in figure 10.1. These results indicate that being able to sense and occupy opportunities one patch away is much better than not being able to reposition at all, but that larger sensing radii provide little

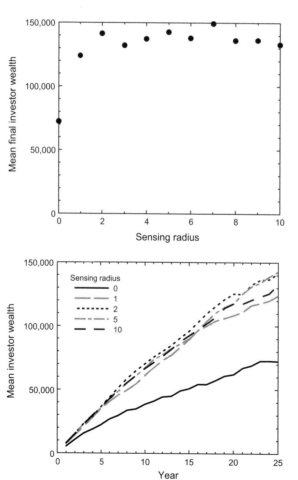

Figure 10.1
Example results of the Business Investor model. Upper panel: how the final mean wealth of the investors varies with their sensing radius. Lower panel: how mean investor wealth changes over simulated time for five values of sensing radius.

additional benefit. Is that because there is not much variation among patches in investor utility, or because all the good business opportunities are already occupied?

10.4.3 Sensing via Networking

What if investors can sense business opportunities that are linked via a network as well as those at neighboring patches? If investors have friends they can call about opportunities in the friends' business neighborhoods, will they be more successful? We can explore such questions by modifying the investor model so that each turtle is linked to some other turtles, from which it can sense information about patches. To do this we need to learn to use link primitives such as `create-links-to` and `link-neighbors`. The Programming Guide section on links will get you started; especially important is understanding the differences between directed and undirected links, and between in-links and out-links.

Let's start by assuming that each turtle is linked to a certain number (e.g., the global variable `number-of-links`, set to 5 for now) of other turtles, chosen randomly. These are directed links: if I am a turtle, I have five links *to* other turtles, from which I sense information about opportunities in their neighborhood. But these turtles I can sense from are not the same other turtles that have links *from* me and get information from me.

How do we create links so that each turtle has links to `number-of-links` other turtles? We can do this in the `setup` procedure where the investor turtles are created, by changing this:

```
; Create the investors
crt initial-num-investors
[
    move-to one-of patches with [not any? turtles-here]
    set wealth 0.0
]
```

to this:

```
; Create the investors
crt initial-num-investors
[
    move-to one-of patches with [not any? turtles-here]
    set wealth 0.0
    create-links-to n-of number-of-links other turtles
]
```

Will this work correctly? Or will it not work because the first turtle created has no other turtles to link with, the second turtle has only one other turtle to link with when it is created, etc.? If you read the NetLogo Dictionary entry for `create-turtles` carefully, you will see that this method *does* work: all the turtles are created first, then they all execute the commands within the brackets, so they do have all the other turtles to link with. (You can confirm this by trying it, then entering this statement in the Command Center: `ask turtles [show count my-out-links]` after hitting the `setup` button. Each turtle will print out the number of links it has to other turtles.)

Now, the only remaining problem is how to let the investors sense information from not only their neighbor patches but also the neighbor patches of the investors they have links to.

We do this by changing how we create an agentset of patches that are potential investments. In the original model we can create this agentset like this:

```
set potential-destinations (patch-set patch-here neighbors
    with [not any? turtles-here])
; Use a temporary test output statement
show count potential-destinations
```

To also include the patches that neighbor the linked turtles, we add this statement:

```
; Add neighbors of linked turtles
set potential-destinations (patch-set
  (potential-destinations) ([neighbors with
    [not any? turtles-here]] of out-link-neighbors))
```

We will concentrate on the complicated parts of this code in the next chapter, but make sure you understand that the primitive `out-link-neighbors` provides an agentset of the turtles that the current turtle has directed links *to*.

Now, when the investors select the patch in the agentset `potential-destinations` that provides the highest utility, just as they did in previous versions, they will sense information from, and consider, the patches surrounding the members of their network. Does this help the investors find better opportunities? We leave that question as an exercise.

10.5 Summary and Conclusions

This chapter should have challenged your ability to teach yourself NetLogo much more than previous ones have. We tried to teach you extremely important techniques such as using `of` and `ask`, writing reporters, and using links, all with little spoon-feeding. If you remain a little unsure about these techniques, don't worry: more practice lies ahead.

Modeling sensing—how agents obtain what information—is a unique and essential part of ABMs. Programming sensing requires model entities to access the variables of other entities. In NetLogo, as in other object-oriented programming languages, it is important to understand the "scope" of variables: global variables can be read and changed by anything in the model, while patch, turtle, and link variables can only be changed by the entity they belong to. (But a turtle can read and set its patch's variables as if they were the turtle's own.) Local variables can only be used by the entity creating them (via the `let` primitive) and only within the procedure or pair of brackets where they are created. But the `of` primitive makes it easy for a model entity to get the value of a variable belonging to another entity, and the `ask` primitive lets one model entity tell another to change one of its variables.

We introduced links in this chapter because they are useful for representing networks. Many ABMs use networks to represent how agents sense information because, in reality, many individuals (especially people) are connected to others in complex ways that cannot be represented spatially. Links make it easy to represent information flow that is not affected by spatial relations or distance.

The sensing concept naturally leads to the next concept we address: adaptive behavior. After agents sense some information, how do they use it to adapt? In chapter 11 we will continue using the Business Investor model to think about ways to model adaptive behavior.

1. Assuming you are in turtle context, write `let` statements that set a new local variable to:

   - The turtle's current size (using its built-in size variable);
   - The label text of the patch that the turtle is on;
   - The minimum size of all turtles;
   - The minimum size of all turtles except this one;
   - The color of the turtle identified by the variable `my-nearest-neighbor`;
   - A turtle selected randomly from those to which this turtle has an out-link;
   - An agentset containing all the turtles to which this turtle has an out-link;
   - The number of turtles on the patch that is one patch right and one patch up from the turtle's patch;
   - A list that contains the value of the variable `wealth` of all turtles on red patches; and
   - The standard deviation in `wealth` of all turtles.

2. Why does this code not work when a turtle wants to identify all the patches within a sensing radius that do not contain any turtles other than itself?

   ```
   let potential-destinations patches in-radius
       sensing-radius with [not any? other turtles-here]
   ```

3. The simulation experiment described in section 10.4.2 looked at how investors' success changed as their sensing radius—the range within which they could sense and consider business opportunities—increased. Design and conduct simulation experiments to address the question asked at the end of that section: why does success seem not to increase as sensing radius increases beyond 2? What are possible explanations, and how can you test whether they are important?

4. Using the model version developed in section 10.4.3, in which investors can sense patches via links to other investors, investigate how the number of links each investor has affects the investors' wealth. In a simulation experiment with 20 investors, vary the number of links per investor from 0 to 15. What happens? Why?

Adaptive Behavior and Objectives

11.1 Introduction and Objectives

The most important reason we use ABMs is that they allow us to explicitly consider how a system's behavior is affected by the *adaptive behavior* of its individual agents—how agents make decisions and change their state in response to changes in their environment and themselves. Therefore, deciding how to model agent decisions is often the most important step in developing an ABM. In this chapter we start thinking about modeling decisions by examining two closely related design concepts: adaptation and objectives. The adaptation concept considers what behaviors agents have and, especially, what decisions they use to adapt to changing conditions. If agents make decisions explicitly to pursue some objective, then the concept of objectives considers exactly what that objective is.

This chapter will be easier to understand if we establish some terminology. We use a set of terms adapted from Grimm and Railsback (2005). By agent *behavior* we refer to something an agent does during a simulation: it makes a decision, uses some resources, moves, produces offspring, etc. (We also refer sometimes to *system behaviors*, which are dynamics of the system of agents.) *Traits* are the rules we give agents for how to behave. Traits can be thought of as the programs that agents execute, and behavior (of the agents and the whole system) as the results of the agents executing those programs as a simulation proceeds. *Adaptive behaviors* are decisions that agents make, in response to the current state of themselves and their environment, to improve (either explicitly or implicitly) their state with respect to some objective. An *adaptive trait* is then the rules we give an agent for making the decisions that produce adaptive behavior. Finally, we refer to *theory* for adaptive behaviors (addressed in chapter 19) as adaptive traits that have been tested and found useful in particular contexts.

The objective of adaptive behaviors is often termed *fitness* in ecology because we assume that the ultimate objective of an organism's behavior is to increase its probability of producing offspring. Economists often use the term *utility* for the objective of human decision-making, but in general we can simply refer to the *objective* of decision-making. If an adaptive trait uses an explicit measure of the agent's objective (or fitness or utility), that measure is called the *objective function* (or *fitness measure* or *utility measure*).

Many decisions can be modeled as a process of identifying alternatives and selecting the one that best meets some objective. But in reality decisions are very often trade-offs: alternatives that provide more of one objective very often provide less of some other objective. Businesses may make decisions to increase profit or market share, but doing so often increases the risk of losses. Animals may seek higher food intake but must consider that eating more can also increase the risk of being eaten. Political parties may act to increase their popularity but at the cost of compromising their core values. Modeling these kinds of trade-off decisions is one of the most challenging, and innovative and exciting, tasks of agent-based modeling.

In this chapter we explore some of the many ways to model decision-making, especially when it involves trade-offs. We focus on theoretical approaches that assume agents explicitly evaluate alternatives by how well the alternatives meet a specific objective function. These approaches are more complex than the simple adaptive behaviors used in many ABMs (e.g., deciding whether to move uphill in the Butterfly model), and they offer great potential for making ABMs interesting and capable. We continue using the Business Investor model developed in chapter 10 because this model includes explicit, objective-based, adaptive traits. The investor model is especially interesting because it illustrates one way (explored even further in chapter 12) that agents can make adaptive trade-offs between objectives such as increasing profit and avoiding risk.

Learning objectives for this chapter are to:

- Understand and use the terminology for adaptive behavior and traits presented in section 11.1.
- Master the concept of, and NetLogo primitives for, subsetting agentsets and identifying the members of an agentset that maximize some objective function.
- Design and analyze simulation experiments that explore how adaptive traits affect the results of an ABM.
- Become familiar with two general approaches for modeling trade-off decisions: maximizing an objective function and "satisficing."

11.2 Identifying and Optimizing Alternatives in NetLogo

Adaptive traits can often use a process of (1) identifying alternatives, (2) eliminating infeasible alternatives, (3) evaluating the feasible alternatives by how well each meets an objective function, and (4) selecting the alternative providing the highest value of the objective. In NetLogo, decision alternatives are often modeled as agents—the patches a turtle could move to, the other turtles that one could interact with, the links that a turtle could follow, and so on. The Business Investor model from chapter 10 represents business opportunities as patches that turtles can move to. Therefore, the decision can be programmed as a process of subsetting an agentset: identifying an agentset of all possible destination patches, eliminating the infeasible ones to produce a subset of feasible patches, evaluating the feasible patches, and identifying the best. Here, we look at how to program adaptive traits in this way, and how in general to identify the particular members of an agentset that meet or optimize some criterion.

11.2.1 Identifying Decision Alternatives by Subsetting NetLogo Agentsets

When alternatives are modeled as agents, NetLogo's built-in agentsets can be used as a set of all alternatives to start the subsetting process. If a turtle is selecting a patch to move to, then the built-in agentset `patches` is automatically a set of all possible destinations. If the turtle can only move to an adjacent patch, then the primitive `neighbors` returns an agentset of such alternatives. Some other primitives that can be used to identify agentsets of alternatives are:

- `turtles`, `links`: all turtles, all links;
- `turtles-at`, `turtles-on`, `turtles-here`: the turtles on a specific patch; and
- `link-neighbors`, `in-link-neighbors`, `out-link-neighbors`: the turtles at other ends of the calling turtle's links.

Now, how can we create a subset of these agentsets with only the ones we are interested in? NetLogo has several primitives that create a new agentset containing only specific members of another agentset. For example, `other` creates an agentset that excludes the turtle, patch, or link executing the statement. The spatial primitives `in-radius` and `in-cone` create an agentset that includes only agents within a specified distance, or distance and angle. So when a turtle uses the statement `let destination-patches patches in-radius 3` NetLogo starts with all the patches, eliminates those more than 3 patches away, and puts the remaining patches in a new, local agentset called `destination-patches`.

(Be aware that creating a new agentset does not create new agents: the new agentset is a new set of references to agents that already exist. A turtle, link, or patch can be a member of several agentsets—just as your friend Thomas can be on your list of friends who let you borrow their car and on your list of friends who like to go to the pub and on your list of friends with whom you exchange Christmas gifts.)

The primitives `with-max`, `max-n-of`, `with-min`, etc., subset an agentset by identifying the members with the highest (or lowest) values of a criterion you specify via a reporter. For example, if you wanted an agentset of the 50 wealthiest investors, you could use

```
let fortune-50 max-n-of 50 turtles [net-worth]
```

where `net-worth` is the turtle variable for wealth.

The most general and important primitive for subsetting agentsets is `with`. The `with` primitive lets you write your own criteria for which agents to keep in the agentset. A simple example is `let potential-destinations neighbors with [pcolor = blue]`. The new local agentset `potential-destinations` is created by starting with the agentset of surrounding patches returned by `neighbors`. Then `with` uses the condition in brackets to exclude any members of this agentset that do not meet the condition of having their patch color blue.

These statements:

```
let potential-mates other turtles with [age > 20]
set potential-mates potential-mates with
    [sex != [sex] of myself]
set potential-mates potential-mates with [not married?]
```

create a local agentset of other turtles that meet three criteria for age, sex, and marriage status. (The boolean variable `married?` is set to true if the turtle is married.) Each of these statements subsets an agentset to exclude members that do not meet one or two criteria. The first statement creates an agentset `potential-mates` by starting with all turtles and using `other` to remove the calling turtle. It then uses `with` to remove any turtles with age not greater than 20. The second and third statements then modify `potential-mates` by using `with` to exclude the turtles that do not meet the criteria for sex and marriage status.

The `and` and `or` primitives let us combine multiple criteria in one subsetting statement. The above three statements could be replaced by this one:

```
let potential-mates other turtles with
    [age > 20 and sex != [sex] of myself and not married?]
```

We could create an agentset of other turtles that do *not* meet these marriage criteria using `or` to exclude turtles that meet any one of the inverse criteria:

```
let non-mates turtles with
    [age <= 20 or sex = [sex] of myself or married?]
```

Remember, from chapter 10, that if you want to merge several agentsets, the primitives `patch-set`, `turtle-set`, and `link-set` can do this.

11.2.2 Optimizing an Objective Function via `max-one-of`

Now we know how to identify a subset of feasible decision alternatives when they are simulated as NetLogo turtles, patches, or links; but how do we identify the alternative that is best with respect to our objective function? The most convenient way to do it is usually via the `max-one-of` reporter, which identifies the agent with the highest value of some other reporter (see the programming note on sensing the members of an agentset in section 10.4.1). When the objective is to avoid something, such as risk, we can of course use `min-one-of` instead.

When you read the NetLogo documentation carefully, you will see that if more than one agent provides the best value of its reporter, then `max-one-of` randomly selects one of them. If you want to use some other way to deal with ties, use the `with-max` (or `with-min`) primitive to get an agentset of agents with highest (or lowest) value of the reporter, and then program a way to select one when several alternatives have the same, best value of the reporter.

NetLogo brainteasers

1. Why is the turtle statement

```
let destination-patches [patches in-radius 3]
```

not the same as this?

```
let destination-patches
    [patches in-radius 3] of patch-here
```

2. If a turtle wants an agentset of patches within radius 3, except for its current patch, why is

```
let destination-patches other patches in-radius 3
```

not equivalent to this?

```
let destination-patches [other patches in-radius 3] of
    patch-here
```

3. If a turtle executes the statement

```
let nearest-neighbors turtles with-min [distance myself]
```

the new variable `nearest-neighbors` will be an agentset containing the calling turtle, not its nearest neighbor. But the statement

```
let nearest-neighbors other turtles
    with-min [distance myself]
```

sets nearest-neighbors to an empty agentset. However, the statement

```
let nearest-neighbors (other turtles)
    with-min [distance myself]
```

sets nearest-neighbors to the nearest other turtle(s). Why do these different statements produce these different results? And: what is the best way to identify the single nearest other turtle?

Answers:

1. in-radius measures the radius from the location of the agent calling it. In the first statement, the turtle might not be in the middle of its patch; in the second statement, in-radius is called by the turtle's patch (patch-here), so the radius is measured from the center of the turtle's patch.

2. Because other is applied to patches (other patches), it only excludes the calling agent if the calling agent is also a patch. In the first statement the calling agent is a turtle, so the turtle's patch is not excluded. In the second statement the of primitive causes patch-here—the turtle's patch—to execute other patches in-radius 3, so now other is called by a patch—and that patch is excluded from the resulting agentset.

3. In the first statement, with-min acts on the agentset other turtles, and the turtle with minimum distance to the calling turtle is the calling turtle itself. In the second statement, the primitive other is applied to the agentset produced by turtles with-min [distance myself], which we just found to be the calling turtle; when an agent applies other to itself, the result is nothing. In the third statement, the parentheses force NetLogo to apply with-min to the agentset other turtles, so we get the closest other turtle. But the best way to get the single nearest other turtle is

```
let nearest-neighbor min-one-of
    other turtles [distance myself]
```

You can test all these kinds of statements using Agent Monitors in NetLogo's View: start a model, select a turtle or patch, open an Agent Monitor to it, and type a test statement into its command window. For example, in a turtle Agent Monitor you can enter

```
ask other patches in-radius 3 [set pcolor blue]
```

You will see that the statement turns all patches within radius 3 blue, including the one containing the inspected turtle. If you then open an Agent Monitor to the turtle's patch and enter

```
ask other patches in-radius 3 [set pcolor red]
```

you see that the inspected patch does not turn red. If, in the turtle's Agent Monitor, you enter

```
show turtles with-min [distance myself]
```

you will see that show outputs the identification number (who) of the turtle being inspected.

In chapter 10 we presented the Business Investor model, which you should have programmed and used to study sensing. This model was also designed as an example in which adaptive decision-making is particularly important and clear: the investor agents make an explicit trade-off between the profits and risks of alternative business opportunities. The "reposition" behavior in this model is a decision of which business opportunity to invest in. This decision is clearly an adaptive behavior, as it is intended to increase the investors' utility (wealth) and it depends on the current state of the investors and their business opportunity environment. The trait used to model this behavior optimizes an objective function, a measure of expected future wealth. The investors identify a set of feasible alternatives—their neighbor patches and their current patch—and select the one that provides the highest value of the objective function.

While this model is simplistic as a business model, it serves as a conceptual example of the many problems in many fields in which agents must make trade-offs among competing and conflicting objectives. The decision-making method used by our business investors is a simplification of classical microeconomic and engineering decision analysis (e.g., Meredith et al. 1973)—agents define an objective function that quantifies how each of several conflicting objectives affects the agent's overall goal, then choose the alternative that maximizes this objective function. For our investors, the overall objective is wealth at the end of a future time horizon; higher profit produces higher wealth, but higher risk increases the chance that all will be lost because the business fails. This general approach has also been developed as decision theory in behavioral ecology (e.g., Mangel and Clark 1986; Houston and McNamara 1999) and for individual-based ecological models (section 7.5 of Grimm and Railsback 2005), and is certainly applicable to many other kinds of problems.

Does this trait succeed in giving the investor agents good trade-off behavior? Do they adapt well when either profits or risks change? Let's do some BehaviorSpace experiments to see what the investors do when we vary the range of profits and risks available to them. Start with the basic version of the investor model described in section 10.4.1.

To conduct these experiments, first

- Create two new parameters via sliders on the Interface. One of these parameters (called `profit-multiplier`) should adjust the annual profit available at each patch: in the `setup` procedure, after setting a patch's state variable for profit, multiply that profit by `profit-multiplier`. Likewise, the second new parameter (`risk-multiplier`) should adjust the patches' state variable for annual risk.

- Now, set up one BehaviorSpace experiment that varies `profit-multiplier` from 0.5 to 1.0 while holding `risk-multiplier` at its baseline value of 1.0; and a second experiment that varies `risk-multiplier` from 1.0 to 2.0 while holding `profit-multiplier` at 1.0.

These experiments examine how investor success and choice of patches vary as we cut the annual profit by as much as half and as we raise failure risk up to double its baseline value. How does the final investor wealth (averaged over all investors and, say, 10 replicates) respond, and how do the investors change their trade-off between profit and risk?

When we vary the range of profits available, the response is very simple. As the range of profits decreases by 50%, the investor's final wealth also decreases by about 50% (right to left in the top panel of figure 11.1). Over this range, investors use the same mean annual risk of about 0.038 and accept the corresponding decreasing profits (lower panel of figure 11.1).

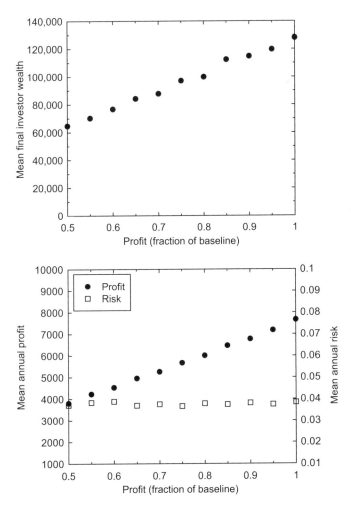

Figure 11.1
Results of an experiment varying the range of business profits. Top panel: the average final investor wealth as profits were varied from 0.5 to 1.0 of baseline. Bottom panel: mean annual profit (left axis) and mean annual risk (right axis) of patches chosen by investors, over all years, for the same scenarios as in the top panel. Dots are the mean of all investors over 10 replicates of each scenario.

The response to increasing risk is a little more interesting. As risk doubles, the average final investor wealth decreases only by about 25% (top panel of figure 11.2). The investors select slightly lower profits so the average risk they select increases by only 72% (from about 0.038 to 0.065), while the available risks increase by 100% (bottom panel of figure 11.2). In chapter 12 we will look more closely at how these agents adapt their investment decisions to profit and risk rates, and their current wealth.

11.4 Non-optimizing Adaptive Traits: A Satisficing Example

The previous section discussed ways to model adaptive decisions as an optimization problem, with agents selecting the alternative that provides the best value of an objective function. But in the real world, people and other agents make decisions without optimizing because it is expensive, or impossible, to sense and analyze all the necessary information about alternatives. Here we discuss an example of the many kinds of theory for such situations.

The trait we explore uses a decision technique called *satisficing*: finding an alternative that is good enough. When we lack information on how well each alternative would meet our

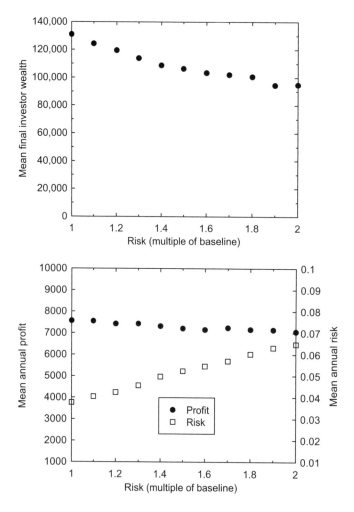

Figure 11.2
Results of varying the range of investment risk. Top panel: the average final wealth as risk varied from 1.0 to 2.0 times the baseline. Bottom panel: mean annual profit (left axis) and mean annual risk (right axis) of patches chosen by investors, over all years. Dots are the mean of all investors over 10 replicates of each scenario.

objective, we can instead seek and choose alternatives that meet our objective well enough. "Well enough" can be defined simply as a threshold value of the objective function that agents are satisfied with; we term this the *satisficing threshold*.

One common application of satisficing is when it is not reasonable to assume that agents can sense information about any alternatives except their current choice, so they must decide whether to switch to another alternative without being able to predict the consequences. An example would be if, in the Business Investor model, the investors cannot obtain information about the profits and risks at patches other than their own. In ecological modeling we sometimes model decisions individuals make, such as where to go on long-distance migrations, without the ability to sense conditions and predict consequences. So how do agents make such decisions and how can we model them?

One general approach to modeling decisions when nothing is known about the alternatives is to assume agents make changes (move, change investments, etc.) more frequently when their objective is not being met well, and change less frequently when their objective is being met. For example, when ecologists model how animals move in search of food, they sometimes use *klinokinesis*—moving faster when food intake is low and moving slower when it is high, and *orthokinesis*—turning less when food intake is low and more when intake is high (as in the Mushroom Hunt model of chapters 1 and 2). A simple version of this approach is to assume agents make a change when their objective function is below a satisficing threshold: if conditions are not good enough to eventually meet one's goal, then one tries something else.

We can use this approach in the Business Investor model. Let's assume now that investors cannot sense the profit or the failure risk at any patch except their own. Now they can no longer simply move to the patch offering highest expected wealth, because they only know expected wealth at their current patch. Instead, let's assume the investors have a satisficing threshold that is a specific rate of increase in their wealth, and that they abandon their patch if it provides a rate of increase less than that threshold. A reasonable value of the threshold is the rate of return they could get from a safe investment such as government bonds. We will use a 5% annual increase in wealth as an example threshold.

If an investor decides to move, how does it decide which patch to move to? We are now assuming investors have no ability to sense information about other patches, so it is reasonable to simply assume they choose a new patch randomly.

To implement this approach to adaptive behavior, make these changes in a new version of the NetLogo file:

- Create, using a slider, a new global variable `wealth-increase-threshold` that defines the satisficing threshold. Use a value of 0.05 (a 5% annual increase in wealth).

- Change the repositioning trait so a turtle moves only if its current utility—now defined as the expected annual rate of wealth increase over the time horizon, as a fraction of current wealth—is less than its satisficing threshold. This trait can be as simple as this:

```
to reposition
  ; Move if expected wealth increase rate is below the
  ; threshold
  ; Potential destinations do NOT include current patch
  if current-utility < wealth-increase-threshold
  [
     let potential-destinations neighbors
         with [not any? turtles-here]
     if any? potential-destinations
         [move-to one-of potential-destinations]
  ]
end
```

- Revise the objective function reporter so it reports the expected change in investment value as an annual rate of increase in wealth. Note that this reporter is now a *turtle* procedure, not a patch procedure, because the turtle no longer needs to evaluate utility at several different patches, only at its own. Therefore, we change the name of this objective function reporter from `utility-for` to `utility-here`.

We can estimate the annual rate of wealth increase by first dividing the expected change in wealth by the investor's current wealth (to get this rate as a fraction of current value), then dividing the result by the time horizon length (which is in years). To prevent division by zero when wealth is zero (on the first tick; any time the investor's business has failed), we can change values of zero to 1.0.

The new reporter can look like this:

```
to-report utility-here
  ; A *turtle* procedure that evaluates its patch
  ; This version uses the satisficing threshold approach

  ; First calculate expected wealth over the time horizon
  if wealth = 0 [set wealth 1.0] ; Prevent divide by zero
  let utility wealth + (profit * decision-time-horizon)

  ; Then factor in risk of failure over time horizon
  set utility utility *
      ((1 - annual-risk) ^ decision-time-horizon)

  ; Now report utility as expected annual rate of wealth
  ; increase, the expected change in wealth over the time
  ; horizon divided by the time horizon length.

  let wealth-increase-rate (utility - wealth) / wealth
  set wealth-increase-rate
      wealth-increase-rate / decision-time-horizon

  report wealth-increase-rate
end
```

■ Add a statement like this to update the turtles' variable for utility after they have finished moving (or staying) and calculating their new wealth:

```
set current-utility utility-here
```

When we use satisficing traits, we should be very curious about how the behavior of the agents and their system depends on the threshold: do agents fare better or worse if they accept lower vs. higher values of their objective as "good enough"? And we should be curious about how good the satisficing decisions are: how close to optimal is the satisficing behavior? This chapter's exercises include looking at these questions.

11.5 The Objective Function

So far in this chapter we have focused on ways to model how an agent makes decisions once an objective function has been defined. But clearly, designing the objective function is an important challenge: agents cannot make good decisions by pursuing the wrong objective. The potential to use more realistic and complex objective functions is an important advantage of ABMs. Economic models often assume businesses and people act only to maximize profit, but other motives are also important to real business owners: keeping a business alive, increasing market share, maintaining good relations with employees and the public, and having fun. Many ecological models assume organisms make decisions only to maximize individual growth, but reproducing and avoiding predation are also very important, and the importance of growth depends on the individual's current state (is it a juvenile or adult? starving or fat?).

So now, let's take a look at the objective function in the Business Investor model and explore some variations. This will give you a feel for how an ABM's results can change when the behavior objective changes.

First, we need to make some changes to the model as described in section 10.4.1 to make the effect of the objective function more interesting. Instead of assuming that all patches provide a positive profit, let's assume that some provide negative net profits: owning these businesses loses money. Therefore, we have a second way for the business to fail: if it loses more money in a year than the investor's wealth at the start of the year, so the investor is out of money. You can implement this change via these steps:

- Change the parameters for annual profit so each patch's profit is drawn from a range of -5000 to 10,000.

- Change the procedure that calculates the investor's change in wealth each tick so that the business fails (the investor has zero wealth) if it reaches a negative value: if the investor's wealth is less than zero, set it to zero.

- In the objective function, if the expected value at the end of the time horizon is negative, set it to zero. This avoids negative utility values, which make simple optimization impossible.

- Create a global variable that counts the number of business failures. Initialize it to zero, and add one to it each time a business fails either due to negative profit or randomly due to risk. Output this variable to an output window or to the Command Center.

Now we can explore some alternative objective functions and their effect on the model. The first alternative is the objective we have been using: maximizing the expected investor wealth over the time horizon, considering both profit and risk. Second, what if we assume investors ignore risk and make decisions only to maximize their profit? We can program this assumption simply by editing the reporter that calculates the objective function, removing the statement that factors failure risk into the function. (You might think that these heedless, greedy investors will make the most money, but you might be surprised.)

A third alternative, at the other extreme of risk-taking, is to assume investors are concerned not about maximizing wealth or profit but about simply avoiding failure as long as they can. The objective function now is the expected probability of surviving the two kinds of failure—due to random risk and negative profit—over the time horizon. This objective also requires only a simple change to implement. Remember that now the investors do not care how much profit they make, but want only to keep expected profit above zero while avoiding risk. Therefore, the part of the objective function based on profit can be set to 1.0 if the expected future wealth is greater than zero and to 0.0 if it is less than zero.

We certainly expect the model to produce different results when each of these objective functions is used. They cover a spectrum from being heedless of risk to being concerned only about avoiding failure. Which do you think produces the highest and lowest investor wealth? The least and most failures? We let you find the answers in the exercises.

11.6 Summary and Conclusions

In this chapter we took our first focused look at what is the most important part of many ABMs: the adaptive traits of the agents. The most compelling reason to use ABMs is because the adaptive behavior of individuals can have important system-level effects. If we fail to capture adaptive behavior adequately, we often will fail to capture how the system is organized and how it behaves. Therefore, it is extremely important to model behavior well. Adaptive traits are so important that we treat them (in chapter 19) as fundamental theory for ABMs.

Here we examined one way to model behavior: as a process of identifying alternatives, evaluating them according to an objective function, and selecting an optimal or good alternative. NetLogo includes primitives that make such objective-maximization traits easy to implement, especially when the alternative actions are represented as NetLogo agents: a patch to move to, a turtle to interact with, a link to follow, etc. This approach assumes that the agent knows what alternatives are available and can sense the information it needs to evaluate the objective function.

In many situations, agents cannot make optimal decisions because they do not have the necessary information or analytical capability. We showed that we can still model decisions quite easily when agents cannot sense information about alternatives or even know what alternatives are available. The example we explored uses the concept of *satisficing*: seeking an alternative that is satisfactory or sufficient, not optimal.

There are many ways of modeling behavior other than the few we illustrate in this chapter. In chapters 15 and 16 we will explore stochastic approaches that cause agents to reproduce patterns of decision-making observed in real individuals. Many other ABMs use rule-based traits that cause agents to reproduce behaviors observed in real individuals (Grimm and Railsback 2005 referred to this approach as "implicit fitness-seeking"; see their section 5.4.); one example is the Flocking model we examined in section 8.4. What is the best approach for your model? We must leave it to you to investigate the literature in your field (and it is almost always profitable to see what people have done in other fields) and perhaps do your own basic research to find appropriate adaptive traits.

11.7 Exercises

1. Write `let` statements that create a local variable containing the following agentsets. Find ways (e.g., test programs, statements typed into the Command Center or Agent Monitors) to show that your statements work correctly.

   - All the blue patches (NetLogo's standard blue color)
   - All the patches colored any shade of blue, from darkest to lightest
   - All the patches with at least one turtle on them
   - All the patches with no turtles on them
   - All the patches with exactly one turtle on them
   - All the turtles in the upper right quadrant of the space
   - In turtle context, all the turtles within a radius of 3 patches
   - In turtle context, all the turtles within a radius of 3, except the calling turtle (the turtle executing the statement)
   - In turtle context, all the turtles that have undirected links with the calling turtle
   - In turtle context, all the turtles that have directed links from the calling turtle
   - In turtle context, all the turtles on patches neighboring all the patches that contain a turtle with a directed link from the calling turtle
   - In patch context, all the turtles within a radius of 3 patches of the patch's center
   - In patch context, all the turtles on patches within a radius of 3 of the patch's center

2. Implement the satisficing trait for repositioning discussed in section 11.4. Check the code carefully in a spreadsheet, especially to make sure you calculate the annual rate of increase in wealth correctly. How does average investor wealth in this version compare to wealth in the original version that assumed investors could sense and optimize utility?

3. With the version of the investor model developed for exercise 2, conduct a sensitivity experiment to see how investor success varies with the threshold rate of wealth increase. Use BehaviorSpace to calculate the mean wealth of all investors at the end of 25-year simulations. What happens to this output when the threshold varies from zero to 0.2 (a 20% annual increase in wealth)? What patterns do you observe in investor movement over time? What do these results tell you about investing? About the model?

4. Implement the Business Investor model version and alternative utility functions discussed in section 11.5. Implement each utility function in a different version of the model (or in separate reporters of the same version) and test it. Use BehaviorSpace to run replicates of each version, and compare the mean and some measure of variability (e.g., standard deviation) in outputs such as the mean investor wealth and the number of failures. What qualitative patterns in investor wealth over time differ among the utility functions? Why?

Prediction

12.1 Introduction and Objectives

Prediction is fundamental to decision-making: when we make even the simplest decision, like whether to walk, bicycle, or drive the car to the grocery store, we automatically anticipate—predict—the consequences of each alternative. If I walk, how long will it take? If I bicycle, will I be able to carry as many groceries as I need? If I drive, will my friends see me and think I'm lazy and environmentally unconcerned? So modeling adaptive behavior often requires modeling prediction. Prediction is a particularly fascinating part of modeling behavior because a prediction is itself a model: we predict the outcomes of our actions by modeling those actions and their consequences in our heads. So we find ourselves modeling how the agent models!

The use of prediction in modeling behavior is quite common, although, as Holland (1995) points out, many models use *tacit* prediction: a simple and often hidden assumption. An ABM may simply assume that its agents seek higher levels of some resource, which conceals the tacit prediction that more resource will increase the agents' ability to meet their objectives. The Business Investor model, in contrast, includes *explicit* prediction: the agents decide which investment to choose by explicitly calculating the expected future value of their investment. Our ABMs will be better designed and understood if we think carefully about what predictions are essential to the agents' behaviors and how best to model those predictions. Even when simple tacit prediction is appropriate for some trait of an ABM, we should make it clear to ourselves and others that we are using prediction and have thought about how to model it.

Our conceptual objective in this chapter is to explore a few ways to model prediction in adaptive traits, and how different prediction models affect outcomes of an ABM. We will continue using the Business Investor model because its adaptive behavior includes an explicit, though simple, prediction. First we will simply explore the effect of changing the time horizon over which agents predict decision consequences. Then we will try modeling how agents develop expectations of how often events occur: instead of simply assuming that investors know their risk of failure, we will assume they update an estimate of that risk from their own experience.

In this chapter we also develop an important modeling skill that has nothing to do with NetLogo: testing and exploring submodels independently before we ever put them into an ABM. Even the seemingly simple traits we have been using so far can produce results that are difficult to understand within the context of the full model. As you move on to more realistic and complex models, it will become even harder to figure out how your detailed assumptions affect the model. To deal with this problem, you will learn to treat key parts of the ABM (adaptive traits being an especially important example) as separate submodels to test and explore by themselves before putting them together into the full ABM.

This technique will require you to program components of your model by themselves in another platform, similarly to how you have been testing your NetLogo software (e.g., in section 6.3.10). Our examples will show how to do this in a spreadsheet, but you could also use statistical and mathematical programming packages such as S-Plus or R, MATLAB, or Mathematica—as long as you can program functions and plot their results over a wide range of inputs.

This chapter's learning objectives are to:

- Develop an understanding that modeling decisions typically requires modeling some kind of prediction of the future consequences of choices, and of how different assumptions about prediction can affect model results.
- Learn the extremely important technique of programming and analyzing key submodels independently to thoroughly understand their behavior.
- Learn to draw contour plots of submodel results.
- Be exposed to one technique—Bayesian updating—for modeling how individuals update their predictions on the basis of what has happened so far.

12.2 Example Effects of Prediction: The Business Investor Model's Time Horizon

The Business Investor model's objective function uses explicit prediction: investors rate alternative business opportunities by their expected wealth at the end of a future time horizon, which they calculate by predicting that the annual profit and failure risk at each patch will be constant over the time horizon. This prediction is accurate for this quite simple model, but it can still be useful even if these rates vary over time. (In reality, we base decisions on inaccurate predictions all the time, and models using inaccurate predictions such as the one investors use here can produce realistic behavior; Railsback and Harvey 2002 provide an example of such a model.)

What is the effect of the decision time horizon in this model? Our intuition tells us that people out to make a quick buck make different decisions than people making a long-term investment; we would expect a short time horizon to cause investors to seek higher incomes but suffer random failures more often. What is the overall consequence of predictions over shorter vs. longer time horizons for the system as a whole, and why?

The obvious way to start addressing this question is with a BehaviorSpace experiment that varies the time horizon and measures how system-level results respond. Let's do this, using the model version produced in section 11.5 that has the annual profit of patches varying from -5000 to 10,000 and investors going bankrupt if either their wealth falls below zero or they experience a random failure. We must now of course make the time horizon a global variable controlled by a slider. We will examine the value of four outputs at the end of the 25-year simulations: mean investor wealth, mean annual profit and risk of patches used by investors (to examine the investors' choices), and total number of failures over the simulation. When we run this experiment we get results similar to those in figure 12.1.

These results seem to confirm our expectation that investors take fewer risks and suffer fewer failures when they base decisions on longer-term predictions. But what exactly causes this shift? How do investor trade-off decisions change as the time horizon changes? And what causes another important pattern in results—that investors move to lower-risk businesses over time as a simulation proceeds, no matter what the time horizon is (figure 12.2)?

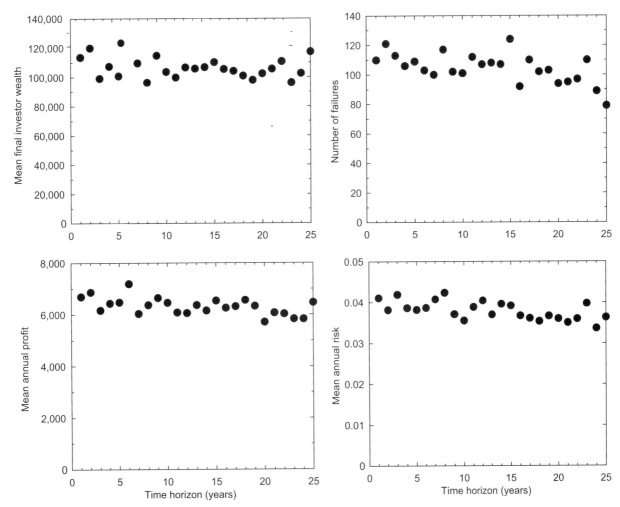

Figure 12.1
Sensitivity of Business Investor model results to the decision time horizon. Each dot is the result at the end of a 25-year simulation. In 25 model runs, the time horizon was varied from 1 to 25 years. Top left panel: mean wealth of investors. Top right: the number of business failures over the model run. Lower panels: mean profit (left) and risk (right) of patches occupied by investors at the end of the model run.

To answer these questions, we need to stop looking at results from the full model and instead analyze just the adaptive trait that the investors use. By isolating the function our investors use to evaluate alternatives, we can understand how the individuals behave without getting tangled up in the full model's complexity.

12.3 Implementing and Analyzing Submodels

The trick we introduce here is an essential one for building and using ABMs productively. We can look at an ABM as a collection of submodels that each simulate one key process, such as an adaptive individual behavior or an environmental process. We can then design, test, and

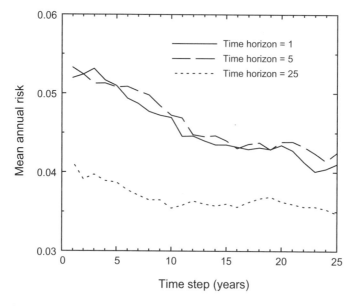

Figure 12.2
Change over time in the mean annual risk of the businesses chosen by investors (in NetLogo, `mean [annual-risk]` of `turtles`, where `annual-risk` is the patch variable for risk).

analyze these submodels separately before putting them in the full model. Analyzing submodels separately is important because it lets us fully understand and test each part of an ABM. If we wait until the full model is assembled to try to understand it, or find any weaknesses or mistakes in it, the job will be much harder or even impossible. And this technique can save a great deal of time and work. Implementing submodels separately before building the full ABM may seem like extra work but actually helps us design and debug the key parts of the ABM quickly and thoroughly, so assembling and testing the full model is much easier.

The way we analyze a single submodel such as the utility function in the investor model is by programming it separately in software that lets us graph and see how it behaves in all possible conditions. The ability to draw contour plots is essential, because contour plots let us see how a submodel behaves as we vary two of its parameters simultaneously. Here, we use Excel because it is widely used (*not* because of its contour plotting capabilities, which are rudimentary), but feel free to use any platform you are comfortable with. (You could even write a special NetLogo program to draw the equivalent of a contour plot on the View.)

Let's implement the original Business Investor model's utility function in Excel. Our goal is to see, via a contour plot such as figure 12.3, how an agent's utility value—expected investor wealth at the end of the time horizon—varies with the two patch variables for investment profit and risk of failure. This plot illustrates behavior because investors always try to move to patches with higher values of utility, or uphill on the contour plot.

To make this contour plot, we need to calculate utility for combinations of profit and risk that include the entire range of both. The range of profit in this version of the model is from -5000 to 10,000, so we can use 16 values that step through this range by 1000. Risk ranges from 0.01 to 0.1, so we can use 10 values stepping by 0.01. The simplest way to calculate utility for all these 160 combinations of profit and risk is in a simple column (figure 12.4).

However, to draw a contour plot Excel requires the values to be in a matrix, so we instead program utility as shown in figure 12.5. This matrix format becomes very cumbersome for more complex submodels, especially when we want to plot how the submodel responds to more than just two variables. Alternatives include doing all the calculations in columns and then manually copying the results into the matrix format, learning Excel's complicated but extremely useful PivotTable and PivotChart features, or using separate plotting software that can import data from

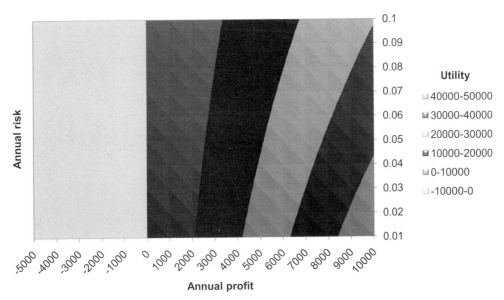

Figure 12.3
Contour plot of investor utility (defined in the "Objective" ODD element of section 10.4.1) when the investor's current wealth is 0 and the time horizon is 5 years. Each band is a range of utility values; the borders between bands are "contour lines" of equal utility. Utility is highest in the lower right corner.

	A	B	C	D	E
1	Time horizon:		25		
2	Current investor wealth:		0		
3					
4	Patch annual profit	Patch annual risk	Expected profit over time horizon	Expected probability of not failing over time horizon	Utility
5	-5000	0.01	=$C$2+(A5*$C$1)	=(1-B5)^$C$1	=MAX(0,D5*C5)
6	-5000	0.02	-125000	0.603	0
7	-5000	0.03	-125000	0.467	0
8	-5000	0.04	-125000	0.360	0
9	-5000	0.05	-125000	0.277	0
10	-5000	0.06	-125000	0.213	0
11	-5000	0.07	-125000	0.163	0
12	-5000	0.08	-125000	0.124	0
13	-5000	0.09	-125000	0.095	0
14	-5000	0.1	-125000	0.072	0
15	-4000	0.01	-100000	0.778	0
16	-4000	0.02	-100000	0.603	0
153	9000	0.09	225000	0.095	21292
154	9000	0.1	225000	0.072	16153
155	10000	0.01	250000	0.778	194455
156	10000	0.02	250000	0.603	150866
157	10000	0.03	250000	0.467	116744
158	10000	0.04	250000	0.360	90099
159	10000	0.05	250000	0.277	69347
160	10000	0.06	250000	0.213	53228
161	10000	0.07	250000	0.163	40739
162	10000	0.08	250000	0.124	31091
163	10000	0.09	250000	0.095	23658
164	10000	0.1	250000	0.072	17947

Figure 12.4
Spreadsheet implementation of the utility function. For each value of profit from -5000 to 10000 (column A), there are 10 values of risk in col. B. The formulas for columns C–E are shown in row 5. Lines 17–152 are not shown.

Excel. Or we can do the analysis in other software that is more adept at contour plotting; for example, you can produce the same contours using the statistical language R via this simple program:

```
# create the profit and risk variables
profit<-seq(-5000,10000,by=1000)
risk<-seq(0.01,0.1,by=0.01)
```

```
# create the array to contour
profit.risk<-expand.grid(profit=profit,risk=risk)

# function to calculate utility
calc.utility<-
  function(current.value,time.horizon,profit.risk)
  {
max(0,(current.value+(time.horizon*profit.risk["profit"]))
* (1-profit.risk["risk"])^time.horizon)
  }

# create an array containing utility values
utility<-
  apply(profit.risk,1,calc.utility,current.value=0,
    time.horizon=5)

# draw the contour plot
filled.contour(profit,risk,
  matrix(utility,ncol=length(risk)),
  col=grey(0:24/24),xlab="profit",ylab="risk")
```

	A	B	C	D	E	Q	R	S
1	Time horizon:		25					
2	Current investor wealth:		0					
3								
4			=$C$2+($C$1*C5)	-100000	-75000	225000	250000	Expected profit over time horizon
5			-5000	-4000	-3000	9000	10000	Patch annual profit
6	=(1-B6)^$C$1	0.01	=IF(C$4>0,$A6*C$4,0)	0	0	42795	47550	
7	0.603	0.02	0	0	0	135780	150866	
8	0.467	0.03	0	0	0	105069	116744	
9	0.360	0.04	0	0	0	81089	90099	
10	0.277	0.05	0	0	0	62413	69347	
11	0.213	0.06	0	0	0	47905	53228	
12	0.163	0.07	0	0	0	36665	40739	
13	0.124	0.08	0	0	0	27982	31091	
14	0.095	0.09	0	0	0	21292	23658	
15	0.072	0.1	0	0	0	16153	17947	
16	Probability of not failing over time horizon	Patch annual risk						

Figure 12.5
Utility function programmed in Excel in matrix format (with columns F–P hidden). Patch annual profit values increase from -5000 to 10,000 in columns C to R; annual risk values increase from 0.01 to 0.1 in rows 6–15. Row 4 calculates the investor value at the end of the time horizon. Column A calculates the probability of not having a failure over the time horizon. The contour plot in figure 12.3 is drawn by selecting cells B5–R15; cell B5 must be empty.

Now that we've programmed the utility function in Excel, let's look at some contour plots and try to understand investor behavior.

At the start of a simulation (or immediately after a failure) the investor's wealth is zero. The utility function, with a time horizon of five years, is illustrated in figure 12.3. This contour is certainly informative: it shows that utility varies strongly with profit but little with risk. Remember that the combinations of profit and risk that fall on the same contour line offer the same utility; this means that a large change in risk is compensated for by a small change in profit. (This is less true at high values of profit.) The investors, moving uphill on this plot, will almost always move toward higher profit (to the right) instead of toward lower risk (downwards). This graph certainly helps explain the results we saw in figure 11.2: as risk increased, the investors accepted only small decreases in investment profit in exchange for lower risk. Also note that any location with negative or zero profit has zero utility: investors have no preference among all the patches with negative profit because they will fail at all of them.

But this plot only represents behavior when investors have zero wealth. How does their decision trait change when they have accumulated, for example, 100,000 units of wealth? When we change the investor wealth and redraw the contours, we get figure 12.6.

We can see several important changes. First, the contour lines are angled quite steeply, meaning that now risk has almost as much effect on utility as profit. An investor gets the same utility from a patch with, for example, annual profit of 10,000 and risk of 0.078 as from a patch with profit of only 1000 and risk of 0.01. Why is risk much more important when the investor has more money? The answer is that a failure is now much more expensive. Remember that a failure causes the investor to lose all its wealth, so the utility function multiplies the probability of not failing by the sum of current wealth and expected income. When an investor has zero wealth and expects an income of up to 10,000, the most it can lose by failing is 10,000. But if

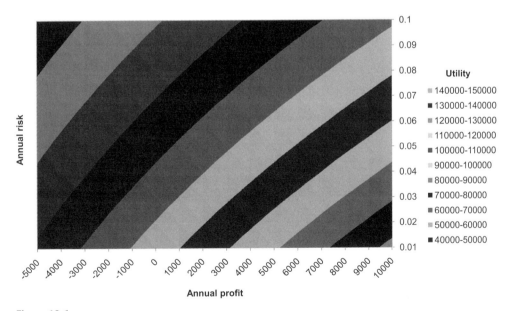

Figure 12.6
The utility function with current investor wealth = 100,000 and time horizon = 5. Utility is highest in the lower right, where risk is lowest and profit highest, and lowest in the upper left.

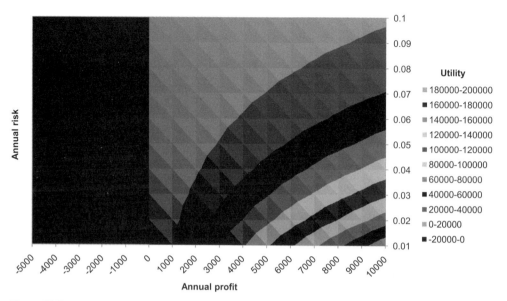

Figure 12.7
Utility function with investor wealth = 0 and time horizon = 25.

the investor has wealth of 100,000 and fails, it loses over 100,000 units. As agents accumulate resources, losing those resources becomes more costly.

The second important change now is that no combinations of profit and risk offer zero utility. Agents can get higher utility by moving from a patch with high risk and positive profit to a patch with negative profit but low risk. Why? Because the risk of failure by going bankrupt is gone: even when losing 5000 per year, an investor cannot lose all of its 100,000-unit wealth over the five-year time horizon. Investors can chose to accept a negative income to reduce their risk of losing their entire wealth. This ability to absorb loss to reduce risk when investors accumulate some wealth explains why investors shift to lower-income, lower-risk patches over time during a simulation (figure 12.2).

Now let's see how this trait changes with the time horizon: how do decisions change if investors predict consequences over much more than five years? If we redraw the contour plot using a time horizon of 25 instead of 5 years (figure 12.7), we see that the contour lines are now curved much more than in figure 12.3. This means that the trade-off between profit and risk changes more as investors go from low to high risk or low to high profit. When profit is low, utility is sensitive to profit, not risk (the contour bands in figure 12.7 are vertical when profit is less than about 1000). But at higher values of profit (moving to the right), the contour bands become more horizontal as utility depends more and more on risk.

Why is risk more important when we change the time horizon from 5 to 25 years? You can understand this question by doing one more simple analysis: plot the risk term of the utility function (i.e., probability of not having a random failure = $(1 - \text{risk})^{\text{time horizon}}$) as the time horizon changes. As the time horizon gets longer, the probability of surviving it without a random failure gets much lower, so reducing the risk of such a failure gets much more important.

Now, how does increasing the time horizon from 5 to 25 years affect investors that have accumulated a substantial wealth? The contour plot of utility for investors with current wealth of 100,000 (figure 12.8) shows that patches offering profits of –4000 or less have zero utility, whereas when the time horizon was 5 years (figure 12.6) even the patches with most-negative

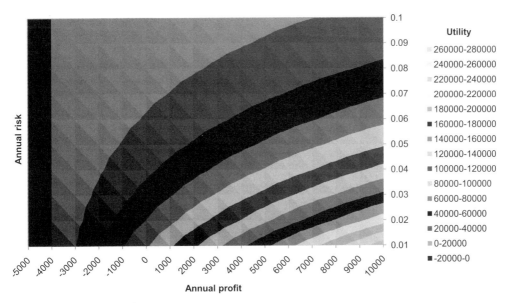

Figure 12.8
Utility function with investor wealth = 100,000 and time horizon = 25.

profits had positive utility values. An annual loss of 4000, over a 25-year time horizon, would use up the investor's current wealth of 100,000 and cause a failure. The longer time horizon therefore makes investors less willing to accept short-term losses in exchange for reduced risk. (This dynamic seems counterintuitive: we would expect long-term investors to be more, not less, willing to accept short-term losses in return for lower risk. But remember that our agents are using a quite stupid prediction: that the investment profits and risks they choose today will persist for the whole time horizon, even though they actually get an opportunity to change their investment each year. How would their behavior change if instead they used a more sophisticated prediction, that better combinations of profit and risk are likely to become available in the future?)

12.5 Modeling Prediction Explicitly

Probabilities and probability theory can be used to model prediction. If, for example, you knew that the probability per year of a business failing is 0.1, then you could reasonably predict that the business will not fail next year, but you could also predict (by applying some theory and assumptions such as that failures are independent over time) that it will fail within 10 years (the probability of not failing over a period of 10 years is 0.35). In the investor model, the agents do not use probability to predict *whether or not* a failure occurs over the time horizon, but instead use the failure risk to evaluate *how likely* they are to go bankrupt over the time horizon. So far, we have assumed that the investors simply know what this failure risk is, whereas in reality investors typically have limited information and limited ability to predict the frequency of unanticipated failures. What if we wanted to be more realistic and model *how* investors might estimate the risk of failure from reasonably available information? For example, let's assume that investors predict the future failure risk of a patch from (a) the mean rate of failure over all patches (e.g., some historical statistic on how many businesses fail per year) and (b) the patch's history: how frequently businesses have failed at the patch since the start of the simulation.

It turns out that Bayesian statistical theory provides a simple way to do this, and has been used often to model prediction in decision-making (e.g., Valone 2006). (You do not need to understand the theory—just accept the method and equations we provide. Technically, we are using conjugate priors of a binomial distribution, which are modeled as a beta distribution; see, e.g., section 2.4.3 of Stauffer 2008.) First, let's assume that our investors start with knowledge of the mean failure risk across all patches and its variance among patches (although we could just as easily assume they have inaccurate information, e.g., by guessing wildly). We distributed risk values uniformly over a range of 0.01 to 0.1, so the mean risk R is about 0.055 and its variance V is about $(0.1 - 0.01)^2/12 = 0.0007$ (this is the equation for variance of uniformly distributed values).

We calculate the two parameters of a beta distribution as

$$\alpha = (R^2 - R^3 - RV)/V$$

$$\beta = (R/V)(1 - R)^2 + (R - 1)$$

Now, if we assume our investors update their estimate of a patch's failure risk each time step by considering how many failures have occurred there so far, we can model this update using Bayesian methods. This is very easy: investors can calculate an updated prediction of annual risk for a patch by adding one to α if there was a failure at the patch, or adding one to β if there was *not* a failure, and then calculating the new estimated risk of failure (F_e) as

$$F_e = \alpha/(\alpha + \beta)$$

This is a simple and mathematically justified theory for how investors update their estimate of a patch's risk, but what kind of behavior will it produce? For example, if a patch has a risk of 0.09, compared to the mean risk of 0.055, what will the estimated risk F_e be if there has been one failure in 10 years of being occupied? How rapidly does F_e vary over time as failures do or do not occur? Will it converge to a relatively accurate value over our 25-tick simulations? To answer these questions, we return to the trick we learned in the preceding section: programming and analyzing the trait by itself. We can, for example, program it in a spreadsheet (figure 12.9).

Programming and analyzing this trait will be an important exercise for this chapter. It turns out that there are important things you need to learn about the trait before using it!

12.6 Summary and Conclusions

Adaptive traits in ABMs almost always include some form of prediction, but modelers often do not think about prediction or even recognize it in their own models. We include prediction in our design concepts because even the simplest decisions involve prediction, and assuming our agents make predictions allows us to design powerful decision-making traits.

There are many ways of modeling prediction. In our Business Investor model, agents simply predicted that the investment profit and risk they currently sense will persist unchanged over a time horizon. Other more complex and potentially useful techniques include statistical models (regression; the Bayesian updating technique presented in section 12.5) to extrapolate predictions from recent experience. Stochastic techniques (including the probability distributions we discuss in chapter 15) can also be used to predict how often, how many times, or how long until some future event happens. But we have seen from the Business Investor model that even an adaptive trait using extremely simple prediction can produce complex behavior.

The technique of programming and analyzing submodels separately, introduced in this chapter, is essential. Whenever you need to add a trait that is not extremely simple, you will save time and frustration by programming it in a spreadsheet or similar platform, calibrating it

	A	B	C	D	E	F
1	Max risk of patches:		0.1			
2	Min risk of patches:		0.01			
3	Mean risk of patches:		0.055			
4	Var of risk:		0.0007			
5	Actual risk of this patch:		0.08			
6						
7	Tick	Random #	Failure?	alpha	beta	Estimated risk
8	0			4.2	71.8	0.055
9	1	0.8764311	=IF(B9<$C$5,1,0)	=D8+C9	=IF(C9<1,E8+1,E8)	=D9/(D9+E9)
10	2	0.6104156	0	4.2	73.8	0.054
11	3	0.5549806	0	4.2	74.8	0.053
12	4	0.5125167	0	4.2	75.8	0.052
13	5	0.8787160	0	4.2	76.8	0.052
14	6	0.2022973	0	4.2	77.8	0.051
15	7	0.6505170	0	4.2	78.8	0.050
16	8	0.6587123	0	4.2	79.8	0.050
17	9	0.8104717	0	4.2	80.8	0.049
18	10	0.7471529	0	4.2	81.8	0.049
19	11	0.3729113	0	4.2	82.8	0.048
20	12	0.9768573	0	4.2	83.8	0.048
21	13	0.0062956	1	5.2	83.8	0.058
22	14	0.8918113	0	5.2	84.8	0.058
23	15	0.8882637	0	5.2	85.8	0.057
24	16	0.0663809	1	6.2	85.8	0.067
25	17	0.0578934	1	7.2	85.8	0.077
26	18	0.0899576	0	7.2	86.8	0.076
27	19	0.0943157	0	7.2	87.8	0.076
28	20	0.3989991	0	7.2	88.8	0.075
29	21	0.0655407	1	8.2	88.8	0.084
30	22	0.5555867	0	8.2	89.8	0.083
31	23	0.7185315	0	8.2	90.8	0.083
32	24	0.2513982	0	8.2	91.8	0.082
33	25	0.3322592	0	8.2	92.8	0.081

Figure 12.9
Spreadsheet implementation of the Bayesian trait for estimating risk. This code implements an investor's estimate of the risk in one patch as a 25-tick simulation proceeds. Notes: The zero tick (row 8) provides the initial estimate of risk; cells D8 and E8 contain the equations for α and β shown in the text, where R is cell C3 and V is cell C4. For subsequent ticks, columns D and E contain the updated values of α and β, and column F contains the resulting estimate of risk. Column B contains random numbers that are compared (in column C) to the patch's actual risk to determine whether a failure occurs each time step.

if necessary (covered in chapter 20), and exploring its behavior under all possible conditions. You are very likely to find and fix problems that would otherwise be difficult to diagnose and understand, and to have a much better idea of how your model behaves and why. Documenting your independent test will greatly increase the confidence of your clients that you really understand what your ABM does and that its results—and your interpretation of them—are robust and dependable.

12.7 Exercises

1. In the Business Investor model, what if the investors made decisions to maximize their expected wealth at the end of the 25-year simulation, e.g., because they plan to retire after

25 years? Instead of always calculating their utility over a time horizon of 5 years, the investors use the number of years remaining until they retire as the time horizon. First, analyze the model's utility function by itself to develop a hypothesis for how you expect agent decisions to change over time. Then modify the investor model to simulate this new trait. Compare the simulation results to your hypothesis.

2. Program and analyze the Bayesian updating trait for estimating failure risk (from section 12.5) in a spreadsheet. What characteristics of it do you think would cause substantial changes in the investor model? (Hint: look at patches with low risk.) Hypothesize how the full model's behavior would change if you used it.

3. Implement the Bayesian updating trait for estimating risk in the Business Investor model and test your hypothesis (from exercise 2) of how it affects behavior.

Interaction

13.1 Introduction and Objectives

Local interaction is one of the defining characteristics of ABMs. The term *interaction* refers to how agents communicate with or affect each other, such as by exchanging information, competing for resources, helping or fighting each other, or conducting business. We also use "interaction" for how agents affect, and are affected by, their environment; environmental interactions such as consuming and producing resources are very important in many ABMs.

System-level models, in contrast to ABMs, must use the same equations and parameters to represent the effects of interaction on all members of the system. For example, competition among members of a population might be represented as a rate at which success of all members decreases as the number of members increases. In an ABM we can model interactions explicitly as ways that individual agents affect each other and their environment. Consequently, the effects of interaction, and even the kinds of interaction, can depend on the state of the agents (including, often, their location) and on their environment. Interaction in ABMs is often *local*—each agent affects only a few nearby others and only its local environment—whereas interaction in system-level models is *global*: all members of the system affect all other members.

Interaction can be modeled as *direct interaction*, in which agents directly affect each other, by exchanging information or resources, trying to eat or steal from other, etc. But ABMs also often represent *mediated* interaction, in which agents interact indirectly via some mediating resource. Competition for a shared resource, and signaling by producing or using something (a chemical; information), are kinds of mediated interactions.

Our objectives in this chapter are to review kinds of interaction often used in ABMs, and the NetLogo techniques for programming them. We introduce a new model that uses both direct and mediated interactions, and lets us see what happens when interactions are global instead of local.

Learning objectives for chapter 13 are to:

- Understand how to model and program competition among agents for a shared resource.
- Master how to program direct interactions, when one agent makes another agent execute some behavior.

- Understand the difference between local and global interactions and how to model each.
- Learn how to program long-term interactions among turtles using two techniques: creating a NetLogo link between interacting turtles, and giving turtles a variable that contains the other turtle(s) it interacts with.
- Understand how NetLogo's lists can be used to model memory.

13.2 Programming Interaction in NetLogo

Interaction occurs even in the simple models we have already built and explored and in fact is almost impossible to avoid in ABMs. One especially common kind of *indirect* interaction is competition for a resource. In NetLogo, it is especially easy to program competition among turtles for a resource that belongs to the patches. The turtles "consume" the resource simply by reducing the value of the resource variable for the patch they are on. If the resource is regenerated, regeneration is modeled as a patch action.

The Cooperation model in the Social Science section of NetLogo's Models Library provides a good example of this approach: cows compete with each other for a resource—grass—which is a patch variable. One way that the cows in this model eat grass is via this procedure:

```
to eat-greedy  ;; turtle procedure
  if grass > 0
  [
    set grass grass - 1
    set energy energy + grass-energy
  ]
end
```

Here, the cows first check to see if there is any grass available in their patch; if so, they consume one unit of it. This unit of grass increases the cow's energy. Because grass is a patch variable, cows can be affected by the other cows on their patch: if too many other cows have eaten there, then there is no grass left. The patches regenerate grass, depending on how much has been consumed by cows, in their procedure grow-grass.

(When modeling competition for patch resources in this way, it is important to think carefully about scales: over the model's time step, do the real things represented by turtles actually compete with each other over an area about the size of the model's patches? If the Cooperation model had a one-day time step and patches 5 meters in width, it would exaggerate the strength of competition because real cows do not graze all day on only a 5 × 5 meter area.)

Let's make up an example of *direct* interaction among agents: turtles with a behavior that depends on a variable happiness of their nearest neighboring turtle. Each turtle can have its own variable neighbors-happiness that holds the happiness of the nearest other turtle; this variable is set like this:

```
; First, identify the nearest other turtle
let nearest-turtle min-one-of other turtles
    [distance myself]

; Now get its happiness and store it in my variable
set neighbors-happiness [happiness] of nearest-turtle
```

And if we want our turtle to interact with its neighbor by changing the neighbor's value of neighbors-happiness, we can do it like this:

```
; Set the neighbor's neighbors-happiness to my happiness
ask nearest-turtle
    [set neighbors-happiness [happiness] of myself]
```

(But this last statement could be a problem because the nearest neighbor of my nearest neighbor might not be myself!)

The key primitives here are ones we already know (e.g., from section 10.3). The primitive of provides the value of a reporter for an agent, which can just be one of the agent's variables; ask is how one agent makes another agent do something; and myself is how we refer back to the agent initiating the interaction in an ask statement (i.e., the "calling" or "asking" agent).

Now we will try these techniques in an example model. Interactions are especially important when agents compete with each other, so as our example we introduce: the Telemarketer model.

13.3 The Telemarketer Model

Let us visit the formerly remote land of Wasellya, which has developed very rapidly, so suddenly all its citizens have telephones. Naturally, the "invasion" of telephone technology is followed rapidly by an invasion of people tempted to start a telemarketing business. Is this a good business risk? How many telemarketers will stay in business, for how long? How does the average life span of a telemarketing company depend on how many are started? Here is the ODD description of a simple model for this problem; you should implement the model.

13.3.1 Telemarketer Model ODD Description

Purpose

The real purpose of this model is to illustrate several kinds of interaction: how to model them and what their effects are on system behavior. While the modeled system is imaginary and none too serious, it does represent a common ABM scenario: a system of agents that compete for a resource and grow or fail as a result. We will use this example system to look at how several measures of system performance—the changes over time in the number of telemarketers in business, and the median time that telemarketers stay in business—depend on how they interact.

Entities, State Variables, and Scales

This model has two kinds of entities: potential customers (households with telephones), which we represent as patches, and telemarketers, which are turtles. The model is spatial: we represent the territory over which a telemarketer contacts potential customers, and the distance between telemarketers and potential customers, in arbitrary units equal to one patch size. World wrapping is turned off, so the space does have borders.

Potential customers have only one state variable in addition to their location. This is a boolean variable that represents whether they have been called by a telemarketer during the current time step. (In the NetLogo program you can use the patch's color for this variable: the patch is one color if it has not yet been called, and a second color if it has been called.)

The telemarketers have state variables for their (floating point) location in space, and for size, which represents how many staff, dialing machines, telephone lines, and other resources they have. Telemarketers also have a state variable for their bank balance, the amount of money they have to pay expenses and grow.

The model runs at a weekly time step. Simulations are 200 time steps long. There are 40,401 potential customers (the patches in a NetLogo world with the origin in the middle and maximum x- and y-coordinates of 100).

Process Overview and Scheduling

The model includes four actions executed in the following order each time step.

First, the patches reset their state variable for whether they have been called by a telemarketer this week to "no" (i.e., set `pcolor` to the color representing "not called yet").

Second, the telemarketers all do their sales calls. Each marketer makes all its calls before the next marketer does its calls. A telemarketer calls a number of potential customers; this number increases with telemarketer size. The potential customers are within a radius of the telemarketer that increases with the marketer's size. (In this early stage of Wasellya's development, the cost of phone calls increases steeply with their distance.) Each potential customer buys something if it has not been called previously in the time step, then sets its variable for whether it has been called to "yes" (i.e., changes its `pcolor`). If that variable was already "yes," the potential customer hangs up on the telemarketer and buys nothing. (Wasellyans are not very sophisticated yet, but they have short tempers.)

The order in which the telemarketers do this second action is obviously important because the first marketer to call a customer makes a sale, whereas marketers who call the same customer later in the week make no sale. We will explore this issue in chapter 14, but here we use NetLogo's default scheduling, which is that the order in which telemarketers execute the action is randomly shuffled each time step.

Third, the telemarketers do their weekly accounting. Income is determined from the number of successful calls. Costs of business increase linearly with size. The bank balance state variable is updated by adding sales income and subtracting costs. If the bank balance is negative, the telemarketer goes out of business and is removed from the model. The telemarketer increases its size if its bank balance is high enough.

The fourth action is observer updates: outputs such as the number of telemarketers still in business, a histogram of their size, and the total number of sales, are reported.

Design Concepts

Basic principles. The basic concept explored in this model is competition among agents for resources that are limited but renewed.

Emergence. The primary model results we are interested in are patterns in the number of telemarketers in business over time as a simulation proceeds. (The size distribution of telemarketing companies—how many big ones and how many small ones there are, over time—is another potentially interesting result.) These results emerge from how many telemarketers are initialized, how they compete with each other, how many customers there are and how they respond to calls, and how the telemarketers turn their income into growth.

Adaptive behavior. The telemarketers adapt to their sales success by deciding whether to grow in size, and how much to grow. However, this behavior is modeled very simply and rigidly: any bank balance above a minimum is automatically spent to increase in size. Potential customers have one simple behavior: when called, they decide whether to make a purchase by whether they have already been called during the time step. There are no adaptive trade-off decisions.

Prediction. The telemarketer's adaptive behavior is based on an implicit prediction that if sales are higher than needed to maintain a minimum bank balance, then increasing in size will further increase income. (Note that this implicit prediction is different from—and perhaps not as smart as—predicting that growth will be beneficial as long as income is *increasing*. Is it smart to grow when income is high but decreasing?)

Sensing. The telemarketers sense whether each potential customer has already bought something on the current time step.

Interaction. There are two kinds of interaction: between telemarketers and potential customers, and among the telemarketers. The telemarketers interact directly with potential customers by communicating to find out whether the customers will buy, and then by making the customers change their state to indicate that they bought during the current time step. The telemarketers' interactions with each other are mediated by the resource they compete for: customers. When the territories of telemarketers overlap, customers of one telemarketer are no longer available as potential sales for other telemarketers.

Stochasticity. There are three stochastic elements of the model. First, telemarketers are placed at random locations, uniformly distributed, when the model is initialized. Second, telemarketers select the customers they call randomly from all the potential customers (see the Telemarketer sales submodel, below). Third, the order in which telemarketers make their calls is randomly shuffled each time step, to avoid bias from the advantage that first callers have.

Observation. The primary outputs of interest can be observed via a plot of the number of telemarketers still in business. However, to understand how the system is working it is also useful to have a histogram of telemarketer sizes (figure 13.1). The number of telemarketers in business is output to a file each time step so that statistics on how long telemarketers stay in business (e.g., their median life span—the time at which half have failed) can be calculated at the end of a simulation.

Initialization

The model is initialized with 200 telemarketers, each with size 1.0 and bank balance of 0.0, and a random location. (For display, these are given random colors and drawn as circles.) The potential customers have their variable for whether they have been called during the week initialized to "no."

Input Data

This model uses no time-series inputs.

Submodels

Telemarketer sales. A telemarketer's potential customers are the patches within the marketer's territory, which is a circle around the marketer's location with radius equal to 10 times the square root of the telemarketer's size. The maximum number of calls a telemarketer can make each step is 100 times its size. (Therefore, unless the territory overlaps the space's edge, there are about 3.14 times more potential customers than calls.) If the number of potential customers is greater than the maximum number of calls, then the telemarketer randomly chooses this maximum number of potential customers to call. If instead the number of potential customers is less than the maximum number of calls (possible for marketers on the edge of the space), then all potential customers are called. The number of successful sales is the number of potential customers called who have not previously been called by a different telemarketer on the same time step (i.e., they still were the color representing "not called yet").

Telemarketer accounting. The weekly cost of business is equal to the telemarketer's size times 50. (Money units are arbitrary.) The income from sales is 2 times the number of successful

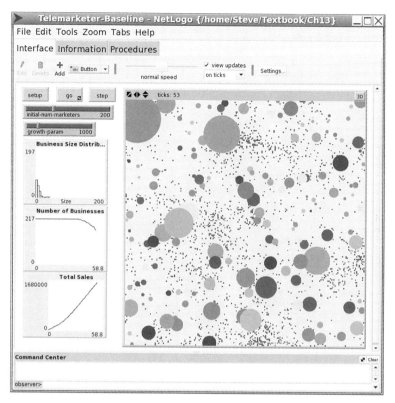

Figure 13.1

An implementation of the Telemarketer model. Telemarketers appear on the View as circles; patches are dark if they have not been called and light if they have. The "Business Size Distribution" histogram shows how many telemarketer businesses there are of each size range. "Number of Businesses" shows the number of telemarketers still in business. "Total Sales" is the total sales of all marketers on the current time step.

sales. The bank balance is updated by adding the income from sales and subtracting the cost of business.

A parameter *growth-param* determines how rapidly telemarketer businesses grow. If the new bank balance is greater than *growth-param*, then the telemarketer converts all but *growth-param* of this balance to increase in growth. For every 1 unit of bank balance above *growth-param*, the telemarketer's size increases by the amount of a parameter *money-size-ratio*, which is 0.001. The value of *growth-param* is 1000, so, for example, if the new bank balance is 1500, then bank balance is set to 1000 and the telemarketer's size increases by 0.5 (which is [1500 − 1000] × 0.001).

If the new bank balance is less than zero, the telemarketer goes out of business and is removed from the model immediately.

13.3.2 Analysis of the Telemarketer Model

The first thing to do after you program this model is of course to produce test output and analyze it to confirm that you have programmed its submodels correctly. Also make sure you understand a key relationship of the model: how the number of sales that a telemarketer must make each week to avoid going out of business changes as the telemarketer's size increases.

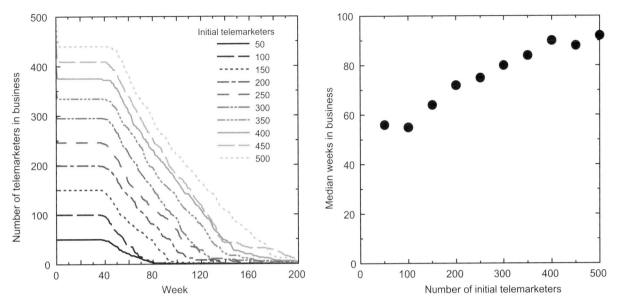

Figure 13.2
Example results from the Telemarketer model. Left panel: the number of telemarketers in business over time, for 10 values of the initial number of marketers. Right panel: median number of weeks that telemarketers stay in business vs. initial number of marketers.

Then you can conduct BehaviorSpace experiments to see how the model responds to two parameters that affect the intensity of interaction among the telemarketers: the number of initial telemarketers, and *growth-param*, which has a baseline value of 1000. How do these parameters affect the number of telemarketers in business over time? Does the number still in business at the end of the simulation vary? Produce graphs similar to those in figure 13.2.

Some of your results may be surprising: you might expect more initial telemarketers to produce stronger competition and more failures sooner, but the right panel of figure 13.2 shows the median survival time of telemarketers *increasing* with the number of telemarketers. You should also note some important qualitative patterns from the Interface. First, the telemarketers that grow rapidly at first never survive very long. Second, the telemarketers who survive and dominate at the end tend to be located at the edge of the space. Why do you think these patterns occur? Are they related?

13.4 The March of Progress: Global Interaction

Let us now imagine that technology progresses quickly in Wasellya and soon the telephone system is so modern that there is no barrier to calling anywhere in the country. Hence, telemarketers no longer compete with each other locally but instead *globally*: all marketers have all patches as potential customers. (Technology making interactions less local and more global certainly seems to be a familiar story.) We can simulate this very easily in NetLogo by getting rid of the restriction that potential customers are within a certain radius. Now the telemarketers simply draw their calls randomly from among all patches. (The model is now *nonspatial*: the locations of telemarketers and customers do not matter.)

How does the switch from local to global interaction affect the system's behavior over time? We leave that question to you as one of the exercises.

13.5 Direct Interaction: Mergers in the Telemarketer Model

No doubt you have noticed that the agents in the Telemarketer model are rather fragile: as competition increases they go out of business quite rapidly, and the larger they are, the easier it seems to be for them to fail. One way that businesses try to avoid such fragility is via mergers: if a company is at risk of failing it might look for a larger company to cover its losses in exchange for a share of future profits. Mergers are a form of direct interaction: agents identify each other as partners and make transactions with each other.

We will now model mergers as a parent-subsidiary relationship: a telemarketer can make itself a subsidiary of one (and only one) parent company, but parent companies can have more than one subsidiary. A very simple version of mergers can be added to the Telemarketer model by adding these assumptions to the weekly accounting submodel:

- If a telemarketer's bank balance falls below zero, and it does not already have a parent company, it tries to avoid going out of business by merging with another telemarketer. It selects a new parent company randomly from all the other telemarketers that have (a) a larger size and (b) a bank balance larger than the telemarketer's deficit ("deficit" is $-1 \times$ bank balance when the balance is negative). If there are no such potential parent companies, the telemarketer goes out of business.
- When a parent company is found, an amount equal to the subsidiary's deficit is transferred from the parent company's bank balance to the subsidiary's, so the subsidiary's bank balance becomes zero.
- If a telemarketer has merged with a parent company, then half of its net income (income from sales minus operating costs) each week is transferred to the parent's bank balance instead of being added to the subsidiary's bank balance. This transfer occurs even when net income is negative, and happens before the bank balance is updated and the following steps executed.
- If a telemarketer's bank balance falls below zero and it already has a parent company, it tries to avoid going out of business by getting money from its parent. If the parent company's bank balance is larger than the telemarketer's deficit, then the amount of the deficit is transferred from the parent to the subsidiary. If the parent company's bank balance is inadequate to cover the deficit, then the subsidiary telemarketer goes out of business.
- When a telemarketer goes out of business, any parent or subsidiary companies do *not* go out of business. (Making parents and subsidiaries go out of business is surprisingly complex to program because the networks of parent-subsidiary relationships can be quite long and complex.)

You should find it relatively easy to program this change, with the hints provided in the following programming notes.

> **Programming note:** Tracking relationships and long-term interactions via variables and via links
>
> To program persistent direct interactions such as the telemarketers' parent-subsidiary relationships, each agent must keep track of which other agent(s) it has a relationship with. In the Telemarketer model, subsidiary telemarketers need to "remember" which other

telemarketer is their parent company so they can send money to it or get money from it. (Parent companies do not need to know who their subsidiaries are.)

In typical object-oriented programming languages, this kind of persistent relation is programmed by giving agents a variable that contains the identity of the agent it interacts with (this variable would be a list of agents if it interacts with more than one). We can use this approach with the Telemarketer model by creating a turtle variable called parent, and initializing it to nobody when the turtles are created. Then when a telemarketer needs a parent company it can identify one via code such as

```
let turtles-to-merge-with
    other turtles with [size > [size] of myself and
    bank-balance > (-1 * [bank-balance] of myself)]

if any? turtles-to-merge-with
  [
    set parent one-of turtles-to-merge-with
    merge-with parent
  ]
```

where merge-with is a procedure that codes the interaction of the subsidiary with the parent once they form the relationship. Then the telemarketer uses the variable parent for all interactions with its parent company. For example, it can send its parent its weekly share of profits via code such as this (which assumes net-profit is a local variable, so it is not necessary to use [net profit] of myself):

```
if parent != nobody
[
  ask parent [set bank-balance bank-balance +
             (0.5 * net-profit)]
  set net-profit (net-profit * 0.5)
]
```

NetLogo's links provide a powerful alternative way to program interaction relationships that you should become familiar with. Instead of turtles needing a variable for their parent company, we create a link from the subsidiary to the parent. For example, to identify and merge with a parent, we can use

```
let turtles-to-merge-with
    other turtles with [size > [size] of myself and
    bank-balance > (-1 * [bank-balance] of myself)]

if any? turtles-to-merge-with
  [
    create-link-to one-of turtles-to-merge-with
    merge-with one-of out-link-neighbors
  ]
```

This code creates a link from the current turtle to its new parent. Note that the primitive out-link-neighbors returns an agentset of all neighbors connected via outgoing links, so we must use `one-of` to convert this agentset (which here never contains more than one turtle) to a turtle. Now, turtles send profits to their parent via

```
if any? my-out-links
[
  ask out-link-neighbors
    [set bank-balance bank-balance + (0.5 * net-profit)]
  set net-profit (net-profit * 0.5)
]
```

Using links may seem a little clumsier than the parent variable alternative in this example, but can be much easier when agents interact with multiple other agents. One advantage of using links is that we can see them in the View, so we can observe how many and how complex they are. Another advantage is that links themselves can have variables and therefore contain information about the relationship they represent; for example, a link could record the tick on which it was created.

NetLogo makes both of these approaches especially easy to use by automatically destroying and cleaning up the parent-subsidiary relationship when one of its members is removed from the model. When an agent executes the `die` primitive, any links to or from it are removed automatically. If we use a variable to hold a turtle (here, the parent company), the variable is automatically reset to `nobody` if that turtle dies. (This kind of cleanup can be a major headache in other programming languages!)

Does giving the telemarketers this merger ability stabilize the system? When you program and test the change, then reanalyze the model, you should find results such as those in figure 13.3: mergers seem to slow down, but not stop, the steady failure of telemarketers until only one is left.

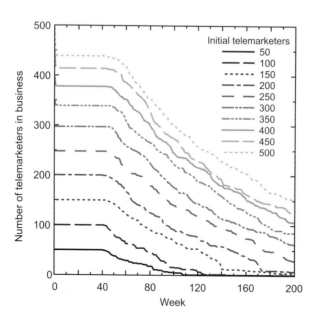

Figure 13.3
Example results from the Telemarketer model with mergers: compare to the left panel of Figure 13.2.

As you can imagine, the citizens of Wasellya soon tire of all the telemarketers and start to wonder whether telephones are such a good thing. What if the potential customers start to resist by remembering which telemarketers have called, and not buying from telemarketers who have already called a certain number of times—can we add this to the model? Memory is often an element of decision-making, and in fact this new process is a simple adaptive trait of the customers that affects their interaction with telemarketers: they decide whether to accept marketing calls by remembering how many times the telemarketer has already called them.

Let us start with the first version of the Telemarketer model (as described in section 13.3) and make these changes:

- Patches have a state variable for how many calls they tolerate from the same telemarketer. This variable is set for each patch when the model is initialized, by drawing a random integer between 5 and 20, and then does not change.

- Patches remember all calls, whether or not they result in a sale. When a telemarketer calls, the patches check whether the number of past calls from the same telemarketer (not including the current call) exceeds their tolerance variable. If so, the customer hangs up and the telemarketer does not make a sale. The customer also refuses any other marketers for the current time step.

Lists are often a useful way to program memory: when some event occurs, information representing the event is added to a list. Later, the list can be examined to see when or how many times an event occurred. NetLogo provides lists that we can use in this way; reread the Programming Guide section on lists and become familiar with the list primitives.

(NetLogo's lists differ from agentsets in two ways that are essential for using them as memory. First, a list can contain multiple entries for the same agent; turtle 13 might be both the fourth and the fifteenth item on a list. Agentsets, however, only keep one entry for each member; if you add turtle 13 to an agentset that already contains it, nothing happens. Second, we can know and use the order in which agents occur on a list—a list's order does not change unless you specifically change it, e.g., by sorting. An agentset is randomly shuffled each time `ask` is used on it, so we never know its order.)

To program memory of how many times a potential customer has been called, we first need to give the patches a variable that holds their memory list. Then, in `setup`, we need to initialize this variable by telling NetLogo that it is a list—unlike other types of variable, a list must be explicitly defined as a list before it can be used. After defining two patch variables called `tolerance` (the number of calls they tolerate before rejecting a telemarketer) and `caller-memory-list`, the following statements in the `setup` procedure initialize these variables:

```
ask patches
[
  set pcolor green

  ; Set tolerance between 5 and 20
  set tolerance 5 + random 16

  ; Initialize the list of callers
  set caller-memory-list []
]
```

The third set statement simply tells NetLogo that `caller-memory-list` is a list that is currently empty. Now, each time a telemarketer calls a patch, the patch will add that telemarketer to this list as a way of remembering the calls.

The code to both remember which telemarketers have called and determine whether to accept a new call can be in one simple reporter, in patch context. This boolean reporter takes a telemarketer as a parameter and returns "false" if the patch refuses any more calls from the telemarketer. The code uses the list primitive `filter` to create a new temporary list that contains only the calls from this particular telemarketer. The length of this temporary list, obtained via the primitive `length`, is therefore the number of calls from the telemarketer.

```
to-report i-accept-call-from? [a-marketer]
  ; First see how many times the caller appears on the
  ; memory list. Do this by getting length of a temporary
  ; list that contains only this caller
  let num-calls length filter
              [? = a-marketer] caller-memory-list

  ; write a-marketer  ; Test outputs
  ; show num-calls  ; Use this only with very few turtles!

  ; Then remember this call by adding it to the memory
  set caller-memory-list fput
      a-marketer caller-memory-list

  ; See if number of previous calls is equal to or below
  ; the tolerance. If so, report "true" to accept the call
  report num-calls <= tolerance ; Report true or false

end
```

Note that this code also uses the list primitive `fput` to add new calls to the beginning of the memory list; the Programming Guide section on lists explains why it is best to add new elements to the beginning of a list instead of to the end.

Programming note: Logical conditions are reporters

In the code for the reporter `i-accept-call-from?`, the statement `report num-calls <= tolerance` probably looks like something is missing. You might have expected this instead:

```
ifelse num-calls <= tolerance
[report true]
[report false]
```

But the logical condition `num-calls <= tolerance` is actually a boolean reporter that returns a value of either true or false. In fact, the NetLogo Dictionary tells us that the primitives `if` and `ifelse` must be followed by a boolean reporter. Because our homemade reporter `i-accept-call-from?` is also boolean, its `report` statement must return either true or false. So we can simply report the value returned by the logical condition `num-calls <= tolerance` directly instead of using the `ifelse` statement.

Now, a telemarketer can interact with each of its buyer patches and make a subset of them that still accept its sales call (return "true" when asked to execute their `i-accept-call-from?` reporter) using a statement such as

```
set buyers buyers with [i-accept-call-from? myself]
```

After programming your patches to remember telemarketer calls in this way, you can test your program by opening an Agent Monitor to a patch. It should at first show the memory list as an empty list, and then, as execution proceeds, show which turtles are added to the list.

How does adding the limited tolerance for telemarketers change both quantitative and qualitative model results? For example, do telemarketers tend to fail at smaller sizes than in the first version? These questions are left as exercises.

13.7 Summary and Conclusions

Interaction is one of the defining, unique characteristics of ABMs: modeling individual agents instead of the whole system lets us represent how each agent interacts, often locally but sometimes globally, with the other agents and its environment. ABMs therefore let us examine the often-important effects of interactions on system behavior. Model interactions can be direct, with one model entity affecting another by changing its state; or they can be indirect, with agents affecting each other by affecting some shared resource.

Many kinds of interaction can be programmed in NetLogo using the same primitives—especially, `of` and `ask`—that we use commonly in sensing. Competition for a shared resource can be programmed very easily by making the resource a patch variable, with turtles reducing the resource variable of their patch. Some kinds of interaction involve relationships: the same pair (or group) of agents interact with each other repeatedly. NetLogo's links are useful for modeling such relationships.

The Telemarketer model provides an example of both direct interaction (the telemarketers change the state of the customers; telemarketers form merger relationships with each other) and indirect interaction (telemarketers compete for customers). We introduced the technique of modeling memory by using lists that contain elements that represent past events. Here, the events were interactions of a customer with telemarketers, but the same technique can be used for memory in many situations. NetLogo's list primitives make it easy to extract certain kinds of events (e.g., interactions with a particular agent) from a memory list.

It should have been clear from the Telemarketer model that when there is interaction, especially competition, the order in which events are executed in the model could have strong effects. The telemarketer that calls a customer first makes a sale, but other telemarketers that call later do not make a sale; so we expect that we would get quite different results if we changed the order in which telemarketers make their calls. This is a question of *scheduling*, and in fact our next chapter uses the same model to look at how scheduling of events affects model results.

13.8 Exercises

1. The first paragraph of section 13.3.2 says it is important, before analyzing the Telemarketer model, to understand how the number of sales that a telemarketer must make

to stay in business changes as its size increases. Draw a graph showing this relationship, with separate lines for several values of `growth-param`.

2. In the analysis of the first Telemarketer model (section 13.3.2), note a counterintuitive result: the median life of telemarketer businesses *increases* as the number of marketers increases, instead of decreasing because of greater competition. Why? How can you show whether it is a result of companies growing too fast? (Hint: look at the parameters that control how rapidly the telemarketers grow.)

3. Can you develop a more successful adaptive trait—deciding when and how much to grow—for the telemarketers? First, define what you mean by "successful"; then try some alternatives and describe how you evaluate them.

4. The second version of the Telemarketer model (section 13.4) is nonspatial: telemarketers interact with each other globally instead of locally. How does this affect model results such as those shown in figure 13.2?

5. Implement the merger version of the model (section 13.5), using NetLogo links to represent the parent-subsidiary relations among telemarketers. Use test output to a file or the Command Center to verify that (a) telemarketers correctly select a parent company that meets the criteria (bigger size; sufficient bank balance), (b) subsidiaries correctly send half their profits to their parent, and (c) when a subsidiary has a negative bank balance it tries to get the deficit from its parent. How does the addition of merger interactions affect results such as those shown in figure 13.2?

6. Implement the model version in which customers remember telemarketer calls, as described in section 13.6. How does this change in how customers interact with telemarketers affect model results?

7. Governments have tried to control telemarketers via "do-not-call" lists. One approach requires each telemarketer to keep a list of customers who have asked them to never call again. Another approach is the national do-not-call list: customers can place themselves on the list, and telemarketers are not allowed to call anyone on the list. Choose one of these two approaches and design and implement a version of it. Assume that some fraction of customers ask to be on the do-not-call lists.

8. Can you modify the model implemented in exercise 6 so that customers only remember the last five telemarketer calls they have received?

14

Scheduling

As you built and used models in previous chapters, you probably found yourself thinking that the order in which model events occur could affect the results you get. The question is how to *schedule* the actions in our model—in what order should events be executed, and how do we make NetLogo execute them in the order we want? Scheduling is an important model design consideration; in early versions of the ODD model description protocol it was one of the design concepts that the other chapters of part II address, but now scheduling is part of the Overview element of ODD (figure 3.1): to provide a meaningful overview of a model's processes, we must specify the order in which they are executed.

Scheduling refers to how we model time. In the real systems we model, things happen continually and concurrently. One way that models simplify reality is by simplifying how time is represented: instead of many agents doing many things at once, we represent what agents do as discrete events that happen in a particular order that we must specify. You might think that we simplify time this way because computers can only execute one piece of code at a time, but that is not as true as it used to be: computers with multiple processors or "threads," now common, can in fact execute several pieces of code at once. The most important reason to simplify how time is represented is to greatly reduce the complexity of models and their results. It is possible to write computer programs so that many agents execute many actions simultaneously, but this makes it much harder to understand what happened and why.

For this chapter, learning objectives are to:

- Master the basic concepts of modeling time: what an "action" is, what discrete time steps are and what assumptions their use makes, and what continuous time scheduling is.
- Know exactly what the primitives `tick` and `ticks` do and how to use them to track time; and how to stop your model when you want it to.
- Develop an understanding of how the order in which actions are executed can affect model results, and how execution ordering can be used to simulate systems in which biases or hierarchies among individuals do and do not occur.
- Understand methods such as synchronous updating for dealing with events that, in reality, occur more frequently than the model's time step.
- Learn the important primitives `repeat`, `loop`, and `while`.

Representing time in a simulation model is much like modeling space. We can represent spatial information as *discrete*: broken into distinct areas within which characteristics do not vary. NetLogo's patches are a discrete representation of space. Spatial information can also be *continuous*: it can vary continuously over distance, no matter how small the distance. NetLogo represents turtle locations as continuous: a turtle in patch 5,10 can be at location 4.8325,10.4171. Likewise, we can represent time as discrete, by assuming that actions occur at regular time steps (ticks, in NetLogo). For example, if we use daily time steps we are simplifying reality by saying that model's state jumps forward by one day at a time, so all events in the model represent what happens from one day to the next. We can also represent time as continuous, by specifying the exact time at which each event occurs. Just as NetLogo allows us to combine discrete patches with continuous spatial locations, it allows us to have a model that represents most actions via time steps while also including some events that occur at specific times between ticks.

The difference between modeling space and time is that when we use discrete time, and assume that events occur only at regular time steps, we still must specify the order in which those events are executed on each time step. Determining this order is the central scheduling issue for most ABMs. Before we explore it, we first need a specific definition for what it is we are scheduling.

14.2.1 Actions

In section 3.3 we first defined the term *action* for the elements of a model's schedule: the schedule defines which actions of the model are executed in which order. An action has three components that specify:

- Which model entities or agents are to execute something: a specific turtle, all patches, the observer, etc.;
- The code that the entities or agents are to execute: a procedure, a NetLogo primitive, a block of code within an ask statement, etc.; and
- If more than one entity or agent executes the code, the order in which these entities execute.

The go procedure of a well-organized NetLogo program is simply a sequence of actions such as

```
ask patches
[
  diffuse pheromone 0.3
  set pcolor scale-color blue pheromone 0 10
]
ask turtles [move]
update-output
```

The first of these statements is an action that specifies (a) which model entities: all the patches, (b) what code to execute: the two statements (diffuse and set pcolor) within the brackets, and (c) in what order: we know that ask randomly shuffles the agentset each time it executes. The second of these actions is similar except that the turtles execute a procedure called move. The third action only affects one model entity, the observer, and tells it to execute a procedure called update-output.

Therefore, to understand how time is represented in a NetLogo program we can start by looking at the go procedure to see what actions are in it and what order they are in. The go procedure is itself an action: when you execute it by pushing its button on the Interface, you are telling one object, the observer, to execute the statements in go.

14.2.2 Discrete Time Steps

The common and (usually) simple way to model time in an ABM is via discrete time steps. All of our example models so far have assumed that all events occur at regular time steps, with one "tick" (one execution of the go procedure) representing a unit of time such as a minute, day, or week. Using time steps simplifies a model by forcing us to follow a specific temporal resolution. If we use daily time steps, then we design the model by thinking about what happens from day to day, so we condense everything that happens within a day into daily events. If our agents move, then we can decide how big the patches are and design the movement algorithm by thinking only about how far an agent might move in a day; there is no need for detail on any moving around from hour to hour.

While discrete time steps are a common and simple way to represent time, they do force us to make decisions that can have strong consequences. When we assume that model events all happen once per time step, we must then decide the order in which the model's action are executed each tick. The order in which actions are executed sometimes affects a model's results strongly (but sometimes it does not). Typically, it is fairly clear what order most actions should be in: if agent behaviors depend on environment variables that change over time, then we need to schedule the environment update before agent behaviors. If we want output that represents the state of the model at the end of a time step, then output should be scheduled last. In subsequent sections of this chapter we discuss some issues of how execution order affects model results and what can be done about it.

In NetLogo we put all the main model actions in the go procedure and use the tick primitive to keep track of time steps. Each time go executes, we simulate one time step and the built-in variable ticks increases by one when tick is executed. Here is one good way to schedule model actions and observer updates in go:

```
to go
  tick
  if ticks > run-duration [stop]

  ask patches [patch-update-procedure]

  file-open "TestOutputFile.csv"
  ask turtles [first-turtle-procedure]
  file-close

  ask turtles [second-turtle-procedure]

  output
end
```

The first action each time step is the tick primitive, which increments the value of the built-in variable ticks and tells the View to update (if the View is set to update "on ticks"). Therefore, the first time step executed is tick 1: ticks is reset to 0 when reset-ticks is executed in the setup procedure. But be aware that the View will now represent the model at the start, not the end, of the tick number that it displays.

The second action is the observer checking whether it is time to stop. In this example, the model stops when the number of time steps we want to execute (a variable called run-duration) has been executed. We discuss such stopping rules in section 14.2.6.

The tick statement and the stopping rule come at the start of go instead of at the end for two reasons. First, by putting tick at the beginning of go, we make sure the tick counter will be accurate when used in any of the subsequent actions—it will equal 1 on the first time step, 2 on the second, and so on. The second reason is to get complete output from BehaviorSpace: when BehaviorSpace runs experiments, it writes its output only at the very end of the go procedure. Therefore, if we stop the model at the end of go, we will not get BehaviorSpace output for that last tick. With the recommended arrangement, to run a model for 200 time steps, we can set run-duration to 200 and BehaviorSpace will produce 201 lines of output that show the model's state before the first time step and then at the end of steps 1–200.

One important action does not appear in the go procedure and is instead scheduled on the Interface tab. That action is updating the View (section 9.2). When using time steps, it is normally best to update the View "on ticks" so you know that it displays the model at the start of a time step. The speed controller on the Interface is actually a scheduling device too: when you turn up the speed, NetLogo just reduces how often it schedules a View update, letting the model run for several ticks before waiting for the View to update.

14.2.3 Execution Order, Biases, and Hierarchies

Why does ask randomly shuffle an agentset every time it executes? The reason is that many models are like the Telemarketer model: an agent that executes first has an advantage (or disadvantage) over agents that execute later. In the Telemarketer model, if the same telemarketers always got to go first, they would always have first access to the limited supply of customers and a distinct advantage over the telemarketers who always had later access to customers. Hence, our simplification of time—assuming that each telemarketer makes its weekly calls all at once—could introduce a bias. The ask primitive provides one way to reduce this bias, by shuffling the order in which agents execute so none get to always go first.

The ask primitive is therefore designed to avoid biases due to execution ordering—but sometimes the systems we model *are* biased: the big guys seem to get to go first and have greater access to resources. These biases often produce the "positive feedbacks" that can make complex systems complex. In business, bigger companies may have more capability for marketing, or research, or setting standards that others must follow (e.g., Blu-Ray vs. HD-DVD disk media). In ecology, bigger individuals often can defend and dominate better habitat, or get more mates. To represent such hierarchies in a model, it can be useful to execute some actions in a specific order: the agents that are dominant in some way get to go first when executing some action.

NetLogo offers a somewhat complicated (but well-documented: see the Programming Guide section on lists) alternative to ask that executes an action on an agentset, in order of some agent variable. For example, suppose we want the Telemarketer model to assume that bigger telemarketers have advantages over smaller ones, such as having faster dialing machines and more persuasive sales staff. We could represent this hierarchy by assuming the telemarketers execute their sales calls in order from largest to smallest, according to their size variable. To program this change, replace this action:

```
ask turtles [do-sales]
```

with this one:

```
foreach sort-on [size] turtles
[
  ask ? [do-sales]
]
```

There are two important primitives used here. `sort-on` creates a new list by sorting another list or agentset (here, the agentset `turtles`). `foreach` executes the block of code within its brackets once for each item on a list, with the character ? representing the current list member. Unlike `ask`, `foreach` does not shuffle the list. Therefore, this code goes through the sorted list and causes each turtle on the list to execute its `do-sales` action.

Programming note: `Ask-concurrent`

NetLogo provides another alternative for scheduling actions in a way intended to avoid biases due to the order in which agents execute: the `ask-concurrent` primitive. While `ask` causes one agent at a time to execute the code within its brackets, `ask-concurrent` starts all the agents and they take turns executing little bits of their action. For example, in the telemarketer model's `go` procedure you could replace

```
ask turtles [do-sales]
```

with

```
ask-concurrent turtles [do-sales]
```

which would cause all the telemarketers to take turns stepping through their `do-sales` procedure a bit at a time. This might seem like just what we want: all the telemarketers are making their sales calls at once.

But if you try using `ask-concurrent` in this way, you might find that it produces very strange results that are in fact mathematically impossible. The total sales each tick (number of patches that bought from telemarketers) can far exceed the number of patches in the model! Why? Telemarketers look at a patch and consider it a sale if it has not already been called by another telemarketer during the current time step. After recording the sale, the telemarketer tells the patch to indicate that it has now been called. However, with `ask-concurrent`, all the telemarketers are doing this at once, so several of them can look at the same patch at the same time. Before the first telemarketer can tell the patch that it has now been called, several other telemarketers might call the same patch and also consider it a sale. The extent to which this occurs depends on how much code there is between the statement where a telemarketer senses that the patch is a sale and the statement where it tells the patch not to accept more calls—the more statements there are between these two events, the more time there is for other telemarketers to sense the same patch.

The problem with `ask-concurrent` is that it does not let us control or even know the exact order in which events are occurring. This loss of control makes models difficult to understand and vulnerable to unforeseen and undetected errors—an extreme example of how important scheduling can be. We recommend you not use it unless you think very carefully about the potential consequences.

14.2.4 Synchronous vs. Asynchronous Updating

In the Telemarketer model, the reason that the order in which telemarketers execute their sales action affects model results is that whenever a telemarketer makes a sale, the customer stops being available to other telemarketers. This is an example of *asynchronous updating*: as soon as an agent affects the environment, the environment is updated so that the next agent to execute experiences a different environment. Another way to avoid the effect of execution order, besides randomizing execution order each tick the way `ask` does, is to use *synchronous updating*: updating the environment all at once, after all the agents have executed their actions that depend on it.

Synchronous updating clearly would be a problem in the telemarketer model because its agents compete very strongly for a resource—uncalled customers—that they deplete rapidly. Not updating which customers have been called until all telemarketers have executed their calls would essentially remove competition from the model. However, synchronous updating can be useful in situations such as when the agents are affected by, but do not alter or deplete, environmental variables, or if they deplete a resource at a rate that is slow compared to the time steps.

Synchronous updating can be implemented in at least two ways. One is to execute one action in which all the agents sense the environment (and update their state variables accordingly), and then a later action in which they update the environment as a consequence of their behavior. The NetLogo Library model Life (under the Computer Science—Cellular Automata category) provides an example of this approach. A second way is to represent the environmental resource being updated using two variables: one for the old state, which agents use as a basis of their behaviors; and one for the new state, which agents update. At the end of a tick, the old state is replaced by the new state. In the chapter 14 section of this book's web site is an implementation of the bird breeding synchrony model of Jovani and Grimm (2008; see exercise 11 of chapter 5) that uses this second synchronous updating approach.

14.2.5 Discrete Events in Continuous Time: The Mousetrap Example

Time steps are not the only way to schedule events in a simulation model. Another approach is "continuous time" simulation, in which unique events are modeled as separate actions that are executed at their own unique times. Instead of assuming that the same actions are executed over and over again each time step, the actions are created and scheduled (i.e., given a time at which to execute) by the model itself. Events can be executed at any point in time, not just at integer time steps. The Telemarketer model could be written, for example, so that whenever a telemarketer calls a customer, it schedules its next call for 10 minutes in the future if it makes a sale and 0.5 minutes in the future if the customer hangs up on it.

Continuous time simulation is not supported extensively by NetLogo, but it can be used. The primitive `tick-advance` can be used instead of `tick` to move the tick counter ahead by a fractional amount of time, or several ticks, instead of by exactly one tick. (See the Tick Counter section of the Programming Guide and the GasLab models in the Models Library.) As an example, examine the following NetLogo code for a simple version of the Mousetrap model, which illustrates chain reactions such as nuclear fission (you can find videos of physical mousetrap models of fission on the Internet). Each patch represents a spring-loaded mousetrap with two balls sitting on it. If a ball lands on a patch, it triggers the mousetrap, which sends its two balls into the air. These balls land on other patches, which may or may not have already been triggered. The time that the balls are in the air increases with the distance they travel. Here, we assume that they land on another patch randomly chosen within a radius of 5 patches, and that their travel time is stochastic, averaging 0.1 ticks per patch-width traveled. Each tick represents one second.

```
patches-own [trigger-time] ; The non-integer time at which
                           ; trap triggers

to setup
  ca
  ask patches
  [
    set pcolor yellow      ; Yellow means trap has not triggered
    set trigger-time -1 ; Initialize trigger time to
                        ; before model starts
  ]
  reset-ticks
end

to pop
  show trigger-time ; Test output- trigger times should
                    ; always increase

  set pcolor red    ; Show the snap
  wait 0.05         ; So we can see things happen on the View
  set pcolor black ; Black means the trap has triggered

  ; Send 2 balls in air, determine where and when they land
  ask n-of 2 patches in-radius 5
  [
    if trigger-time < 0 ; Patch only triggers if it
                        ; has not already
    [
      ; Set the time when the ball will land on and trigger trap
      set trigger-time ticks +
          (random-float 0.2 * distance myself)
    ]
  ]
end

to start
  ; Set off one trap to start the action
  ask one-of patches [pop]

  ; Now keep stepping time forward to next time a trap is
  ; triggered, as long as there are any balls in the air
  while [count patches with [trigger-time > ticks] > 0]
  [
    ; Find the triggered trap that has next trigger time
    let next-patch-to-trigger min-one-of patches with
                [trigger-time > ticks] [trigger-time]
    let time-til-next-trigger
        [trigger-time] of next-patch-to-trigger - ticks
```

```
  ; Move time forward to when next trap triggers and trigger it
  tick-advance time-til-next-trigger
  ask next-patch-to-trigger [pop]

  plotxy ticks count patches with [trigger-time > ticks]
]  ; while
end
```

Programming note: Using `repeat`, `loop`, and `while` to execute code repeatedly

NetLogo has three primitives that cause a set of statements to be repeated within a procedure. (This is called "looping" by programmers.) `repeat` simply repeats the statements within its brackets a specific number of times. The number of times the statements are to be repeated can be a constant, as in `repeat 10 [ask turtles fd 1]`. Or the number of repeats can be an expression, as in `repeat count customers [ask one-of customers [buy-something]]`: the `repeat` primitive starts by evaluating `count customers` to determine how many times to repeat the statements in brackets.

`loop` is like `repeat` except that the number of times to repeat the statements is not specified before the loop starts. Instead, you must include a `stop` or `report` statement in the loop to stop it (unless you want to learn for yourself that "Tools/halt" on the NetLogo menu is how to stop an endless loop).

The primitive `while`, which we use in the Mousetrap code, also repeats a set of statements inside its brackets without specifying in advance how many times to repeat the statements. Instead, `while` repeats until a boolean expression becomes false. An example is `while [any? turtles with [pcolor = blue] [ask turtles [fd 1]]`. This statement tells the turtles to step forward one unit, repeatedly, until no more turtles are on blue patches. The boolean expression inside the first set of brackets is reevaluated to determine whether to go again *after* the code inside the second set of brackets is repeated each time. Hence, in this example, the turtles will not stop moving forward the instant the last turtle steps off a blue patch; instead, all turtles will step forward, then `while` will check whether any turtles are on a blue patch and decide whether to execute `ask turtles` again.

(In most programming languages, looping statements such as these are very important and are introduced at the beginning of a course. In NetLogo we deal mostly with agentsets, and looping over the members of an agentset is hidden within the `ask` primitive.)

After entering the Mousetrap code in the Procedures tab, you need only put buttons for the `setup` and `start` procedures, and a plot, on the Interface (figure 14.1). Note that the `start` procedure here is quite different from a typical `go` procedure. First, we only execute it once, not repeatedly via a "forever" button. Once started, the model creates and schedules its own actions: when one patch triggers another, it tells the triggered patch when it should execute the `pop` procedure that triggers its mousetrap. The `while` loop in `start` simply moves the tick counter ahead to the time at which the next patch needs to "pop." Instead of us telling NetLogo what tick to stop at, the `while` loop just stops when there are no more patches to execute.

You can set the "view updates" switch on the Interface to "continuous" to see each mousetrap trigger as it occurs. Or you can set the updates to "on ticks" to see how many triggers happen each simulated second; the View is updated each time the value of ticks reaches the

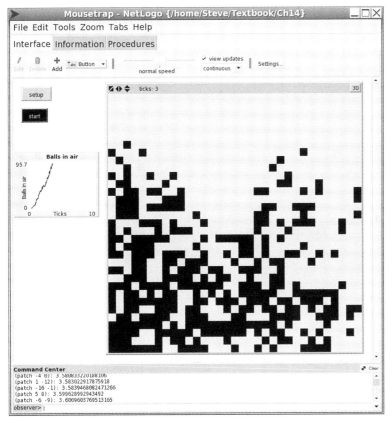

Figure 14.1

The Mousetrap model. Patches represent traps and turn black when they have been triggered. The Command Center displays the time (fractional tick) at which each trap is triggered. In this run, the first trap triggered was near the lower left corner.

next whole number. Looking at both of these ways to update the View illustrates the key use of continuous time simulation: to represent how the number and frequency of actions changes over time as a consequence of what is happening in the model, and to keep track of the exact order in which such events occur. If we used a time-step model, representing the timing and sequence of trap triggers accurately would require extremely small time steps. (In the NetLogo Library's Mathematics section is a Mousetrap model that cannot represent such detailed characteristics of the system: it ignores variation in ball travel time and assumes that each ball is in the air for exactly one tick.)

14.2.6 Temporal Extent: Stopping Rules

Modeling time also includes specifying the extent of time that we want to simulate. This means we need a "stopping rule" that tells NetLogo when the model is done. In most of our example models we simply specify the number of ticks that we want to execute. However, other criteria can be used to stop models; we tell a model to run until some particular condition occurs. For example, instead of

```
tick
if ticks > run-duration [stop]
```

you could use

```
tick
if count turtles with [color = blue] = 0 [stop]
```

The Mousetrap model stops when there are no more balls in the air—no more patches waiting to execute their pop procedure. When we use a condition such as this to stop a model, the number of ticks it takes to finish can be an important model result.

14.3 Summary and Conclusions

The main point of this chapter is that we need to think about how we represent time, and about the order in which the actions in our model are executed. NetLogo makes it very easy to use discrete time steps, and this approach is often best. However, it is important to understand what using time steps implies: that the model can only represent things that happen as time jumps from one step to the next, that things that happen each step are assumed to happen concurrently (all at once, at the time step); but the order in which these "concurrent" events are actually scheduled as actions executed by the computer can have major effects on results.

When designing a typical time-step model, there are several scheduling questions that we typically need to think about. First, what are its main actions—what entities in the model (environment, agents, observer) do what behaviors on each time step? Second, in what order should these main actions be executed? Models often start each time step by updating the environment, then the agents execute their behaviors, and finally output is produced. Third, for actions executed by multiple entities (all the patches, all the turtles, etc.), in what order do the entities execute their behavior? If the entities do not interact, the execution order may not matter; but if they do interact, we can use different scheduling to represent non-hierarchical systems (e.g., by using ask, which shuffles the order in which agents execute a behavior) or hierarchical systems (by executing the agents in a particular order). Fourth, we usually need to specify a stopping rule: the number of ticks we want to execute or the condition (e.g., all turtles are dead, or wealthy, or self-actualized) when we want the model to stop. In a well-organized NetLogo model, the answers to all these questions are found in the go procedure.

When the order in which agents execute an action has strong effects that seem undesirable, there are a few tricks that *may* be useful. One is to use synchronous updating, so that all agents base their behaviors on information sensed at the same time. Another is to break the schedule's actions into smaller actions that are ordered differently.

But there are alternatives to time step simulation. Continuous time simulation can be more appropriate when actions cannot be usefully approximated as happening at regular ticks and when the exact order in which events occur is especially important. With this approach to scheduling, the model itself determines what actions are to be executed at what times, and these times may be continuous, not discrete. Compared to some other agent-based modeling platforms, NetLogo does not provide extensive tools for continuous time, but our Mousetrap model illustrates one way to do so using the tick-advance primitive.

1. The basic version of the Telemarketer model (as described in section 13.3) is scheduled so that the order in which telemarketers do their sales calls is randomized each time step by using `ask`, but it could be sorted so that execution is from largest to smallest telemarketer. Which seems most realistic, if all telemarketers have the same equipment and expertise? Which seems most realistic if bigger telemarketers have better equipment and nearly always reach customers earlier than smaller telemarketers? When telemarketers execute in sorted order, how are the results different from those in figure 13.2, qualitatively and quantitatively?

2. The description of the Telemarketer model with mergers in section 13.5 is ambiguous about whether mergers happen via asynchronous updating (each telemarketer decides whether to merge as soon as it finishes its weekly accounting) or via synchronous updating (mergers are conducted after all telemarketers have finished their weekly accounting). Try both approaches; what effect do they have on results?

3. Modify the basic version of the Telemarketer model to use `ask-concurrent` instead of `ask` for the action in which telemarketers do their calls. Are there strong effects on results?

4. Program the Mousetrap model described in section 14.2.5. Explain why a random number is used in determining how long each ball is in the air (i.e., the trigger time of the patches that the balls land on)—the reason is not just to represent unpredictability in the flight of each ball. Add a plot that accurately shows the number of mousetraps remaining untriggered over time. Finally, design and conduct experiments that quantify model results such as the time of peak activity (when the most balls are in the air), the time at which the last trap triggers, and the fraction of all traps that get triggered. How do these results vary as you change the maximum distance that balls can fly (from its standard value of 5), or the number of balls per trap (e.g., from 1 to 5 instead of just 2)?

5. Can you program the Mousetrap model to show on the View which patch triggered which other patch—that is, the path that each ball takes?

6. Compare the Mousetrap model in NetLogo's library to the one you built for exercise 4. Modify your model's scales, parameters, and initialization to make them as similar as possible to the library model: set the World to the same size, set the maximum distance that balls can fly to the same value, set the travel equation so that mean travel time equals the library model's travel time of one tick; and make the model start by triggering the center trap. What are the key differences remaining between the two models' formulations? What qualitative and quantitative differences (in results such as the maximum number of balls in the air, or the times at which activity peaks and ends) are there in the results produced by the two models?

Stochasticity

In modeling, the word "stochastic" describes processes that are at least partly based on random numbers or events. Stochastic processes therefore produce different results each time a model executes because the random events or numbers are different each time. The word "deterministic" is used as the opposite of stochastic, describing processes that are modeled with no randomness so they produce exactly the same result each time they are executed. Some people assume that ABMs, and even simulation models in general, are all highly stochastic. That is often a mischaracterization: it implies that all the important dynamics in ABMs result from random numbers, but many ABMs have relatively few stochastic processes and their most interesting and complex results arise from unique individuals executing deterministic behaviors in a variable environment.

We have already used stochastic processes: many of our example models use random numbers provided by NetLogo (also called "*pseudorandom* numbers; section 15.3). Here, we present more of the standard techniques for using stochastic processes in simulation modeling. In overview, when we represent some source of variability in our model as stochastic—as a sequence of random numbers—we do three things:

- Select an appropriate statistical distribution for the random numbers;
- Select the distribution's parameters, or model how the parameters vary during the simulation; and
- After the model is built, use replication to understand how much of the variability in its results is due to its stochastic processes.

This chapter uses some basic probability and statistical concepts that beginning modelers may not be familiar with yet. These concepts are quite important for simulation modeling, so you should make the effort to understand them at least at the introductory level presented here.

The learning objectives for this chapter are to:

- Understand why and how modelers use random numbers, and how we determine the effect of stochasticity on our model results.

- Know what a random number distribution is, and the difference between continuous and discrete distributions.
- Know the properties of uniform, normal, Bernoulli, and Poisson distributions, and how to draw random numbers from these distributions.
- Understand use of random generator seeds to control the random numbers produced by NetLogo.
- Learn the very common technique of modeling agent traits stochastically by using data to parameterize random number distributions.
- Practicing "defensive programming" to catch and correct run-time errors.
- Become aware that there are many "recipes" in the literature to use when you must write your own code for numerical methods such as random number distributions.

15.2 Stochasticity in ABMs

15.2.1 Uses of Stochasticity

Why are random numbers and stochastic processes so widely used and important in simulation modeling? The most general answer to this question is that we often want to represent some kind of variability, but without representing all the detail needed to explain *why* that variability occurs. Simplification is the most essential characteristic of modeling, and one way to simplify variables or processes that change over time or space is to assume that they are random. Then, instead of having to model what causes those variables or processes to change, we simply draw random numbers to make them change.

One of the most common uses of stochasticity is to assign initial values to variables, especially when initializing a model. For example, an economic model might treat households as agents, with each household having a variable for net worth. When the model starts, we need to give each agent a unique but realistic net worth. Let's say we have some data that indicate the net worth values of real households follow a normal distribution with a mean of $223,000 and a standard deviation of $83,400. We could initialize the model by giving each agent a net worth drawn randomly from a normal distribution with this mean and standard deviation. Now we have produced a realistic distribution of net worth values without all the trouble of figuring out why net worth varies in this way among households.

A second common use of stochasticity is to model processes that produce variable outcomes when we do not want to represent those processes in great detail. To model some process in a way that produces realistic results, we can parameterize a stochastic process with data on the rate or frequency of real events. Imagine a model (of crop or livestock production, or forest fires, or tourism) that depends on when it rains. Instead of modeling the weather, which is extremely complex, we can simply examine some weather data. These data would tell us, for our study site, what percentage of days in the past 10 years have had rainfall (perhaps broken out by month). Now we can treat this observed *frequency* of rainfall as a *probability* of it raining on any day in our model. On each simulated day we can get a random number between zero and one; if this number is less than the probability of rainfall, then our model experiences rain.

Treating observed frequencies as probabilities in a stochastic process can also be used to model agent behavior. An anthropological model might simulate how often a prehistoric family moves to a new location. If there are archeological data indicating how frequently real families moved, then the observed frequencies could be used as the probability of a model family moving on any particular time step.

The above two examples of stochastic process models are very simple and leave out "contingencies" (how probabilities depend on other variables) that could be important. For example, you might think that even prehistoric people must have been smart enough to move not just at random times, but instead when they had a good reason to. In general, it is best to start our models as simply as possible, but sometimes we clearly need to add some of these contingencies. In many places, the probability of it raining on one day depends very much on whether it rained the day before, and there are statistical methods to estimate the daily probability of rainfall in a way that captures this dependency. (Whether rainy days tend to follow each other may not be important for a model of crop production, but it might for a tourism model!) Likewise, there are a variety of methods to make stochastic models of agent behavior more sophisticated. Grimm et al. (2003) used a probabilistic model of when marmots disperse from their home territory that uses field data to estimate how the probability of dispersal changes as the marmot gets older. Statistical techniques such as logistic regression can be used to model how the probability of some event varies with one or several variables. The model we build in chapter 16 illustrates some of these techniques.

15.2.2 Analysis of Stochasticity via Replication

When we use stochasticity to represent variables or processes and behaviors in a model, we raise the question of how random our results are. If an ABM is very simple and has only one major environment variable that is modeled stochastically, will the results just be entirely random? Or if our ABM uses many stochastic processes, will it always produce nearly the same results because the effects of its parameters and inputs are swamped by all the randomness? When we vary the inputs or parameters of a model, how much of the change in results is due to randomness? These questions are the reason we use replication, a technique we introduced with the concept of simulation experiments in chapter 8. When we use replication in a simulation experiment we simply run the model multiple times, with the random numbers being the only thing that changes (e.g., figure 8.3).

It is very important to understand what replication does and does not tell us about a model: the results from a replication experiment only describe the effects of stochasticity in our model, and nothing else. Replication does not tell us how much uncertainty there is in a model, because uncertainty comes not just from stochastic processes but also from uncertainty in parameter values, inputs, and assumptions. Hence, we should not use replication as a descriptor of a model's reliability or how well the model fits observations. Variability from a replication experiment should not be compared to variability in the real system being modeled *unless* we are willing to assume that the stochastic processes in the model represent all sources of variability in the real system that are not otherwise represented in the model. And remember that statistical tests that depend on the number of replicates (e.g., t-tests) must be used carefully because the number of replicates is arbitrary: a change too small to be scientifically meaningful (compared to all the sources of uncertainty in a model and in data on the real system) can be "statistically significant" if we run enough replicates.

But replication is appropriate, and essential, for comparing different versions of a model to each other. When we change parameter values, equations, assumptions, or input data and want to know what the effect is, we must run the model enough times to distinguish the effect of the change from the model's stochasticity. The best way to "analyze" such replication experiments is often to simply choose a reasonable number of replicates (e.g., 10 for a simple model) and graph the range of results as we did in figure 8.3.

15.3.1 Terminology

Now let's look in detail at how to program stochastic variables and processes in NetLogo. You already have some basic experience with stochastic processes, having used the primitives `random` and `random-float` to do things like draw random numbers from a certain range. Let us start by defining some terms. At this point you should read the NetLogo Programming Guide section on "Random Numbers"; otherwise, you are unlikely to understand this.

First, modelers often talk about *pseudorandom* numbers as well as random numbers. *Pseudorandom* describes the random numbers we use in computer programs. These numbers are not truly random because they are generated by a mathematical algorithm. In this book, as in the NetLogo documentation, we just use the simpler term "random number" even though we really use pseudorandom numbers.

The computer program that generates random numbers is called, amazingly, a *random number generator*. The random number generator in NetLogo uses a sophisticated algorithm rigorously tested to ensure that it produces results that very closely resemble a truly random process. However, we can reproduce exactly the same sequence of pseudorandom numbers by setting the generator's *seed*. Setting the generator's seed is an important technique because it lets us eliminate the random numbers as a source of variability among model runs. From now on, we assume that you understand (from the Programming Guide) random number seeds and the NetLogo primitives that control them.

We do not use NetLogo's random number generator directly but instead use primitives called *random distributions*. This term is potentially confusing because in statistics, the distribution of a random variable is a mathematical function describing how likely different values of the random variable are to occur. A uniform distribution, for example, means that all values between a minimum and maximum are equally likely to occur; for a variable following a normal distribution, values closer to the mean are more likely to occur. In NetLogo and other simulation platforms, though, a random number distribution is a primitive or procedure that produces random numbers that have known statistical properties. If you repeatedly execute the NetLogo primitive `random-normal 100 20` you will produce a series of floating point numbers that are normally distributed with a mean near 100 and standard deviation near 20. (In the Command Center, enter `repeat 100 [show random-normal 100 20]`.)

15.3.2 Important Random Number Distributions

Selecting appropriate random number distributions can be a key part of designing a model. If you use random numbers for any but the simplest purposes, you should make yourself familiar with the basic probability theory behind common distributions so you understand their assumptions and use them appropriately. This information is available from statistics and probability texts and on-line sources such as Wikipedia. Here we briefly describe only the distributions you are most likely to use.

Modelers typically use (and NetLogo includes) two distinct types of distributions: continuous and discrete. Continuous distributions produce random numbers that are real numbers in the mathematical sense: they include fractions and can be any value over some range. Discrete distributions produce results that can only have certain discrete values such as true vs. false, or integers.

Continuous Distributions

A uniform real number distribution (the NetLogo primitive `random-float`) simply provides random floating point numbers between a specified minimum and maximum, with all values within that range equally likely. Especially important is a uniform distribution with minimum 0.0 and maximum 1.0; in fact, when modelers talk in general about a "random number" they usually are referring to this distribution. Such "random numbers" are useful for modeling whether an event occurs from a parameter that represents the probability of the event occurring. If, for example, we assume that the probability (between zero and one) of a turtle dying on each tick is held by a global variable called `mortality-prob`, we can simulate whether each turtle dies by having turtles execute this each tick:

```
if random-float 1.0 < mortality-prob [die]
```

A normal distribution (the primitive `random-normal`) produces a sequence of random floating point numbers that create the familiar bell curve, with 68% of the numbers falling within a range of plus-or-minus one standard deviation of the mean, and 95% of numbers within two standard deviations of the mean. To use this distribution, we must specify the mean and standard deviation. If we want to give each patch a value for some variable called *food-level*, and our field data indicate that food levels of real patches seem normally distributed with a mean of 5.0 and standard deviation of 1.5, we can use

```
ask patches [set food-level random-normal 5.0 1.5]
```

While a normal distribution produces values more likely to be closer to its mean, it is important to understand two of its other characteristics: (1) unlike a uniform distribution, there is no limit on what range of values a normal distribution can produce, so (2) when you draw many values from a normal distribution, it is quite likely that a few will be so far from the mean that they are nonsensical for the model. In the above example, a negative value of `food-level` probably makes no sense and could cause mathematical errors or absurd behavior in the model. The probability of drawing a negative value of `food-level` is (from normal distribution theory) 0.000429, which seems small. However, this probability is one chance in 2330, and if the model has 2500 (50×50) patches then drawing at least one negative value is very likely.

What can you do about the possibility that random normal values are occasionally nonsensical? The answer is a technique that software professionals use all the time: defensive programming. "Defensive programming" means writing your software so it constantly watches for problems, and when a problem happens the software either corrects it or tells you about the problem and stops. Write your NetLogo program so it checks for nonsensical values from `random-normal` and does something about them. For example, you can specify a minimum value, such as 0.5, by using this modification of the above statement:

```
ask patches [set food-level max
             (list 0.5 random-normal 5.0 1.5)]
```

Or you can have your program simply draw another random number if it obtains a negative one. A `while` loop is useful for this; in the following code, we draw the random number inside a `while` loop that repeats if the random number is not positive:

```
ask patches
[
  let my-rand -1.0
```

```
  while [my-rand <= 0.0]
  [
    set my-rand random-normal 5 1.5
    ; show my-rand ; Test output
  ]
  set food-level my-rand
]
```

But be aware that if you use these tricks you are changing the distribution of values actually used in your model; the mean of the values you keep and use will be different from the mean you specify in the random-normal statement. (Another way to avoid negative values is to use a log-normal random number distribution instead of a normal distribution; see exercise 7.)

NetLogo includes primitives for two other continuous distributions, exponential and gamma, which we do not discuss here.

Discrete Distributions

Discrete random number distributions are important because ABMs are inherently discrete, integer models. Many of the processes we model use true-false choices (does a turtle execute some behavior or not? does it die or not?) and integer variables: the number of agents that do something (die, move, etc.), the number of patches a turtle can sense, the number of new turtles created by reproduction, and so on.

We have already used one discrete distribution in this chapter without knowing it. The Bernoulli distribution returns a value of either false or true, for one trial of an event when the probability of that event being true is provided as a parameter. Above, we used a uniform real distribution to model when turtles die:

```
  if random-float 1.0 < mortality-prob [die]
```

This mortality model is actually a discrete process: the turtle either lives (mortality = false) or dies (mortality = true), and the probability of mortality is the parameter *mortality-prob*. Another way to program this is, therefore, using the Bernoulli distribution:

```
  if random-bernoulli mortality-prob [die]
```

The random-bernoulli reporter randomly returns either a true or false value, with the probability of returning "true" equal to its parameter (here, mortality-prob).

There is only one problem with the above NetLogo statement, which is that NetLogo does not have a random-bernouilli primitive. But you can easily fix that by writing your own:

```
  to-report random-bernoulli [probability-true]
    ; First, do some defensive programming to make
    ; sure "probability-true" has a sensible value
    if (probability-true < 0.0 or probability-true > 1.0)
    [
      user-message (word
      "Warning in random-bernoulli: probability-true equals "
          probability-true)
    ]

    report random-float 1.0 < probability-true
  end
```

This example also includes some defensive programming. The Bernoulli distribution's parameter is a probability value, and probabilities must by definition be between 0.0 and 1.0. By testing whether the parameter `probability-true` is within this range, we can detect any code or logic errors that give this probability parameter an unexpected value.

NetLogo does have primitives for two integer distributions. The primitive `random` is a uniform integer distribution: it produces integers that are evenly distributed over a specified range. This primitive assumes that one end of the range is always zero, so if we want a turtle to create a number of offspring that is uniformly random between 2 and 6, we would tell the turtle to

```
hatch 2 + random 5
```

The Poisson distribution models the number of times that some event occurs over a specified interval (usually, an interval of time or space), when the average rate of occurrence (mean number of events per unit of time or space) is known and when the events are independent of each other (the occurrence of one event does not depend on when the previous event occurred). NetLogo's `random-poisson` primitive takes a parameter that is equal to (a) the average rate at which the event occurs (mean number of events per unit of time or space), multiplied by (b) the interval length: the number of units of time or space being modeled. (This parameter is not explained well in the NetLogo Dictionary.) For example, if we know that patches sprout new turtles at an average rate of 0.4 times per tick, then we can model the number of turtles that a patch sprouts each time step by telling the patches to

```
sprout random-poisson 0.4
```

The code `random-poisson 0.4` produces an integer number of sprout events per tick, when the mean number of sprouts is 0.4 per patch per tick. Therefore, a patch would sprout zero turtles on most ticks, one turtle on fewer than 40% of ticks, and would only rarely sprout two or more. (Keep in mind that using the Poisson distribution assumes that the sprouting events are independent over time and patches, which means that it does not take time to recover between one sprout and the next, that sprouting does not become more probable as the time since last sprouting increases, and that sprouting in one patch is not more or less likely if sprouting happens in adjacent patches. The distribution can still be useful even when these assumptions are not completely true.)

If instead we want to model how many turtles are sprouted by all the patches, we can still use `random-poisson`. Now the mean occurrence rate is still 0.4 sprouts per patch per tick, but we treat each patch as an interval in space. The Poisson distribution parameter is now the average rate of sprouting (0.4) times the number of intervals being modeled, which is `count patches`. So we can model the total number of sprouts per tick in all patches as

```
let number-sprouts random-poisson (0.4 * count patches)
```

There is one other integer distribution you should be familiar with, even though NetLogo does not have a primitive for it. The *binomial distribution* models the number of events that occur when there are n chances for an event to occur and the probability of each event occurring on a given chance is P. Another way to think of the distribution is this: if you execute the Bernoulli distribution n times, how many "true" values do you get?

An example application of the binomial distribution is modeling how many patches sprout a turtle at one tick, when the probability of a patch sprouting one or more turtles is the variable `sprout-probability`. If NetLogo had a binomial distribution, it would take two parameters: the probability of a "true" on any trial, and the number of trials. Then you could program this process as

```
let number-patches-that-sprout
    random-binomial sprout-probability count patches
```

Sometimes it is not clear whether to use a Poisson or binomial distribution. The key distinction between these two distributions is that the binomial distribution applies to problems where there are a number of true vs. false trials, and the number of true events cannot be greater than the number of trials. The Poisson distribution is used to model the number of times some event occurs in a certain interval of space or time, when it is possible for the number of events per interval to be greater than one.

Programming note: Numerical recipes and NetLogo extensions—don't reinvent the wheel again!

Suppose you find yourself in the not-too-unlikely situation where your model really needs a negative binomial random number distribution (which provides random numbers distributed like a normal distribution—except they are integers) but, alas, there is none in NetLogo. By now you are a confident programmer and think you could write your own, if you just knew the algorithm.

Luckily, computer scientists have written handy reference books telling us exactly how to program many, many numerical functions like random number distributions. Search your library for books with titles including terms such as "numerical recipes," "computational methods," and the kind of method you are looking for, such as "random number generation." You will find algorithms and example code for programming all sorts of functions: statistics, geometry (is a point x,y within some polygon? It's not a simple problem but there is a very elegant method!), chemical reactions, etc.

But before you even look for an existing recipe, look carefully to see whether what you need has already been implemented in NetLogo. Very generous and talented programmers have already written sets of NetLogo procedures to do a variety of handy things such as importing spatial data from geographic information systems, storing information in arrays and hash (lookup) tables instead of just lists, and "profiling" your program—showing what parts of it take up the most computer time. These additions are called "extensions." There is a whole Extensions section of the NetLogo documentation, and additional ones are listed on the NetLogo web site (see also section 24.5). You can also search NetLogo's Models Library and its Code Examples section, and even the Internet, for NetLogo code that does what you need.

So if you need a method that is not in NetLogo, do not give up and do not start from scratch! First look to see if there is already an extension or other code that does what you want. If not, just program your own using a published algorithm, so your code will much more likely be correct and efficient, and finished sooner.

15.3.3 Random Number Generator Seeds

As you know from reading the NetLogo random number documentation, you can control the random numbers in your model with the primitive random-seed. If you place this statement

```
random-seed -260409600
```

at the top of the setup procedure, your model will always produce exactly the same results (unless you change some other part of the model) because the random number generator will produce exactly the same sequence of random numbers. (For modeling success, it is important for the seed to be a lucky number. The above seed is the Unix date code for the birth date of the wife of one author.) But changing the seed to any other integer will produce different random numbers.

When running simulation experiments, be aware that BehaviorSpace does not override any `random-seed` statements in your program. If your model uses `random-seed` as shown here, and you tell BehaviorSpace to produce 10 replicates, all 10 "replicates" will be identical. If you want to produce replicate simulations, remove any such `random-seed` statements.

What if we want one part of a model to be random, but consistent among model runs (including BehaviorSpace replicates), while the rest of the model varies stochastically each time the model runs? For example, we might want a patch variable *resource-level* initialized to random values from a normal distribution with mean of 100 and standard deviation 20, but we want these initial *resource-level* values, and only them, to be exactly the same each time the model runs. Doing this will allow us to see how much variation in model results is due to stochasticity other than the initialization of *resource-level* values. To do this, we can use statements such as these in `setup`:

```
to setup
  with-local-randomness
  [
      random-seed 131313
      ask patches [set resource-level random-normal 100 20]
  ]
  ...
end
```

or instead:

```
to setup
  random-seed 131313
  ask patches [set resource-level random-normal 100 20]

  random-seed new-seed
  ...
end
```

You can figure out how these statements work by reading the NetLogo Dictionary entries for the primitives `with-local-randomness`, `random-seed`, and `new-seed`.

15.4 An Example Stochastic Process: Empirical Model of Behavior

In section 15.2.1 we discussed the use of observed data and stochastic processes to model processes such as agent behaviors. Now let's look at an example, by modifying the Business Investor model that we first built in chapter 10. Agents in this model have one important adaptive trait: they decide each time step whether to switch to one of the other investment opportunities (patches) available to them. These investment alternatives may provide higher or lower profit and may have higher or lower risk. In the baseline version of this model (described in section 10.4), the agents make this decision by selecting the alternative that maximizes an objective function that represents expected investor wealth at the end of a time horizon.

Table 15.1 Observed Frequencies of Investor Decisions

Investor alternative	Number of times available	Number of times chosen	Frequency chosen when available
Change to investment with higher profit and lower risk	115	93	81%
Change to investment with higher profit and higher risk	2665	116	4.4%
Change to investment with lower profit and lower risk	2589	120	4.6%
Change to investment with lower profit and higher risk	4668	0	0%
Do not change investments	5000	4671	93%

Now, imagine that instead of relying on this theoretical approach to modeling investor behavior, we had data from real investors making similar decisions. How could we use these data in an empirical (observation-based) trait? To be specific, let's assume we have the observations from 5000 separate decisions by real investors—a seemingly large data set—reported in table 15.1. (These "observations" are actually results from two runs of the investment model from section 10.4.)

These observations provide information on the *frequency* that investors chose the four different alternatives. For example, of the 115 times that investors had available to them new investments with both higher profit and lower risk, they chose such an investment 93 times, so they chose this alternative, when it was available, with a frequency of 81%.

To use these data in a stochastic model of the decision, we can make the assumption that the observed frequencies are equal to *probabilities* that agents will choose the four alternatives when available. Thus we assume that investors will change to an investment with higher profit and lower risk with a probability of 0.81 when at least one such investment is available, to an investment with high profit and higher risk with probability 0.044, and so on.

While 5000 observations seems like a big data set, you may already be thinking that a number of questions remain unrepresented by the data. For example, what did the investors do when more than one alternative was available to them—if they could move to either higher profit and higher risk, or lower profit and lower risk, how often did they choose one over the other? Because we have no data on this question, we will have to make up an unsupported assumption for our model. We also have learned nothing about how the *magnitudes* of investment risks and returns affect the probability of switching: we might strongly suspect that investors would be more likely to switch to an investment with higher risk if the increase in profit was big instead of small (or if the increase in risk was small instead of big), but the frequency data do not let us prove or quantify such a relation.

Now you should be able to implement this stochastic decision trait in the business investment model, via statements such as these:

```
; First identify potential neighbor destination patches
let potential-destinations neighbors with
    [not any? turtles-here]

; Identify any alternative patches with higher profit and
; lower risk
set HiProfit-LowRisk-alts potential-destinations with
```

```
      [(annual-profit > [annual-profit] of myself) and
       (annual-risk < [annual-risk] of myself)]

  ;  Decide whether to move to one, using global
  ;  probability parameter
  if any? HiProfit-LowRisk-alts
    [
      if random-bernoulli P-move-to-higher-profit-lower-risk
      [
        ;  Now move there
        move-to one-of HiProfit-LowRisk-alts
        stop  ; Turtles can only move once per tick!
      ]
    ]

  ;  Repeat for higher profit and higher risk, etc.
```

(You will need four such blocks of code for the different situations: whether there are any alternatives available with higher profit and higher risk, with lower profit and lower risk, etc.)

Now, you might think that this model should perfectly reproduce observed behavior of real investors. But does it? We leave analysis of this stochastic model to the Exercises.

15.5 Summary and Conclusions

Agent-based models are not necessarily stochastic: very interesting and complex results can emerge from models that are completely deterministic (using only equations, logic, etc. with no randomness). Yet, very often, we want some form of variability in our model but do not want all the detail to explain where the variability came from. Stochastic processes are for representing such variability.

When we design a stochastic element of a model, we need to select an appropriate random number distribution. To select a distribution, we first look at whether we need the random numbers to be floating point numbers, integers, or boolean values. Then, do we want the numbers to be uniformly distributed over a discrete range, or perhaps normally distributed with some values more likely than others and an extremely broad range of possible values? Do we need the values to be positive? Then we look at the theoretical assumptions behind the available distributions to choose the most appropriate. The continuous distributions that many people are most familiar with, especially the normal distribution, are not the most useful in ABMs because the processes we model are often integer or boolean, not real-number; or they have limits to the range of usable values—negative sizes or ages often make no sense, but they are likely to be drawn from a normal distribution.

Sometimes a distribution can be useful even when its theoretical assumptions are not completely met. The Poisson distribution, for example, can often be useful for providing random integers even when its assumption of independence among events is not realistic. Whenever you aren't completely sure what numbers a distribution will produce, take 10 minutes to write a test code (in NetLogo or a spreadsheet, etc.) and investigate.

Once we have selected an appropriate distribution, then we need to assign its parameters. Each distribution has its own kinds of parameters: a uniform distribution requires the minimum and maximum of the range of values it can produce, a normal distribution requires a mean and standard deviation, and so on. These parameters are often treated as constants that are estimated from data: the frequency with which events have been observed in real systems is assumed to represent the probability of their occurring in the model, or the mean and standard deviation of field measurements of some variable are used as the parameters of a normal distribution in the model.

However, it is very common and useful for the parameters of a statistical distribution to be modeled—the parameters can vary as a function of whatever is going on in the ABM. For example, the probability of an agent executing some behavior can vary with its state variables such as size, age, or location. Logistic regression is one technique for determining from data how the probability of an event is related to one or several variables.

Sometimes we want to turn off some or all of the randomness in a model, by keeping the random numbers from changing every time we run the model. We can do this by setting the random number seed. Via the `with-local-randomness` primitive, NetLogo even lets us switch randomness off and on in different parts of a model.

In this chapter we introduced the technique of defensive programming: putting checks in the code to see if variables have impossible or unrealistic values. This technique is critical for keeping models from producing nonsense without our realizing it. Defensive programming can be particularly appropriate when working with stochastic processes, and especially when using distributions like the normal distribution that can occasionally produce "outlier" values.

There are potentially useful random number distributions that are not built into NetLogo. Some of these (the Bernoulli, binomial, and log-normal distributions, for example) are fairly trivial to program. You should be aware, however, that efficient, tested algorithms for other distributions—and many other numerical methods—are readily available in the computer science literature. Whenever you need a method that seems even remotely likely to have been programmed before, you should check for a NetLogo extension or "numerical recipe" so you do not have to design it yourself.

15.6 Exercises

1. Write a simple NetLogo procedure that draws 1000 values from the distribution `random-float 100` and writes them to a file. Analyze the results to see how variable the results were. How many numbers were between 0 and 10, between 10 and 20, etc.? How close were the mean and standard deviation to the theoretical values for this kind of distribution? (What is the equation for the standard deviation of this kind of distribution?) Repeat the experiment several times to see how variable its results are.

2. Repeat exercise 1, but this time draw random integers from `random-poisson 0.8`. This distribution models, for example, how many offspring a turtle produces per tick when the average rate at which turtles reproduce is 8 times in 10 ticks. How many times did this distribution produce values of 0, 1, 2, etc.? How close are the mean and standard deviation to the theoretical values?

3. Analyze the stochastic variability in results of the original Business Investor model (as described in section 10.4), specifically in the effect of investor time horizon on mean final investor wealth as shown in the upper left panel of figure 12.1. First, repeat the simulation experiment using 10 replicates for each time horizon scenario. What are the means and standard deviations in results among these replicates for each time horizon? Second, determine how much of this stochastic variability is due just to the randomness in how the model was initialized. Use NetLogo's random seed primitives to make the `setup`

procedure produce exactly the same patch and turtle states every time the model is initialized. What other parts of the model are stochastic? How do the standard deviations in model results change now that the same initial conditions are used each time?

4. Write a NetLogo program that counts how many times it draws values from the normal distribution discussed in "Continuous Distributions" in section 15.3.2 (mean = 5; standard deviation = 1.5) until it obtains a negative value. What answers do you get when you run this program 10 times or so? (Hint: this program can take as few as four statements, three of which are `to go`, `reset-ticks`, and `end`. The third statement uses the `while` primitive.)

5. Write a random binomial distribution as a NetLogo reporter. It will have two parameters: the probability per trial of a "true" event, and the number of trials; and it will report the number of true events. (Hint: first type in the `random-bernoulli` reporter from "Discrete Distributions" in section 15.3.2.) Repeat the experiment in exercise 2 using your random binomial distribution instead of `random-poisson`. How are results different? Why? Compare the difference in results between the binomial and Poisson distributions when the parameter is 0.08 instead of 0.8.

6. In NetLogo's Simple Birth Rates library model, should the procedure `grim-reaper` be replaced with a Poisson or binomial distribution? Why? Do it, so `grim-reaper` looks like this:

```
to grim-reaper
  let num-turtles count turtles
  if num-turtles < carrying-capacity [stop]
  let death-rate
      (num-turtles - carrying-capacity) / num-turtles
  show death-rate
  ask n-of (random-binomial death-rate count turtles)
          turtles [die]
  ; OR ??
  ask n-of (random-poisson (death-rate * num-turtles) )
          turtles [die]
  end
```

7. The log-normal distribution has several advantages over the normal distribution for many modeling applications: it can model distributions of relatively small numbers while producing only positive values, and it is skewed so that values between zero and the mean are more common than values above the mean. Write a NetLogo reporter that provides a log-normal distribution. You can do this by creating a normal distribution that returns the logarithm of the variable you are modeling. The trick is calculating the mean and standard deviation of this normal distribution. If X represents the variable that has a log-normal distribution, and μ and σ are the mean and standard deviation of X, then a log-normal distribution of X can be programmed as `exp(random-normal M S)`, where: $M = \ln(\mu) - (\beta/2)$, $S = \sqrt{\beta}$, and $\beta = \ln[1 + (\sigma^2/\mu^2)]$. Write a test procedure that calls your `random-lognormal` reporter 1000 times and writes the results into a list, then histograms the list to let you see how the results are distributed.

8. Program the stochastic version of the Business Investor model described in section 15.4. Use the frequencies reported in table 15.1 as probabilities of investors choosing each

decision alternative when it is available. Make sure investors cannot move more than once per tick. Include an output file showing how often turtles move to all combinations of higher vs. lower investment return and risk—do they reproduce the frequencies of choosing these alternatives used to parameterize the stochastic decision trait? How well do investors perform, in comparison to the original version of this model? What patterns are different? Why?

Collectives

16.1 Introduction and Objectives

Agent-based models represent a system by modeling its individual agents, but surprisingly many systems include intermediate levels of organization between the agents and the system. Agents of many kinds organize themselves into what we call *collectives*: groups that strongly affect both the agents and the overall system. This chapter is about how to model such intermediate levels of organization.

If we need to include collectives in a model, then how do we do it? One way is to represent collectives as a completely emergent characteristic of the agents: the agents have behaviors that allow or encourage them to organize with other agents, the collectives exist only when the agents organize themselves into collectives, and the collectives are given no behaviors or variables of their own. But collectives can also be represented as a separate, explicit type of entity in an ABM with its own actions: the modeler specifies how the collectives behave and how agents interact with the collectives.

If we wanted to model collectives as a separate kind of agent, how could we program them in NetLogo? So far we have focused on models that have only two kinds of entities: patches and turtles (and occasionally, links). Fortunately, NetLogo provides *breeds*: a way to model multiple kinds of turtles or links. Breeds let us do many things, including modeling collectives as explicit entities.

In this chapter your learning objectives are to:

- Learn what collectives are and several ways of modeling them.
- Become expert at using NetLogo breeds to represent kinds of turtles.
- Develop experience with stochastic modeling by programming and analyzing a new example model, of African wild dogs that live in packs.
- Become familiar with logistic functions, which are particularly useful for representing probabilities and functions that vary nonlinearly between zero and one.

16.2 What Are Collectives?

Cooperation among individuals—often, in small groups so that cooperators can be distinguished from non-cooperators—has important advantages, so it is not surprising that systems of all kinds include groups of cooperating individuals. In biology, cells of different types form organs and other structures to perform functions that the whole organism depends on. In many animal and human societies, family and social groups are very important to the behavior, survival, and reproduction of individuals. Economic systems contain many networks and organizations through which individuals and firms cooperate.

We have already looked at several models that include structures that can be considered collectives. In chapter 8 we looked at NetLogo's Flocking model, in which flock-like collectives emerge from individual behaviors. These flock collectives do not really affect this model's individuals, because emergence of flocks is the only thing the model simulates. But in many species of real birds and fish, and in detailed ABMs of these species, the collective flocking or schooling behavior is believed to strongly affect the individuals, especially by reducing their vulnerability to predators and increasing their ability to migrate successfully (Partridge 1982, McQuinn 1997, Huse et al. 2002). In chapter 13 we built a version of the Telemarketer model in which telemarketers formed "mergers," collectives that let agents share financial resources and potentially stay in business longer.

In the Flocking and Telemarketer models, the collectives arose only from the behavior of the agents. There were no traits in the model for how flocks or merged businesses behave, only traits for the individual agents.

A more common approach is to add the collectives to the model explicitly as another type of agent. A collective agent typically has its own state variables (especially, a list of the individuals in the collective) and traits (often including rules for how individuals are added and removed from the collective). The individuals often have a state variable for which collective they belong to and traits for how the collective affects them. This approach of modeling collectives explicitly often adds the least complexity: we can ignore all the individual-level detail that, in reality, produces the behavior of the collective. We can think of collectives and their traits as a simplified model of their individuals.

16.3 Modeling Collectives in NetLogo

When we need to model collectives explicitly with their own state variables and traits, how do we do it in NetLogo? For that matter, what do we do whenever we need more kinds of agent than just patches, turtles, and links? There are several alternatives.

16.3.1 Representing Collectives via Patches or Links

If the individual agents in a model are represented as turtles, sometimes it makes sense to represent collectives of turtles as patches. This approach can work, for example, when a collective is a group of individuals that occupy the same space. A population of animals or people that live in family or social groups that each occupy a territory (or business, or town) can use the patches to represent the groups. Patch variables can describe the collective's characteristics (e.g., the number of members, perhaps broken out by age, sex, etc.), and the primitive `turtles-here` provides an agentset of member turtles. The patch's traits could include rules for collective behaviors such as adding and removing individuals. (In chapter 19 we introduce a model that uses patches to represent a social group.)

NetLogo's links might seem natural for modeling collectives because we can think of a collective as a group of individuals that are linked in some way. However, NetLogo does not have a built-in way to give variables or rules to a *network* of linked turtles—only to the turtles or links. So links are not easily used to represent collectives with their own behaviors.

16.3.2 Using Breeds to Model Multiple Agent Types

NetLogo provides *breeds* for modeling different kinds of turtles or links. Here, we discuss only turtle breeds; link breeds are quite similar. Breeds are equivalent to *subclasses* in an object-oriented programming language: they let you create several specialized kinds of turtles that have different characteristics from each other, while still sharing some common variables and traits. The details of how to define and use breeds are explained in the NetLogo Programming Guide. Here are the key points:

- If you want to use breeds other than just turtles, you must define them at the top of the program.
- Turtles have a built-in state variable `breed`. If you create a turtle using `create-turtles` (or `hatch` or `sprout`), its `breed` variable is set by NetLogo to "turtles." Using `create-<breeds>` creates a turtle of the specified breed: `create-wolves` creates a turtle with its value of `breed` equal to "wolves."
- You can change what breed a turtle is by simply using `set breed`.
- Each breed can have its own unique state variables, defined via a `<breeds>-own` statement.
- Each breed still has the built-in variables (`xcor`, `ycor`, `color`, etc.) that NetLogo provides for all turtles.
- The statement `turtles-own` can still be used to define variables that all breeds use.
- If you have defined the breeds "wolves" and "sheep," you can use `ask wolves [...]` or `ask sheep [...]` to have only wolves or only sheep do something. But `ask turtles [...]` still causes all turtles of all breeds to do something. Likewise, you can use `mean [age] of sheep` and `mean [age] of turtles` to get the average age of just sheep and of all turtles.
- Several of the agentset primitives have breed versions. For example, `sheep-here` reports an agentset of sheep on a patch, while `turtles-here` reports all turtles on the patch.
- Breeds do not have their own NetLogo contexts (see section 2.2); all are treated as being in turtle context. Hence, if you have sheep and wolf breeds, a wolf can try to run a procedure that you wrote for sheep. However, if a wolf tries to use a variable that was defined only for sheep (e.g., `sheep-own wool-thickness`) NetLogo will produce a run-time error.

Different breeds can be very different (e.g., sheep that eat grass and wolves that eat sheep), but they can also be only slightly different versions of the same agent. For example, in the Telemarketer model of chapter 13 it appeared that telemarketer agents that grew rapidly tended not to survive long. One way to investigate whether that observation is true would be to create two breeds of telemarketer ("growers" and "savers") that differ only in the parameter controlling how rapidly they expand, and include both in the same simulation. Likewise, we could analyze the effects of mergers in the Business Investor model by letting a breed with a trait to form mergers compete against a breed that lacks the trait.

We focus here on using breeds to model collectives. We can represent individuals as one breed, while representing collectives as a separate breed with completely different variables and traits. To make this work, we must keep track of which individuals belong to each collective, and which collective each individual belongs to. The trick is giving each individual a variable (e.g., `my-group`) that contains the collective it belongs to, and giving each collective a list or agentset of all the individuals that belong to it (`my-members`). The hard part is remembering to always keep these variables up to date. Each time an individual leaves or joins a collective, or is created, we need to remove it from or add it to its collective's agentset of members. And each time an individual joins or leaves a collective, we need to update the individual's variable for which collective it belongs to. (We illustrate ways to do this in section 16.4.2.) Forgetting to update these collective membership variables is a common mistake that can be hard to find.

16.4 Example: A Wild Dog Model with Packs

Some of the clearest examples of collectives are populations of animals that form family or social groups. These animals typically cooperate within their collectives to obtain food and other resources. Mating and reproduction can be tightly regulated within the collectives: a group's dominant ("alpha") male and female mate, but suppress reproductive behavior by other members. Therefore, the characteristics of a collective strongly affect the fitness (survival, growth, reproductive potential) of its members. Yet the behavior and fate of the individuals strongly affect their collective: the reproductive output of its alpha couple depends on how other members decide whether to remain or seek better opportunities elsewhere, and whether individuals leave or die determines the size and structure (e.g., number of young vs. old members) of the group.

Our example is a model of African wild dogs (the species *Lycaon pictus*), which are endangered and carefully managed in a number of nature reserves in South Africa. (This example is loosely based on the management model of Gusset et al. 2009, but distinctly different.) African wild dogs usually live in packs that include an alpha male and female, the only individuals that reproduce. After pups grow into adults, they decide whether to stay in their natal pack in hopes of become an alpha when their parent dies, or to "disperse": leave the pack in hopes of forming their own new pack. The dispersers travel in groups of one or more siblings of the same sex until they meet a group of the opposite sex from a different pack, with whom they may form a new pack.

The model of Gusset et al. (2009) is a tool for evaluating management to reduce the risk of extinction for the dog population of one particular reserve, the Hluhluwe-iMfolozi Park in South Africa. Here, we consider two hypothetical management actions (which are not necessarily realistic for this park). One is to increase the "carrying capacity" of the reserve, the maximum number of dogs that there are resources to support. Carrying capacity could be increased by increasing the size of the reserve or by reducing the competition for prey from other predators such as lions. Increasing the size of the reserve could, however, reduce the ability of disperser groups to find each other and form a new pack. The second management action is attempting to reduce the mortality of dispersers, which is high.

Detailed observations of the wild dogs have been collected for many years, so we have extensive empirical information to use. These observations include the following:

- The frequency with which packs produce pups decreases as the total population of dogs increases. This relationship appears to reflect competition with other dogs over the park's resources. It is reasonable to expect the relationship to change—reproduction becoming more frequent at higher dog populations—if more resources were available.
- Subordinate adults (defined as being two or more years old and not the alpha male or female) often leave their pack as dispersers. Adults that are the only non-alpha of their sex leave their pack 50% of the time, each year. When two or more subordinate adults of the same sex are in a pack, they always leave as a same-sex group of dispersers.
- When two disperser groups of opposite sex and originating from different packs meet, they usually—64% of the time—join to form a new pack.
- The annual mortality rate varies with dog age and is much higher for dispersing dogs than for those in a pack.

This information tells us several things about how the model can be structured. First, it is clear that we need to model the individual dogs because many processes depend on individual variables such age, sex, and social status. Second, it is clear that not one but two kinds of collectives are important: packs and disperser groups. Therefore, we will need three breeds: dogs, packs, and disperser groups.

Something else that is clear from this information is that stochastic traits would be very useful. The behaviors of dogs and the collectives they form are complex, but frequencies and rates observed in the field can be used as probability parameters for stochastic traits. Reproduction, dispersal decisions, and mortality can all be modeled as stochastic events with probabilities based on observed frequencies.

Finally, this information and our two hypothetical wild dog management actions help clarify the model's purpose. The management actions could change the reserve's carrying capacity and therefore the relation between the total number of dogs and the probability of each pack reproducing each year. Increasing the size of the reserve could reduce the probability that disperser groups encounter each other. And other management actions could reduce the probability that dispersing dogs die before forming a new pack.

16.4.1 Model Description

Here we provide a description of our simplified Wild Dog model. Because we discussed the system and model extensively above, we provide here only the overview and initialization parts of the ODD description, with added detail so you can program it. (Completing the ODD description is exercise 2.)

Purpose

The purpose of the model is to evaluate how the persistence of a wild dog population depends on (a) the reserve's carrying capacity, (b) the ability of dispersing dogs to find each other, and (c) the mortality risk of dispersing dogs. Measures of "persistence" include the average number of years (over a number of replicate simulations) until the population is extinct, and the percentage of simulations in which the population survives for at least 100 years.

The model's purposes as a NetLogo exercise are to demonstrate the use of breeds to represent collectives and to illustrate stochastic modeling techniques.

Entities, State Variables, and Scales

The model includes three kinds of agent: dogs, dog packs, and disperser groups. Dogs have state variables for age in years, sex, the pack or disperser group they belong to (to keep track of which dogs belong to which pack), and social status. The social status of a dog can be (a) "pup," meaning its age is less than one; (b) "yearling," with age between 1 and 2; (c) "subordinate," meaning age is greater than 2 but the dog is not an alpha; (d) "alpha," the dominant individual of its sex in a pack; and (e) "disperser," meaning the dog currently belongs to a disperser group, not a pack.

Dog packs have no state variables except for a list (or, in NetLogo, an agentset) of the dogs belonging to the pack. Disperser groups have a state variable for sex (all members are of the same sex) and an agentset of member dogs.

The time step is one year. The model is nonspatial: locations of packs and dogs are not represented. However, its parameters reflect the size and resources of one nature reserve.

Process Overview and Scheduling

The following actions are executed once per time step, in this given order.

1. Age and social status update:

- The age of all dogs is incremented. Their social status variable is updated according to the new age.
- Each pack updates its alpha males and females. If there is no alpha of a sex, a subordinate of that sex is randomly selected (if there is one) and its social status variable set to "alpha."

2. Reproduction—Packs determine how many pups they produce, using these rules:

- If the pack does not include both an alpha female and alpha male, no pups are produced.
- Otherwise, the probability of a pack producing pups depends on the total number of dogs in the entire population at the current time step, N. (N does not include any pups already produced in the current time step.) The probability of a pack reproducing, P, is modeled as a logistic function of N, and the parameters of this logistic function depend on the carrying capacity (maximum sustainable number of dogs) of the reserve. P has a value of 0.5 when N is half the carrying capacity and a value of 0.1 when N equals the carrying capacity. The carrying capacity is 60 dogs. (See the programming note below on "Parameterizing and programming logistic functions.")
- If the pack reproduces, the number of pups is drawn from a Poisson distribution that has a mean birth rate (pups per pack per year) of 7.9. Sex is assigned to each pup randomly, with a 0.55 probability of being male. Pup age is set to 0.

3. Dispersal—Subordinate dogs can leave their packs in hopes of establishing a new pack. These "dispersers" form disperser groups, which comprise one or more subordinates of the same sex that came from the same pack. Each pack follows these rules to produce disperser groups:

- If a pack has no subordinates, then no disperser group is created.
- If a pack has only one subordinate of its sex, it has a probability of 0.5 of forming a one-member disperser group.
- If a pack has two or more subordinates of the same sex, they always form a disperser group.
- Dogs that join a disperser group no longer belong to their original pack, and their social status variable is set to "disperser."

4. Dog mortality—Mortality is scheduled before pack formation because mortality of dispersers is high. Whether or not each dog dies is determined stochastically using the following probabilities of dying: 0.44 for dispersers, 0.25 for yearlings, 0.20 for subordinates and alphas, and 0.12 for pups.

5. Mortality of collectives—If any pack or dispersal group has no members, it is removed from the model. If any pack contains only pups, the pups die and the pack is removed.

6. Pack formation—Disperser groups may meet other disperser groups, and if they meet a disperser group of the opposite sex and from a different pack, the groups may or may not join to form a new pack. This process is modeled by having each disperser group execute the following steps. The order in which disperser groups execute this action is randomly shuffled each time step:

- Determine how many times the pack meets another disperser group (variable *num-groups-met*). *Num-groups-met* is modeled as a Poisson process with the rate of meeting (mean number of times per year that another group is met) equal to the number of other disperser groups times a parameter controlling how often groups meet. This meeting rate parameter can potentially have any value of 0.0 or higher (it can be greater than 1) but is given a value of 1.0. The following steps are repeated up to *num-groups-met* times, stopping if the disperser group selects another to join.

- Randomly select one other disperser group. It is possible to select the same other group more than once.
- If the other group is of the same sex, or originated from the same pack, then do nothing more.
- If the other group is of the opposite sex and a different pack, then there is a probability of 0.64 that the two groups join into a new pack. If they do not join, nothing else happens.
- If two disperser groups do join, a new pack is created immediately and all the dogs in the two groups become its members. The alpha male and female are chosen randomly; all other members are given a social status of "subordinate." The two disperser groups are no longer available to merge with remaining groups.

Initialization

The model is initialized with 10 packs and no disperser groups. The number of dogs in each initial pack is drawn from a Poisson distribution with mean of 5 (the Poisson distribution is convenient even though its assumptions are not met in this application). The sex of each dog is set randomly with equal probabilities. The age of individuals is drawn from a uniform integer distribution between 0 and 6. Social status is set according to age. The alpha male and female of each pack are selected randomly from among its subordinates; if there are no subordinates of a sex, then the pack has no alpha of that sex.

Programming note: Parameterizing and programming logistic functions

A logistic function produces an output that varies from 0.0 to 1.0 (or 1.0 to 0.0) in an "S" shape, as its input increases linearly. Logistic functions are particularly useful for modeling how the probability of some event depends on one or several variables, because (a) they produce values always in the probability range of 0.0 to 1.0, and (b) they reproduce how probabilities can be very low over a wide range of conditions, very high over another range, and change sharply between those two ranges. For example, figure 16.1 illustrates how the probability of a dog pack reproducing depends on the total dog population. Over a range of low populations, the probability of reproducing is high: once the population is below about 20, competition for resources is unimportant and the probability of reproducing is close to 1.0. At high population values the probability of reproducing approaches zero: once competition is too high for reproduction, it does not matter how much higher

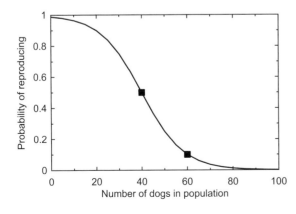

Figure 16.1
Example logistic function, relating probability of dog pack reproduction to dog population.

it gets. In an intermediate population range—about 20 to 60—it is much less certain what will happen as probability of reproduction decreases sharply from high to low.

Here is one way to define and program logistic functions. First, identify two points on the logistic curve. In figure 16.1, the square markers indicate two points: where the population ($X1$) is 40 and the probability ($P1$) is 0.5, and where the population ($X2$) is 60 and $P2$ is 0.1. (It does not matter if $P1$ is higher or lower than $P2$.)

Next, calculate four intermediate variables:

$$D = \ln[P1 / (1 - P1)]$$
$$C = \ln[P2 / (1 - P2)]$$
$$B = (D - C) / (X1 - X2)$$
$$A = D - (B \times X1)$$

Now you can calculate P for any value of X as:

$$P = Z / (1 + Z)$$

where:

$$Z = \exp[A + (B \times X)]$$

The intermediate variables A–D need only be calculated once (as long as the shape of the logistic function does not change), so variables A and B can be global variables and calculated during setup.

The parameters for logistic functions can be estimated from data using logistic regression, or they can be simply guesstimated to provide reasonable probabilities over a range of X values. In all cases, you should program and graph logistic functions (e.g., in a spreadsheet) before using them in your model, to make sure the parameters produce reasonable results.

16.4.2 Implementing the Model

There are several new challenges to programming this model in NetLogo. First, we need to start by defining the breeds and their state variables.

```
breed [dogs dog]
breed [packs pack]
breed [disperser-groups disperser-group]

dogs-own
[
  age
  sex
  status
  my-pack
  my-disperser-group
]

packs-own [pack-members]
```

```
disperser-groups-own
[
  sex
  group-members
]
```

Then we can proceed with defining the global variables as usual. Now, when writing the setup procedure, how do we create the initial packs and their dogs? One way is to create each pack, and as it is created have each pack create its dogs:

```
create-packs initial-num-packs
[
    ; set a location and shape just for display
    setxy random-xcor random-ycor
    set shape "box"

    ; create the pack's dogs
    let num-dogs random-poisson initial-mean-pack-size

    hatch-dogs num-dogs
    [
      ; first, set display variables
      set heading random 360
      fd 1

      ; now assign dog state variables
      ifelse random-bernoulli 0.5
          [set sex "male"]
          [set sex "female"]

      set age random 7
      set-my-status  ; a dog procedure that sets
                     ; social status from age
      set my-pack myself   ; set dog's pack to the one
                           ; creating it
    ] ; end of hatch-dogs

    ; Initialize the pack's agentset that contains its dogs
    set pack-members dogs with [my-pack = myself]

    ; show count pack-members   ; Test output – off for now

    ; now select the alpha dogs
    update-pack-alphas   ; a pack procedure to give the
                         ; pack 2 alphas

] ; end of create-packs
```

Now we have created the initial packs, and each pack has created its dogs. The pack has an agentset of its dogs (pack-members) and each dog has a variable for its pack (my-pack). Keeping these variables up to date is how the model keeps track of which dogs are in which pack.

The organization of the go procedure should be quite clear from the detailed model description provided above. In the following example go procedure, some actions (updating dog ages and social status; mortality of disperser groups) are programmed right in the go procedure because they are extremely simple.

```
to go

  tick
  if ticks > years-to-simulate [stop]

  ; First, age and status updates
  ask dogs
  [
     set age age + 1
     set-my-status
  ]

  ask packs [update-pack-alphas]

  ; Second, reproduction
  ask packs [reproduce]

  ; Third, dispersal
  ask packs [disperse]

  ; Fourth, mortality
  ask dogs [do-mortality]

  ; Fifth, mortality of collectives
  ask packs [do-pack-mortality]
  ask disperser-groups [if count group-members = 0 [die]]

  ; Sixth, pack formation
  ask disperser-groups [do-pack-formation]

  ; Finally, produce output
  update-output

end
```

Now let's look at another example of how collectives are formed: code for how subordinate dogs decide whether to leave their pack and how they form a disperser group when they do leave. Note that this dispersal decision is modeled as traits of packs, not of individual dogs. Also note that, while the disperse procedure must make four separate checks for whether single or multiple males or females disperse, they can all use the same generic procedure create-disperser-group-from to form their new disperser group. This procedure takes as its input an agentset of the dogs that disperse (e.g., my-subordinates with [sex = "female"]).

```
to disperse ; a pack procedure

  ; First, identify the subordinates and stop if none
  let my-subordinates pack-members with
      [status = "subordinate"]
  if not any? my-subordinates [stop]

  ; Now check females
  if count my-subordinates with [sex = "female"] = 1
  [
      if random-bernoulli 0.5
      [
        create-disperser-group-from
              my-subordinates with [sex = "female"]
      ]
  ]

  if count my-subordinates with [sex = "female"] > 1
  [
    create-disperser-group-from
          my-subordinates with [sex = "female"]
  ]

  ; And check males
  if count my-subordinates with [sex = "male"] = 1
  [
    if random-bernoulli 0.5
    [
      create-disperser-group-from
            my-subordinates with [sex = "male"]
    ]
  ]

  if count my-subordinates with [sex = "male"] > 1
  [
      create-disperser-group-from
            my-subordinates with [sex = "male"]
  ]
end ; to disperse

to create-disperser-group-from [some-dogs]
  ; a pack procedure
  ; "some-dogs" is an agentset of the dispersers

  ; First, create a disperser group and put the dogs in it
  hatch-disperser-groups 1
```

```
[
  ; Set disperser group variables
  set group-members some-dogs
  set sex [sex] of one-of some-dogs

  ; Display the group
  set shape "car"
  set heading random 360
  fd 2

  ; Now the disperser group sets the variables of the
  ; dispersing dogs
  ask some-dogs
  [
    set my-disperser-group myself
    set status "disperser"
    set color green

    ; and display them in a line from the disperser group
    move-to my-disperser-group
    set heading [heading] of my-disperser-group
    fd 1 + random-float 2
  ] ; end of ask some-dogs

] ; end of hatch-disperser-groups

; Finally, remove the dispersers from their former pack
let dogs-former-pack [my-pack] of one-of some-dogs
ask dogs-former-pack
  [set pack-members pack-members with
      [status != "disperser"]]

end ; to create-disperser-group-from
```

With these examples and hints, you should be able to program (and test!) the full Wild Dog model.

16.4.3 An Example Analysis: Effect of Disperser Group Meeting Rate

Now that you have programmed the Wild Dog model, let us look at how it could be used to address management decisions. Some potential actions to manage reserves for wild dog populations could affect the rate at which disperser groups meet each other and form new packs. For example, some reserves have fences to keep dogs from leaving; adding or removing fences could increase or decrease the rate at which disperser groups meet. How important is this meeting rate to the persistence of the total dog population?

In the model, the meeting rate is represented as a parameter (called, e.g., *disperser-meeting-rate*) that represents the mean number of times per year any pair of groups meet each other. To investigate the importance of how easily dispersers meet each other, let's do a sensitivity experiment that varies the value of *disperser-meeting-rate* over a wide range.

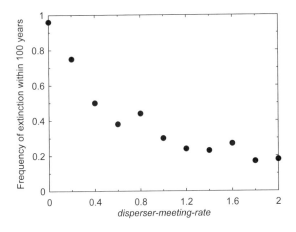

Figure 16.2
Response of extinction frequency in the Wild Dog model to the rate at which pairs of disperser groups meet.

What output of the model should we look at? Remember that the model's purpose is to study how management affects persistence of dog populations, and one measure of a simulated population's persistence is how frequently it goes extinct within a certain number of years. We will look at how frequently the dog population goes extinct in hundred-year simulations: how many times, out of 500 replicate simulations, there are no dogs alive after 100 years. (We use more replicates than in previous experiments because time to extinction is a particularly variable output.)

You should easily be able to set up a BehaviorSpace experiment for this analysis. It will vary the value of *disperser-meeting-rate* over a range from, for example, 0.0 to 2.0. If you set BehaviorSpace's reporter for measuring runs to `count dogs`, its output at the end of each run will tell you whether the dogs went extinct. From this experiment, you should be able to produce a graph that looks similar to figure 16.2.

This analysis indicates that increasing the rate at which disperser groups meet above the baseline value (1.0) would not benefit population persistence immensely: the frequency of extinction decreased by about 40% as *disperser-meet-rate* was increased from 1.0 to 2.0. However, *reducing* the ability of dispersers to meet is predicted by the model to have stronger negative effects: as *disperser-meet-rate* was reduced from 1.0 to 0.0, the frequency of extinction tripled. This analysis could indicate to managers that they should avoid actions that make it harder for disperser groups to meet, while not assuming that improving disperser meeting would make the population substantially less likely to go extinct.

16.5 Summary and Conclusions

In many agent-based systems, collectives—groups of individuals that strongly affect the fate and behavior of the individuals and the dynamics of the system—are important. One way to model collectives is as an emergent trait of the individuals: individuals are modeled in such a way that they can form collectives, and the collectives are strictly an outcome of the model. The NetLogo Flocking model is an example of this approach. Modeling collectives as emergent from individuals is especially useful when the purpose of the model is, as with Flocking, to explain how collectives form. But this approach is likely to be overly complicated if the collectives have important behaviors of their own.

Collectives

For systems in which collectives have important and even moderately complex behavior, it is typically necessary to model the collectives explicitly as a separate kind of entity with its own state variables and traits. In NetLogo, one way to model collectives explicitly is to assign the properties and traits of collectives to patches. This approach can work well when the collectives each occupy a piece of some landscape, and in models that are nonspatial so patches can represent only the collectives. In other models, though, the breeds feature of NetLogo is especially useful for representing collectives because it lets us create multiple kinds of turtle (or link).

Collectives are often essential in ABMs, but their use has a cost in addition to programming. Because collectives are an additional level of organization, we now need to model and analyze how the properties and behavior of collectives and individuals affect each other, as well as how the system affects both collectives and individuals. How do we know whether our ABM captures these multilevel dynamics adequately? How do we know whether our traits for individuals and collectives contain the right processes and produce the right kind of dynamics? These are questions for the next stage of the modeling cycle, when we take the first version of a model that we've programmed and start analyzing and revising it, to learn about the model and, finally, about the system that we are modeling.

16.6 Exercises

1. Program the logistic function in a spreadsheet (or other software) and reproduce the graph in figure 16.1. Produce graphs for $P2 = 0.9$ instead of 0.1, and for $X1 = 20$ instead of 40.

2. Write a complete ODD description for the Wild Dog model. Is the information in section 16.4.1 sufficient?

3. Program the Wild Dog model and test your program. How can you test such a stochastic model?

4. With your Wild Dog model, analyze these outputs: (a) the percentage of 100-year simulations that produce extinction (as in figure 16.2) and (b) the median time to extinction of the population (which could require simulations longer than 100 years, if extinction often takes more than 100 years). How do these outputs vary with the disperser meeting rate parameter? The carrying capacity? The annual risk of mortality for dispersing dogs? Which of these variables seem important to the population's persistence?

5. Some species of dogs include "transients": a type of individual that no longer belongs to a pack or disperser group but can join a pack to replace an alpha that has died (i.e., Pitt et al. 2003). Add transients as another "breed" in the Wild Dog model, using these traits:

   (a) All members of disperser groups become transients if their disperser group has not joined another to form a pack in its first year.

   (b) At the start of their first year as a transient, transients seek packs that lack an alpha of their sex (after packs replace their missing alphas with subordinates if they can). Each transient has a probability of 0.5 of finding and joining each such pack. Transients execute this search in randomized order.

   (c) If a transient fails to find a pack to join, it remains a transient.

   (d) Transients have an annual mortality probability of 0.5.

   How do transients change model results?

6. The Wild Dog model treats replacement of dead alphas as a pack trait. Instead, try to design a way to model this behavior as an individual dog trait: each dog must determine whether it becomes an alpha. What additional assumptions, variables, and processes do you need to (or want to) add? You can do this exercise in words instead of in NetLogo.

Pattern-Oriented Modeling

". . . mere prediction is not enough. The researcher should also be concerned about the generative mechanisms behind his results" (Heine et al. 2005)

Introduction to Part III

17

17.1 Toward Structurally Realistic Models

Now that we have introduced you to agent-based modeling, NetLogo, and important concepts for designing ABMs, part III will focus on a more strategic level. Part II was spent in the engine room of the ship RV *Agent-Based Modeling* learning how to set up the engine of an ABM and keep it running, via design, implementation, and testing. Now, in part III, we leave the engine room and head for the bridge to work on getting somewhere with our vessel: we want to learn about how real agent-based complex systems work.

Remember from chapter 1 that we use models as simplified and purposeful representations to answer research questions and solve problems. A "simplified representation" means that we use only a limited and preferably small number of variables to represent the real system. The reason is obvious: more variables make a model more complex so that we would need more time and resources to build, test, understand, and communicate the model. Models in general and ABMs in particular should therefore be as simple as possible and use the most parsimonious representation that still captures key characteristics and behaviors of a real system. We refer to such models as "structurally realistic": they include key structural elements of a real system's internal organization, but they are certainly not "realistic" in the sense that they include everything we know about the system. By "internal organization" we mean the structures and processes that, in the real system, produce the system's characteristic properties and behaviors.

In chapter 1 we also learned that we keep models simple by using the question or problem addressed with the model as a *filter*. Aspects of the real system (entities, state variables and attributes, processes, spatial and temporal resolution and extent) should be included in the model only if we, or the experts we are working with, consider them absolutely essential for solving a particular problem.

The problem with ABMs is that the question or problem that a model addresses is often inadequate as a filter for deciding what must be in models: by itself, the problem does not provide enough information to ensure that our models are structurally realistic and really capture the essence of the system's internal organization. Part III is about the solution to this problem: *pattern-oriented modeling* (POM). POM is the use of patterns observed in the real system as the additional information we need to make ABMs structurally realistic and, therefore, more general and useful, scientific, and accurate.

227

We can illustrate the basic concept of POM with a historical example from ecology. Numerous population models have been developed to explain the pattern of strong cycles in the abundance of small mammals in boreal regions (lemming population booms and crashes are a famous example). All these models can "solve" the problem they were designed for: they can produce population cycles. However, they all make different assumptions about the key mechanisms causing the cycles (Czárán 1998). Therefore, it is very likely that some or even most of the models are not structurally realistic but instead wrong about the mechanisms causing population cycles. They reproduce the cyclic pattern for the wrong reasons!

The cyclic pattern, as the general research question or problem that these models address, is a filter (explaining a strong pattern like this is a very common research question). A filter separates things, such as models that do and do not reproduce the cyclic pattern. Obviously, the filter "cyclic pattern" is not enough to identify the internal organization of these populations: it is too easy to make a model produce cycles. We obviously need more filters that are better at excluding unrealistic mechanisms.

The basic idea of POM is to use *multiple* patterns to design and analyze models (Grimm et al. 1996, 2005; Grimm and Berger 2003; Wiegand et al. 2003; Grimm and Railsback 2005, chapter 3). Each pattern serves as a filter of possible explanations so that after using, say, three to five patterns we are more likely to have achieved structural realism. An unrealistic model might pass one filter, but probably not also the second and third one. POM thus uses *multicriteria assessment* of models using multiple patterns that have been observed in the real system, often at different scales and at both the agent and system levels.

"But what is so novel about this?" Platt (1964, p. 347) asked this question when he introduced the term "strong inference": "This is the method of science and always has been, why give it a special name? The reason is that many of us have almost forgotten it." Likewise, POM is what scientists do all the time: look for explanations of phenomena by seeking models that can reproduce patterns produced by the phenomena. But in all the complexity of agent-based modeling we easily get distracted from this master plan of science.

The basic idea of POM is not new but fundamental to any science; it is to use multiple patterns as indicators of a system's internal organization. By trying to reproduce these patterns simultaneously we try to decode this internal organization. Using POM for agent-based modeling includes using patterns to design a model's structure (chapter 18), develop and test theory for agent behavior (chapter 19), and find good parameter values (chapter 20).

17.2 Single and Multiple, Strong and Weak Patterns

What exactly do we mean by "pattern"? A pattern is anything beyond random variation. We can think of patterns as regularities, signals, or, as they are sometimes called in economics, "stylized facts": "stylized facts are broad, but not necessarily universal generalizations of empirical observations and describe essential characteristics of a phenomenon that call for an explanation" Heine et al. 2007, p. 583; after Kaldor 1961).

For example, light emitted by an excited atom is not white, which means that it does not include all wavelengths of the visible spectrum but only a few, characteristic wavelengths. Atoms of different elements can be distinguished by their spectra. Thus, the spectra are beyond random variation: they "call for an explanation" because they indirectly reflect the structure of atoms. Our job as scientists is to observe characteristic patterns and "decode" them to understand how a certain system works. For atomic spectra, this process led to quantum theory.

For most systems, however, one single pattern is not enough to decode the internal organization. Multiple patterns, or filters, are needed. Imagine that you need to meet someone you do not know at the airport. Only one "pattern," say, knowing that he is male, is not enough to identify the right person—the pattern "male" only filters out half of all people. But if you also know that the person's age is about 30, and that he will wear a green T-shirt and carry a blue suitcase, you probably will identify him right away.

There are several interesting things to note about this set of patterns used to identify a person at the airport:

- It is small—only four patterns;
- The patterns are simple or "weak," each captured in just a few words and none strongly descriptive;
- The patterns are qualitative, not quantitative;
- The patterns are about different things—sex, age, clothing, and baggage; and
- It is relevant specifically to the problem of identifying the person at the airport—the patterns are useful because they can be detected at a distance, but you would certainly use different characteristics to describe the same person for a different problem (e.g., deciding whether he should marry your sister).

Our point is that a small number of *weak* and *qualitative* but *diverse* patterns that *characterize a system with respect to the modeling problem* can be as powerful a filter as one very strong pattern (e.g., a photograph of the person), and are often easier to obtain.

What do we mean by "patterns that characterize a system with respect to the modeling problem"? These are patterns that we believe, from what we know about the system, to be caused by the same mechanisms and processes that drive the system's behavior relevant to the problem we model. Another way to think about these characteristic patterns is that if your model did *not* reproduce them, you should not trust it to solve your problem. For example, in chapter 18 we discuss a European beech forest model that reproduced one key pattern of real forests (a horizontal mosaic of developmental stages), but was not trusted by beech forest experts because it was not capable of reproducing additional patterns (in vertical structure) that they considered essential.

How do you identify a set of diverse patterns that characterize the system for the problem that you are modeling? This task, like much of modeling, uses judgment, knowledge of the system, and often, trial and error. Some models address striking or "strong" patterns that clearly call for explanation; for example, some semi-arid ecosystems have strongly banded vegetation (figure 17.1). More often, systems exhibit weaker patterns, such as that certain state variables stay within limited ranges or that the system responds in characteristic ways to a variety of changes.

Detecting and agreeing on what constitutes a useful pattern is subjective to some degree. Experts on a system may disagree on whether a certain observation is a characteristic pattern. Thus, as with any assumption underlying model design, the choice of patterns is experimental: we make a preliminary choice, then implement and analyze the model and see how far we can get. Our preliminary understanding is likely to include blind spots, biases, and inconsistencies. But progressing through the modeling cycle will focus us on meaningful and useful patterns. Indicators of having chosen the right set of patterns are that we find mechanisms that reproduce these patterns robustly (chapter 23), and that our model makes secondary predictions that can be used for validation (chapter 18).

Figure 17.1
Banded vegetation pattern as reproduced by the model of Thiery et al. (1995): bands of shrubs or trees (grey and black) alternate with bands that lack such vegetation (white). This pattern is observed in semi-arid landscapes where water runs off in a uniform direction (here, from top to bottom). (Figure produced from a NetLogo reimplementation by Uta Berger, using File/Export/View from the main menu.)

Matching patterns: qualitative or quantitative?

POM includes the step of testing whether a model reproduces specific patterns, so we need to define criteria for when a pattern is matched. Do we need to define the patterns numerically and designate some statistical tests for whether model results match them? At least until the final stages of modeling, it is important not to be rigidly quantitative; otherwise, it is too easy to get trapped in a bog of statistical inference before understanding anything.

We recommend following Platt's (1964) advice: "Many—perhaps most—of the great issues of science are qualitative, not quantitative, even in physics and chemistry. Equations and measurements are useful when and only when they are related to proof; but proof or disproof comes first and is in fact strongest when it is absolutely convincing without any quantitative measurement."

For example, Watson and Crick found their model of DNA's structure using qualitative criteria first; only later did they turn to quantification (Watson 1968). Thus, rigorous quantitative patterns are indeed needed (chapter 20), but only after we use qualitative analysis to exclude inappropriate submodels. Proof usually comes first, quantification second.

17.3 Overview of Part III

In chapter 18, we explain how modeling can be made pattern-oriented by designing a model structure that allows (but does not force!) a set of patterns observed in the real system to emerge in the model. Then (chapter 19) we "zoom in" and learn how to find the most appropriate submodel for the agents' key behaviors, a process we call theory development. We identify useful theory for agent behavior by "hypothesizing" (trying) alternative behavior submodels and "falsifying" (rejecting) those that fail to cause the ABM to reproduce observed patterns.

For models used to support decision-making in the real world, we often need to go one step further and calibrate models to quantitatively match observations and data. ABMs typically include parameters for which good values cannot be determined directly from data or literature. We have to "guesstimate" them, which can be sufficient for qualitative predictions, theory development, and understanding much about the real system. But for quantitative modeling, uncertainties in the guesstimated parameters are often too high. In such situations, we can again employ the idea of using patterns as filters (Wiegand et al. 2004); in chapter 20 we discuss calibration and the use of patterns to narrow the ranges of parameter values.

The topics covered in part III are all key steps of model development and the modeling cycle. For ABMs and complex models in general, these steps are not trivial or simply intuitive. They need to be thought about carefully so we can convince others that our models are in fact structurally realistic and well calibrated. Documentation of these steps is therefore important and a key element of the "TRACE" modeling cycle description format of Schmolke et al. (2010).

Part II was focused on specific parts of a model or its computer implementation. But part III is more strategic and thus addresses full ABMs. The exercises will therefore be more comprehensive and more and more like project work. This means that you are learning to steer and navigate the RV *Agent-Based Modeling* to explore the unknown. Working strategically on full models can be quite complex and sometimes frustrating at the beginning, but it is also exciting—after all, you are finally addressing the very purpose of agent-based modeling: to learn how real agent-based complex systems work.

Patterns for Model Structure

18.1 Introduction

The first very important decision in designing a model is selecting the set of entities and their state variables and attributes that represent the system. We could think of endless things that are in any system we want to model, and infinite variables that characterize each. But in models, we usually—wisely—include only a very few kinds of entities and only a few variables (size, location, etc.) to describe each. Once we have chosen the entities and state variables, we also know what processes need to be in the model: the processes that are important for making the state variables change.

We must use the purpose of a model as a guide for deciding what entities and variables to include. However, a model's purpose can provide too narrow a point of view, leading to a model that is too simple in structure to be tested thoroughly. For example, if our job is to predict the growth of the human population, the simplest model for this purpose could be an exponential growth equation with the population size as its only state variable and one parameter fit to historic population data. But then how would we know whether the model is a sufficiently good representation to project the population for 20 years, or 100 years, into the future? How could we know whether the model is structurally realistic?

Including too few entities and state variables causes models to be *underdetermined*: they do not, due to their simplicity, provide enough kinds of results to be tested adequately. They are too poor in structure and mechanisms (DeAngelis and Mooij 2003). What can we do to make the model richer, but not too rich and complex? We might, for example, know a lot about the age structure of human populations—how many people there are of each age—in different regions. A model that is capable of reproducing historical patterns in both total population size and regional age structures is much more likely to include the key processes of demography than a model focusing only on global population size.

In this chapter we introduce the first task of pattern-oriented modeling: using observed patterns to design a model structure that includes enough to make the model sufficiently realistic and testable but not unnecessary complex. We describe the steps in this task, then illustrate them using a real forest modeling project. Then we provide a second case study, this time from the field of business management, in which observed patterns were used not to design a model but to evaluate the structure of several existing models.

Our goal in this chapter, and in the rest of part III, is to give you an understanding of the overall strategy of POM. Unfortunately, POM is not a simple recipe that quickly leads to miraculous insights. Instead, it is a strategy for doing science with ABMs; it requires work and knowledge of the system being modeled, but should make the process more productive and more fun.

Learning objectives for chapter 18 are to:

- Understand the four steps in using POM to design a model's structure, and the examples of this process that are presented here.
- Be able to structure an ABM from (a) knowledge of the system and problem that the model addresses and (b) observed patterns that characterize the system's processes relevant to the problem.

18.2 Steps in POM to Design Model Structure

Once again, the goal of POM for model structure is to decide what entities—kinds of agents and other things—need to be in a model, and what state variables they need to have, to give the model enough realism and complexity to be testable and useful, while avoiding unnecessary complexity. The following steps are a concise guide to this process.

1. Start by formulating your model (writing a description of it using the ODD protocol), with the purpose of the model as the only filter for designing the model structure. Include the entities, state variables, and processes that seem the minimum necessary to solve the problem you model, given your current understanding or conceptual model of the system. At this stage the formulation should seem too simple to possibly be useful.
2. Identify a set of observed patterns that you believe to characterize your system relative to the problem you model, as discussed in section 17.2. Spend the time necessary with the literature and with experts to really understand what is known about the system and how its processes are believed to give rise to the patterns, especially if you do not know the system well already. Limit yourself to a manageable set of patterns, typically two to six, most often three or four. Often, some patterns are closely linked and not independent of each other; try to find a diversity of patterns, such as responses to several different kinds of change. Try to rank the patterns by how important they are in characterizing the system and problem.
3. Define criteria for pattern-matching: how will you decide whether the model does or does not reproduce each pattern? Start with qualitative criteria such as visual pattern-matching and trends in statistical summary output, though quantitative criteria may be needed in later steps of POM. We discuss pattern-matching more in chapters 19 and 20; for now, the only concern is having some idea what outputs you will use to evaluate whether the model reproduces your characteristic patterns.
4. Now, review your draft model formulation and determine what additional things need to be in the model to make it *possible* for your characteristic patterns to *emerge* from it. Are any new entities or state variables needed so that the patterns could emerge? Are any of the patterns believed to be driven by processes that are not already in the model? Are new variables or outputs needed to observe whether the patterns are reproduced? Whatever you identify—typically, only a few variables and processes—should be added to the model formulation.

Now you are ready to advance through the modeling cycle. Revise the draft formulation to include the changes identified in these steps, then program the model and test the software thoroughly. When you have done this you will be ready for the next stage of POM, testing your ABM and its theory for individual adaptive behavior, the subject of chapter 19.

The story of two models of the spatiotemporal dynamics of European beech forests illustrates the use of observed characteristic patterns to design ABMs—and how the choice of patterns can affect a model's structure and, consequently, its believability and success.

Before the advent of large-scale agriculture, much of central Europe was covered by forests dominated by beech trees. These "climax" forests consist of a mosaic of patches of large, old trees with closed canopy and virtually no understory (foresters use the words "canopy" for the high layers of leaves that block light to lower levels of a forest, and "understory" for woody plants below the canopy); and patches with gaps in the canopy and a mixture of young and old trees. The models we examine were designed to help understand the mechanisms causing this mosaic pattern and, ultimately, decide how big and old forest preserves must be to resemble the "Urwald" (ancient forest).

18.3.1 Before BEFORE and after BEFORE

The first model (Wissel 1992) is a simple grid-based model. Its design focused on reproducing a single pattern observed in natural beech forests: the mosaic of patches in the stages described above. The patch types are believed to follow each other in a development cycle: a canopy that is closed (fully occupied with branches and leaves) opens when old trees die due to storms or age, which lets younger trees establish and grow in a mixture of ages and heights, until the canopy is closed again and most of the small trees die because they no longer get sufficient light. To reproduce this single pattern, the model needed to include only grid cells (patches, in NetLogo) with state variables only for which developmental stage the cell is in and how long it has been in that stage. Trees grow slowly, so the time step is 10 years. The model's main process is the progression of each cell through the cycle of developmental stages. This cycle assumes that cells progress from early colonizing species such as birch, to a mixture of later colonizers, followed by stages dominated by young and then older beech trees. In the final stage of the cycle, a cell contains a single large beech tree, until this tree dies. The model's second process is an effect of one cell on its neighboring cells: beech trees can be killed by too much sunlight on their bark, so when a cell changes from the large beech tree state to open (the big tree dies), the cells to the north (and, to a lesser degree, east and west) are more likely to have their trees die soon.

This model had a mixed reception. It qualitatively reproduces the observed mosaic pattern (figure 18.1), so landscape ecologists, who focus on such patterns, were quite happy with it

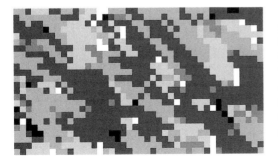

Figure 18.1
View of Wissel's (1992) beech forest model, from a NetLogo reimplementation by Jan Thiele. The shading represents the time since a cell was empty: cells range from white on the first time step after becoming empty to black at the maximum age of a cell, after which the cell's old beech tree will die of age.

(Turner 1993). However, beech forest scientists were not convinced that the model produced these patterns for the right reasons. They questioned several model assumptions and, in particular, believed that the developmental stages were not automatic but instead emerged from processes that occur mainly in the vertical dimension and produce important vertical patterns. Vertical competition for light was believed the most important structuring process: if the forest canopy is closed, too little light reaches the understory, leading to high mortality of smaller trees. If, on the other hand, a large tree dies, the trees of the lower layers receive a pulse of light and can grow, possibly to reach the canopy.

In other words, these forest scientists saw an additional set of patterns: the vertical characteristics of the different developmental stages. The "late" stages have almost exclusively canopy trees and almost no understory. When large trees die, their patch cycles through stages with only small seedlings, then taller saplings, and finally tall trees again. Further, the forest scientists believed they knew *why* most big trees die: they are typically toppled by high wind during storms.

Neuert et al. (2001; Rademacher et al. 2004) developed the second model, called BEFORE, by adding state variables to represent the vertical characteristics of a cell and the processes believed to cause vertical characteristics to change. Hence, BEFORE was designed so that the characteristic vertical patterns of forest stages could emerge from underlying processes, and so that the large-scale spatial mosaic of developmental stages could emerge from the vertical patterns. If these patterns did all emerge in a reasonably realistic way, then forest scientists would have much more confidence in BEFORE as a structurally realistic and useful model than they did in the original model.

Formulating this model structure took several months. It required reading the relevant literature and discussing the structure with beech forest experts. The modelers were unhappy with including the vertical dimension because it made the model considerably more complex and difficult to implement (in the dark days before NetLogo). But the empirical evidence and expert opinions convinced them that to understand the internal organization of beech forests, it was necessary to include vertical structure and mechanisms in the model. (Early individual-based forest models, e.g. Botkin et al. 1972 and their successors, represented vertical competition for light but focused on species composition and often ignored horizontal interactions.)

18.3.2 The Structure of BEFORE

To illustrate the model structure derived from the characteristic patterns, we provide here the "Entities, state variables, and scales" of BEFORE's ODD description, and a short summary of the model's processes. A full ODD description is available through this book's web site.

Entities, State Variables, and Scales

BEFORE has two kinds of entities: vertically compartmented grid cells and individual large beech trees. Trees are tracked as individuals only after they grow out of the "juvenile" stage, explained below. BEFORE is simpler than the original model in one way: it ignores tree species other than beech. In the "Urwald" no other species grow into an opening, so they are unimportant to the problem being modeled.

Horizontally, grid cells are 14.29-meter squares (corresponding to the area covered by one very large beech). Vertically, cells have four compartments that each represent a height class of beech (Figure 18.2). These compartments are separate entities within each cell, with each compartment representing some of the cell's trees:

- *upper canopy*: trees 30–40 m high (maximum height of a beech tree);
- *lower canopy*: trees 20–30 m high;

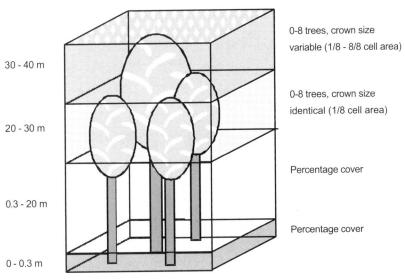

	0-8 trees, crown size variable (1/8 - 8/8 cell area)
30 - 40 m	
	0-8 trees, crown size identical (1/8 cell area)
20 - 30 m	
	Percentage cover
0.3 - 20 m	
	Percentage cover
0 - 0.3 m	

Figure 18.2
The four height compartments of a cell in BEFORE. Individual trees of the two highest compartments are represented as individuals, but beech seedlings and juveniles of the lower two compartments are represented only by the percentage of the cell covered by them. (From Rademacher et al. 2004.)

- *juveniles*: small trees 0.3–20 m high; and
- *seedlings*: trees up to 0.3 m.

These four vertical compartments have the following state variables:

- *Upper canopy* and *lower canopy* compartments have a list of the trees present in the compartment. Zero to eight trees can be present in each of these compartments of each cell. These two compartments each have a state variable calculated from the number and size of trees in them: *F4* for upper canopy and *F3* for lower canopy. *F4* and *F3* are the percentage of the cell area occupied by trees in the compartment, calculated by adding the percentage of cell area occupied by each tree (area occupied by trees is explained below). *F4* and *F3* range from 0 to 100%, in steps of 12.5% (because of how the area of an individual tree is defined, below), and are used for calculating vertical competition for light.
- *Juveniles* and *seedlings* compartments each have one variable, *F2* (juveniles) or *F1* (seedlings compartment), for the percentage of cell area (0–100%) covered by juvenile trees and seedlings (respectively) in these height ranges. In these lower compartments, trees are not represented as individuals but only via *F2* and *F1*. The juvenile compartment also has a state variable for the number of previous time steps in which *F2* was greater than zero.

Individual trees have different state variables in the two canopy compartments:

- *Upper canopy trees*: These are the highest trees, with vertically unrestricted access to light. They have state variables for their age and "crown area": the fraction of the cell's area occupied by the tree. The crown area is quantified in eighths (12.5%) of cell area. A cell could contain, for example, two upper canopy trees with crown areas of 1/8 and 5/8, or a single giant beech with a crown area of 8/8 (the entire cell).
- *Lower canopy trees*: These are rapidly growing trees that can almost reach the upper canopy. But if the canopy is filled by upper canopy trees, the lower canopy trees no longer

receive enough light to grow. Beech is shade-tolerant, so trees can survive in the lower canopy for about 90 years, and if a gap opens in the upper canopy, lower canopy trees can fill it quickly. State variables are age and the time spent in this compartment. All trees of this compartment have the same crown area: 1/8 of cell area.

Note that this model structure is still highly simplified: it neglects, for example, the number of trees in the lowest compartments, the exact location of trees within their grid cell, and anything about tree size except crown area and which height compartment trees are in.

The time step is 15 years. The spatial extent is 54 × 54 grid cells, representing about 60 hectares.

The processes in BEFORE all occur within grid cells and among neighboring cells. The establishment ("birth"), growth, and mortality of trees all depend on vertical competition for light, which is determined by the variables $F2$, $F3$, and $F4$, representing how much of the cell area is covered by trees. Toppling of big trees by wind storms is a stochastic process. Wind-toppled trees can hit and destroy trees in neighboring cells. The opening of gaps in the canopy affects neighboring cells by making their trees more vulnerable to future wind storms, but also by allowing slanted light to reach their lower compartments and promoting growth. (This slanted light interaction among cells was at first not considered important by the forest experts, but it turned out to be essential. This experience reminds us that experts may know their system very well and yet have incomplete or inaccurate conceptual models.)

18.3.3 Consequences of Structural Realism: Validation and Independent Predictions

Because BEFORE was designed using more patterns, and more diverse patterns, it is more complex than the original beech forest model of Wissel (1992)—though still conceptually simple. Was the additional complexity worthwhile? When BEFORE was implemented and tested, it successfully reproduced emergent patterns including the observed mosaic of developmental stages over space, the cycle of developmental of stages over time in a cell, and system-level properties such as the proportion of the forest in the final stage (upper canopy trees) and the cyclic variation of this proportion. The model was thus successfully "verified."

However, BEFORE is rich enough in structure and mechanism to produce output that can be scanned for new patterns that were not considered or even known when the model was formulated, tested, and calibrated. We call such new patterns *secondary* or *independent* predictions. If such predictions are confirmed by new empirical evidence (from the literature, new measurements, or new experiments), we call the model "validated" instead of just "verified"; the process of searching for and testing independent predictions is called *validation*. (However, the terms *verification* and *validation* are not used consistently in the literature.)

We have an intuitive understanding of validation: consider the following mental exercise:

$$F(2, 3) = 10, F (7, 2) = 63, F (6, 5) = 66, F (8, 4) = 96$$

F is a simple mathematical operation combining two numbers, for example 7 and 2, that leads to the given results. What is F, the generative mechanism? (We leave the answer as a fun exercise for you; the key is, of course, to discover patterns in the input-output relationship of F. Be sure you note the multiple kinds of patterns you use to identify the generative mechanism.) Now, how would you test your "theory" of F? By testing whether your theory predicts data not already used, that is, by independent predictions! (You can find these data—$F(9, 7)$—on this book's web site.)

Why is it important to test a model with independent predictions? Because even when we use multiple observed patterns for verifying a model, there is still a risk that we "twist" it to reproduce the right patterns for the wrong reasons, for example by calibration (chapter 20), or by combining submodels that are each unrealistic but by accident work well in combination. The only way to avoid this risk is to search for evidence of the model's structural realism that was not used during model development and verification.

For example, BEFORE was analyzed for patterns in the age distribution of the canopy trees and the spatial distribution of very old and very large canopy individuals ("giants")—but only after BEFORE was designed and calibrated. The two most characteristic of these new patterns were that neighboring canopy trees typically have an age difference of 60 years, and 80% of all very large or old individuals have an individual of the same type within a radius of 40 meters. It turned out that these results were surprisingly close to empirical information from the few remaining natural beech forests in Europe (Rademacher et al. 2001).

In further work (Rademacher and Winter 2003), BEFORE was modified to track of how much dead wood (which provides habitat for many species) the model forest produces. This change required adding a few new deduced state variables (e.g., tree trunk volume, because dead wood is measured as a volume) and processes. (*Deduced* means here that trunk volume was calculated from existing state variables—height, crown area, age—via empirical relationships.) The resulting predictions matched observations quite well.

These experiences indicate that the higher complexity and costs for formulating, implementing, testing, and analyzing BEFORE paid off very well. The Wissel model provided little more than one possible explanation of one pattern. It could not be tested against more than this one pattern because it was too poor in structure. BEFORE, in contrast, took considerably more effort to design and implement but could be tested in a variety of ways, leading to higher confidence in its results and to important insights about the internal organization of real beech forests.

18.4 Example: Management Accounting and Collusion

The main lesson from the beech forest model example is that using a variety of observed patterns to design an ABM's structure produces a model that is more structurally realistic, testable, and believable. Could we not apply this lesson in reverse and evaluate how believable existing models are by how many patterns they were designed to explain?

Heine et al. (2005) discuss patterns in a management problem of large businesses: the decision by central management of how much money to invest in the company's different divisions. This decision is based mainly on information provided by the division managers, who tend to overstate the expected returns from investment in their division because their pay increases with how much of the company's profit is made in their division. Groves (1973) invented a mechanism of distributing investment among divisions in a way that compels managers to predict profits honestly. In this mechanism, a division manager's pay is also influenced by the sum of the profits predicted by the other managers, in such a way that (as can be shown mathematically) all managers maximize their pay by reporting honest predictions.

However, this "Groves mechanism" for eliciting honest management information can be undermined by collusion among division managers. This potential has been examined in a number of empirical studies and several models, some based on game theory and some simulation models. Heine et al. (2005) were concerned about the "ad hocery" of the models: the models' design and evaluation seemed arbitrary instead of structurally realistic. Heine et al. thus scanned the empirical literature for patterns (in their terminology, "stylized facts") to evaluate these existing models. They identified the following general patterns or stylized facts, a list of conditions under which collusion is favored:

1. When the number of potential colluders (number of divisions in a company) is small,
2. Under long time horizons,
3. In uncomplicated settings (low dynamics, low complexity, high transparency, etc.),
4. When the potential benefits of collusion are high,
5. When group composition is stable, and
6. When there are effective ways for colluders to enforce their agreements.

Heine et al. then checked whether the existing models of collusion to circumvent the Groves mechanism referred to these patterns (were capable of potentially reproducing them) and, if so, whether they successfully reproduced and explained them. The first of two game theory models only addressed pattern 1, but contradicted it. The second game theory model addressed, and partially reproduced, patterns 2 and 6. The first of two simulation models partially reproduced pattern 1 but contradicted pattern 6. The second simulation model partially reproduced patterns 2 and 3, and convincingly reproduced pattern 6. Hence, Heine et al. (2005) concluded that only the second simulation model had much credibility for explaining collusion in the management accounting problem; only that model considered enough observed patterns to leave "little room for arbitrary model design or questionable parameter calibration" (Heine et al., 2005, *Abstract*).

This example shows that POM for model structure is useful not only when designing an ABM; the idea that models are more credible and useful when they include entities and state variables that allow them to be tested against multiple patterns (or "stylized facts") is general and powerful. Here, it was used to evaluate existing models instead of designing new ones. And this example shows that POM is applicable to many kinds of models, not just ABMs.

18.5 Summary and Conclusions

The general idea of POM is to use multiple observed patterns as indicators of a system's internal organization. Trying to reproduce such patterns with our ABMs makes us decode the internal organization of the real system. Experienced modelers use patterns in exactly this way all the time—it is not surprising that Heine et al. (2005) used POM in their business management study even without being exposed to examples from other fields—but often the use of patterns is not explicit or systematic.

The point of POM is to consciously think in terms of pattern, regularities, or stylized facts right from the beginning and to select model structure accordingly. For beginners, it can take some time to get familiar with POM and to understand that it is, in fact, quite simple. Like a neophyte detective, you will learn to identify patterns in data and expert knowledge and to focus not just on "strong," striking patterns but also on weak ones: combinations of weak patterns can be very powerful for designing and testing model structure. POM is not a technique, like programming or a statistical test, but a general strategy and attitude, a continuous awareness of the fact that we as scientists are working very much like detectives. And, by being pattern-oriented, we can avoid two extremes: simplistic models that are too poor in structure and mechanisms and therefore likely to reproduce a very few patterns but for possibly the wrong reasons, and models that are too complex because they were designed without the guidance of characteristic patterns (as well as the model's purpose) as a guide to what can be left out as well as what must be included.

1. On the web site section for this chapter is a list of models in which observed patterns were particularly important. Select one or more of the models and identify the patterns that were important in their design. How did the authors search for and document the patterns? Describe how the patterns affected the design of the model, being specific about questions such as these: What scales do the patterns occur at and how did they affect the model's scales? What things were included in the model to make it possible for the patterns to emerge, and what other things were required to be in the model by the problem it addresses? (This difference is sometimes unclear because explaining observed patterns is often the purpose of a model.) Identify any important information that is unclear from the published description of the model and how it was designed.

2. You might know that the periodic table of chemical elements reflects the structure of atoms as explained by quantum physics. But the periodic table was developed long before quantum physics existed. How was this possible?

3. Exercise 1 of chapter 1 was about modeling the checkout queues in a grocery store. Assume your task is to model how the distribution (e.g., mean, standard deviation, maximum) of times that customers spend in queues depends on characteristics of the checkout counters and characteristics of the customers. What patterns can you observe that would help you decide how to structure your model?

4. In your field of research, try to perform a study similar to that of Heine et al. (2005). Identify a general or specific question or problem that has been modeled several times in different ways. Then identify a set of observed patterns, both strong and weak, that you believe to characterize the system for the specific question or problem. Finally, evaluate the degree to which the existing models refer to the patterns: do the models contain the structures that would be necessary to test whether they can reproduce the patterns?

5. Start the process of building your own model of a particular system and problem by identifying patterns for POM. Identify the system and problem that you will model, and how the model's scales, entities, and state variables are determined or constrained by the problem. Then identify a set of observed patterns that characterize the system for your problem. What additional structures are needed for your model to have the potential for these patterns to emerge?

Theory Development

In the previous chapter we focused on model structure. Now we will turn to processes and how to model them. As we design a model's structure and formulate its schedule—the Overview part of the ODD protocol—we identify *which* processes we need without bothering yet about *how* they work. Then, to get the modeling cycle going, we start with a very simple, often obviously wrong, representation of the model's processes. For example, we often just assume that agents make decisions randomly. However, after we have a first implementation of the entire model, we need to unsimplify and come up with sufficiently realistic and useful representations of key processes. How can we do this?

First of all, we do not want to consider all the processes in a model with the same level of detail. ABMs typically include a number of relatively minor processes that can be represented simply or by using submodels from the literature. Results of the model are unlikely to be too sensitive to the details of such submodels. But ABMs also typically have a few key processes, in particular agent behaviors, that we believe are most important. Agent behaviors are especially important because they are the reason we use an ABM in the first place, and also because we are less likely to find a useful way to represent them in the literature. Therefore, the questions usually are: What is the most important agent behavior, and how do we identify or invent a good submodel (*trait*, using the terminology introduced in section 11.1) to represent it?

We call this stage of model design *theory development*, with a *theory* being a trait for some particular agent behavior that has been tested and proven useful for explaining how the system works, in a particular context (see Grimm and Railsback 2005, chapter 4). The goal is to find models of key agent behaviors that are simple—usually, a gross simplification of real behaviors—but complex enough to produce useful system behaviors. This definition of theory has more in common with theory in physics and engineering, where equations or models are not accepted as theory until they have proven useful and general for solving real problems. The idea of theory for complex systems as models of the individual characteristics that give rise to system-level dynamics is not new (e.g., Auyang 1998), and is in fact a common approach for modeling complex physical systems such as tall buildings and electronic circuits.

How do we test models of agent behavior to know which to accept as theory? The answer of course is via pattern-oriented modeling (POM): we accept agent traits as theory when they cause the ABM to reproduce the set of patterns that we chose to characterize the system's dynamics (section 17.2). In this chapter we introduce the use of ABMs as "virtual laboratories" where you test and improve traits for key behaviors. After explaining the general process of theory development, we illustrate it with several examples. Then we introduce a new example model in which the key behavior is the decision by animals that live in a social group of whether to stay in the group or leave to find a new group. Developing theory for this behavior is your exercise.

Theory development is the most unique, important, and academically fertile part of agent-based modeling. How to model key agent behaviors is obviously fundamental to agent-based simulation, and there are many important discoveries to be made. Currently, almost every ABM requires new theory for a new behavior or context, and developing the theory will often be as important as anything else learned in the modeling study. Luckily, we can take advantage of standard and powerful scientific methods, with our ABM as our laboratory, to develop theory efficiently and convincingly. This stage of modeling is, more than any other, about basic scientific discovery.

Chapter 19 learning objectives are to:

- Understand the general concept of developing theory by posing alternative hypotheses, conducting controlled experiments to identify the most useful hypothesis, then revising the hypotheses and conducting more experiments.
- Understand that theory in agent-based science includes models of agent behavior that explain important system behaviors; and that we develop this theory via pattern-oriented testing of hypothesized traits.
- Develop experience with this theory development process.

19.2 Theory Development and Strong Inference in the Virtual Laboratory

We can consider an ABM a "virtual laboratory" where we test alternative traits for key behaviors. We do so by plugging these alternative traits into the ABM and testing how well the ABM then reproduces patterns observed in the real system. Through a cycle of testing and refining traits, perhaps combined with research on the real system to find additional patterns, we exclude ("falsify") unuseful traits and home in on useful ones that we can treat as theory.

Contrasting alternative theories, or hypotheses, is the foundation of what is simply called *scientific method*. In 1964, biophysicist John Platt rephrased this approach and called it *strong inference*. He wondered why some fields of science move forward so much faster than others. His answer was that progress depended largely on how systematically people used this method (Platt 1964, p. 347):

Strong inference consists of applying the following steps to every problem in science, formally and explicitly and regularly:

1. Devising alternative hypotheses;
2. Devising a crucial experiment (or several of them), with alternative possible outcomes, each of which will, as nearly is possible, exclude one or more of the hypotheses;
3. Carrying out the experiment so as to get a clean result;
4. Recycling the procedure, making subhypotheses or sequential hypotheses to refine the possibilities that remain, and so on.

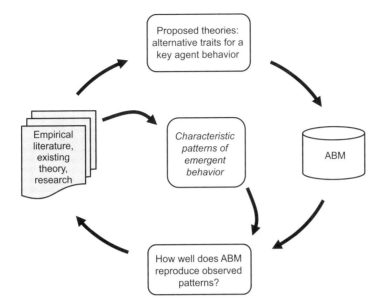

Figure 19.1

The POM cycle for developing theory for an agent behavior. Existing literature on the real system, and sometimes new empirical research, is used to identify patterns that characterize the system behaviors that emerge from the agent trait. The empirical information is also used, along with existing theory, to propose alternative theories for the behavior. The alternative theories are implemented in an ABM and tested by how well the ABM then reproduces the characteristic patterns. Promising theories can be refined by finding or developing more empirical information that provides additional characteristic patterns and suggests improvements to the theory, and then repeating the cycle.

The approach we use to developing theory for agent behavior (figure 19.1) is a subcycle of the modeling cycle, and closely follows Platt's steps:

1. Identify alternative traits (hypotheses) for the behavior;
2. Implement the alternative traits in the ABM, testing the software carefully to get "clean results";
3. Test and contrast the alternatives by seeing how well the model reproduces the characteristic patterns, falsifying traits that cannot reproduce the patterns; and
4. Repeat the cycle as needed: revise the behavior traits, look for (or generate, from experiments on the real system) additional patterns that better resolve differences among alternative traits, and repeat the tests until a trait is found that adequately reproduces the characteristic patterns.

Where do we get alternative theories to test and contrast? Sometimes traits for key behavior have already been proposed by others or identified as the model was conceived. However, often it is not at all clear how to model behavior, so we start with both empirical data and existing theory. Are there sufficient observations of the real world to develop statistical and stochastic traits of the kind we explored in sections 15.2.1 and 15.4? Is there decision theory for the kinds of agent and decision being modeled? Many scientific fields include a subdiscipline focused on behavior of individuals (e.g., cognitive psychology; behavioral ecology) that may offer applicable theory. However, classical theory for how agents make decisions (e.g., businesses maximize profits; animals maximize growth) is often too simplistic to produce realistic behaviors in ABMs, or simply does not apply to the decisions made in a somewhat realistic model. But it is always important to understand existing theory as a starting point, and in some fields quite useful kinds of theory can be adapted to ABMs. Several approaches for modeling behavior are discussed in chapters 4 and 7 of Grimm and Railsback (2005).

In trying to identify traits to test as theory, it is important to remember that your goal is not to reproduce all behaviors that your agents are known to exhibit! If you try to do so, you will never finish. Instead, remember that you seek a trait that is just realistic enough to reproduce the patterns of interest, and make sure you have a clear limit on what patterns you need to reproduce.

You should always start with *null theories*, that is, theories that do not use any specific mechanism or assumption. Typical null theories are to just assume that agents make decisions randomly, and (the opposite) to assume agents always do exactly the same thing. Why start with null theories? One reason is that it allows you to develop and test the rest of the ABM before focusing on behavior traits. Another reason is that it helps you learn how sensitive the output of the ABM is to how you represent the key behavior. You might find that some key patterns emerge robustly even without detailed behavior; then you will have learned that the behavior in fact is not key to explaining the patterns (an example in section 19.3.3 illustrates this nicely). Or you might learn that some important patterns are not reproduced, showing that the theory development exercise is in fact important.

Finally, we remind you of an important technique from chapter 12: exploring and testing traits independently before you put them into the ABM for pattern-oriented evaluation. Your progress in theory development will be much faster if you implement hypothesized traits separately (e.g., in a spreadsheet) and understand them thoroughly before plugging them into the ABM.

19.3 Examples of Theory Development for ABMs

So far in the history of ABMs, there have been only a few studies that explicitly followed the theory development cycle. They include development of theory for how trout decide where and when to feed (Railsback et al. 1999; Railsback and Harvey 2002; Railsback et al. 2005; summarized in section 4.6 of Grimm and Railsback 2005), and the study depicted below in section 19.3.3. But there have been many studies that were in fact pattern-oriented theory development even though they were not described as such by their authors. Here, we highlight a few diverse examples.

19.3.1 Fish Schooling and Bird Flocking

If you did the exercises in chapter 8, then you already have experience doing theory development! In section 8.4, we discussed the simulation experiments that Huth and Wissel (1992) used to find a model of individual behavior that explains how groups of fish form schools. Exercises 4 and 5 of chapter 8 then asked you to contrast some alternative traits for how animals form schools and flocks. This work on fish schools was one of the earliest examples of theory development for ABMs, and is portrayed (along with empirical studies and models that preceded and supported it) in chapter 11 of Camazine et al. (2001).

The fundamental assumptions of fish schooling and bird flocking models are that the individual animals adapt their movement direction and speed to match those of their neighbors, and move toward their neighbors while maintaining a minimum separation distance and avoiding collisions. However, there are important details, especially: *which* neighbors does an individual adapt its movement to? Huth and Wissel (1992) identified two alternative assumptions—potential theories—for how fish adapt their movement. One theory was that fish react to only the nearest other fish; a second was that fish adapt to the average direction and location of several neighbors.

To contrast these alternative theories, Huth and Wissel simulated them and compared the results to several observed patterns. One pattern was simple and qualitative: did the model fish look like a school of real fish? But the other patterns were quite quantitative, using measures of how compact and coordinated real fish schools are: mean distances between fish and their nearest neighbor, mean angle between a fish's direction and the average direction of all fish, and the root mean square distance from each fish to the school's centroid. The theory that fish adapt to the average of several neighbors reproduced these patterns robustly over a wide range of parameter values, while the alternative of following one nearest neighbor did not reproduce the patterns and was hence falsified.

The experiments of Huth and Wissel (1992) considered only the most basic patterns of schooling and focused on fish. However, pioneering studies such as theirs merely started the modeling cycle, and subsequent studies have produced more detailed models, more quantitative data on real animals, and much more refined theory. For example, Ballerini et al. (2008, which cites much of the more recent literature in this field) used extensive high-resolution measurements of starlings to support the theory that these birds adapt their movement to their 6–7 nearest neighbors, regardless of how many close neighbors there are or how close they are. Ballerini et al. then implemented this theory in an ABM and contrasted it to the alternative that birds adapt to neighbors within a certain distance. The patterns used to contrast these two theories concerned how robust the flock is to being broken apart when its birds swerve to avoid a predator. The theory that birds always respond to 6–7 neighbors created a more realistically cohesive model flock than did the alternative that birds respond to all neighbors within a fixed distance. The fixed-distance alternative was therefore "falsified" from being the best trait for flocking—in contexts where flock response to predators is important.

19.3.2 Trader Intelligence in Double-Auction Markets

"Double-auction" markets include agents (people or firms) who are either buyers or sellers, with buyers buying goods from those sellers who ask for a price lower than the buyer's value of the good. Many real double-auction markets exist, including stock markets, and they are governed by rules that serve purposes such as letting buyers know the best available price and sellers know what buyers are willing to pay, determining the exact price at which trades are made (trades occur only when a buyer is willing to pay more than a seller is asking, so the trading price is between these two values), and updating this information as trades are executed and participants adjust their prices and willingness to pay. Duffy (2006) describes a series of experiments, using both ABMs and real people, to develop theory for how traders decide when to buy or sell. The exact theory question is: What assumptions about trader intelligence are sufficient to reproduce patterns observed in double-auction markets operated by real people?

Double-auction markets have been simulated with ABMs many times, and the obvious theory question is how to represent the elaborate ways that people make trading decisions in sufficient detail so that market models produce realistic dynamics. What are realistic dynamics? A number of empirical experiments have used real people to represent buyers and sellers in game-like artificial markets that are simplified in the same ways that the ABMs are, and these experiments have identified patterns such as that people rapidly converge on a very efficient solution, the "equilibrium" price where the number of willing buyers equals the number of willing sellers.

The first agent-based theory experiment described by Duffy (2006) did exactly what we recommend in section 19.2: start with a null theory that is extremely simple. Gode and Sunder (1993) simulated a market with "zero-intelligence" traders: traders decided randomly (within very wide bounds) the price at which they would buy or sell. Hence, traders could make trades that lose instead of make money. However, the market rules that link the highest offer to buy with the lowest selling price were still in effect. As expected, this null theory resulted in wild fluctuation in the market's exchange price. Then Gode and Sunder (1993) made the next step: adding the smallest bit of intelligence by constraining the random trades to exclude those that lose money. Now, the result was surprising: the simulated market rapidly converged, just as the human market did, on an efficient equilibrium price. In fact, these minimal-intelligence trader agents made more total profit than the humans did.

One conclusion from these experiments is that theory for trading decisions in a market ABM might not need to be elaborate at all, because much of the system's behavior emerges instead from the market's elaborate trading rules. However, the results of this experiment triggered so much interest that a number of subsequent studies (also described by Duffy 2006) identified additional patterns and tested whether they could be reproduced using minimally intelligent trader behavior. These studies followed the same protocol of establishing "real" patterns by having people operate a simplified artificial market, and then simulating the same trading rules in an ABM. At least one additional pattern (effects of a price ceiling imposed on the market) was reproduced by the minimal-intelligence trading theory. However, in general these experiments found that as markets are made more complex and realistic, human traders produce less efficient results but minimal-intelligence agents do far worse. And even the original experiment of Gode and Sunder identified one pattern of the human market (low price volatility) that was not reproduced by the ABM.

These experiments provide an excellent example of the pattern-oriented theory development cycle: some very basic patterns were established via empirical research, then simulation experiments were used to find theory just complex enough to reproduce the patterns, and then the cycle was repeated by establishing more patterns—by conducting new empirical experiments that better characterize real markets—and determining how much more "intelligence" needed to be in the theory to reproduce those patterns. As Duffy (2006) said: "Using ZI [zero intelligence—null behavior] as a baseline, the researcher can ask: what is the minimal additional structure or restrictions on agent behavior that are necessary to achieve a certain goal."

19.3.3 Harvest of a Common Resource

Understanding how people make decisions about harvesting a common resource is important for problems ranging from how to manage natural resources (e.g., air quality; public land) to understanding how cooperative societies first started. In these problems, individual people or businesses consume a resource that is shared by many; if individuals maximize their own short-term gain, the common resource is quickly used up. Janssen et al. (2009) used an ABM, results from a laboratory experiment, and POM to study how people make one particular kind of common resource decision.

Like the double-auction market studies described above, Janssen et al. designed an ABM to be tested against data collected on real people but in an artificial, simplified context. They modeled a laboratory experiment in which human subjects harvested a shared resource in a computer game. Players moved an avatar around a resource grid, harvesting tokens that "grew" on each cell; players were actually paid for each token they harvested. To produce negative feedback between harvest rate and resource availability, which makes total harvest much more sensitive to harvest behavior, the time until a cell grew a new token after being harvested was made to decrease with the number of neighbor cells that were also harvested. Several people operated on the same grid, competing for the available tokens and experiencing the consequences of each other's decisions. The exact movements of each participant were recorded. The ABM of Janssen et al. then simulated those movements as a way of quantifying the participants' harvest behavior.

Janssen et al. quantified three patterns that characterized the laboratory experiments with real people. The patterns are the range (mean ± 1 standard deviation) over time of three metrics that depicted (1) the number of tokens available for harvest, (2) the inequality of harvest among the participants, and (3) how often participants changed direction as they moved their avatar through the grid. A pattern was assumed to be matched by the ABM if the ABM's results fell within the observed range.

Janssen et al. (2009) tested three alternative traits for how people moved to harvest the tokens. The first two were null theories embodying two extremes: making decisions for no reason, and always using the same decision. The first trait was for avatars to move randomly, and the second was for them to always move toward one of the nearest tokens. The third trait included some mechanisms observed in the human experiment: it assumed avatars tend to move toward closer tokens, to avoid tokens that are closer to the avatars of other participants, and to prefer continuing in their current direction. The relative strength and exact effects of these three tendencies were controlled by seven parameters, so when testing the third trait Janssen et al. ran their model with 33,000 combinations of values for these parameters.

In their theory development experiments, Janssen et al. found that neither null theory reproduced the first and third observed patterns. The third trait reproduced the second pattern under most parameter combinations, and reproduced the other two patterns for only a few parameter combinations. From this first round through the theory development cycle, Janssen et al. concluded that their second pattern, inequality of harvest, seemed built into the system because it was produced by even the null traits. They also concluded, from the parameter combinations that caused the third trait to reproduce all three patterns, that harvesting tokens close to the avatar and moving in a constant direction seem particularly important parts of the human behavior.

19.4 Exercise Example: Stay or Leave?

Now, we set up an exercise that you can use to try theory development yourself. We describe a problem and a simple ABM that addresses it, providing all the detail except the agents' key trait. We also provide the patterns you will need to develop and test theory for this trait.

19.4.1 Social Groups and the Stay-or-Leave Adaptive Trait

Many animals live in social groups with reproductive suppression. One example is the wild dogs we modeled in chapter 16; other examples are wolves, gorillas, and alpine marmots, which all live in packs or family groups that have "alpha" individuals. The alpha individuals reproduce and prevent lower-ranked individuals from reproducing, so the only way an individual can achieve "fitness"—if we define fitness as reproduction—is to become an alpha. However, all the individuals obtain the benefits of group living, such as more efficient feeding and better ability to avoid predators. (You may have observed similar hierarchies in human groups, with "alpha animals" suppressing the "fitness" of lower-ranked individuals.) This social organization evolved in many species, so it seems to be a general evolutionary response to certain environmental conditions.

In this social organization, subordinate individuals face an important adaptive decision: whether to stay in the group or to leave in hopes of finding another group or empty territory where they can be an alpha and reproduce. If a subordinate individual stays, it enjoys the benefits of the group but can only reproduce if it lives until the current alpha dies (or becomes weak enough to be displaced). But even outliving the current alpha is not enough when there are any higher-ranking (older or bigger) subordinates who will likely become alpha next. If a subordinate leaves, it obtains the possibility of becoming an alpha quickly, but exposes itself to the risks of being alone until it forms or finds a new group, or until it returns unsuccessful to its home territory. ABMs are well suited for modeling populations of animals that live in social groups because their dynamics depend heavily on this adaptive behavior of subordinate individuals.

19.4.2 The Woodhoopoe Example: Observed Behaviors and Patterns

As an example system, let us consider populations of the red-billed woodhoopoe, a common group-living bird of eastern and southern Africa. In a long-term field study, du Plessis (1992) observed reproductive and mortality rates, group sizes, and behaviors. Du Plessis found that each group occupies its own spatial territory and has a strict hierarchy for both males and females. Only the alpha couple reproduces. The rank of subordinates seems determined mainly by age: if an alpha animal dies, the oldest subordinate of its sex replaces it. But subordinate adult woodhoopoes (one year old or more) sometimes undertake so-called "scouting forays": they leave their territory and explore the surrounding area. Usually they return to their home territory, but if they find a territory lacking an alpha of their sex they stay and become the alpha. During the scouting foray, the risk of being killed by predators is considerably higher than in the home territory.

Neuert et al. (1995) used these observations to build an ABM, and we will now describe a simplified version of it. The key behavior for which we need theory is the decision by subordinates of when to make scouting forays. We will use the simplified ABM as a virtual laboratory to test alternative theories, putting them in the model and seeing how well it then reproduces several key patterns.

We also provide three patterns that a theory for scouting forays should cause the ABM to reproduce. (These patterns are not directly from the observations of du Plessis 1992.) The first pattern is the characteristic group size distribution explained in figure 19.2. The second pattern is simply that the mean age of adult birds undertaking scouting forays is lower than the mean age of all subordinates, averaged over the entire simulation. The third pattern is that the number of forays per month is lowest in the months just before and just after breeding, which happens in December.

19.4.3 Description of the Woodhoopoe Model

Here is the ODD description of a simplified ABM of woodhoopoe populations, except that its key adaptive trait, the scouting foray decision, is left for you to design as exercise 19.2. (Because the submodels are so simple, we here include them in the "Process overview and scheduling" element of ODD.)

Purpose

The purpose of the model is to illustrate how the dynamics of a population of group-living woodhoopoes, and the dynamics of its social groups, depend on the trait individuals use to decide when to leave their group. The model provides a laboratory for developing theory for the woodhoopoes' scouting foray trait.

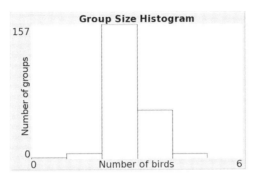

Figure 19.2
Characteristic woodhoopoe group size distribution, illustrated as a NetLogo histogram. The histogram shows how many social groups there were (y-axis) for each number of birds per group (x-axis). The distribution includes results at the end of month 12 of each year, over years 3–22 of a 22-year simulation. Only adults (age > 12 months) are counted. The number of groups with zero adults (first bar) is very low, often zero. The number of groups with only 1 adult (second bar) is low, and the number of groups with 2 adults is highest. As the number of adults increases above 2, the number of groups becomes rapidly smaller; here there are no groups with more than 4 adults.

Entities, State Variables, and Scales

The model entities are territories and birds. A territory represents both a collective—a social group of birds—and the space occupied by the group (territories can also be empty, though). Territories are represented as a one-dimensional row of 25 NetLogo patches, "wrapped" so that the two ends of the row are considered adjacent. The only state variables of territories are a coordinate for their position in the row and a list of the birds in them. Birds have state variables for their sex, age (in months), and whether they are alpha. The time step is one month. Simulations run for 22 years, with results from the initial two "warm-up" years ignored.

Process Overview and Scheduling

The following actions are executed in the given order once per time step. The order in which the birds and territories execute an action is always randomized, and state variables are updated immediately, after each action.

1. Date and ages are updated. The current year and month are advanced by one month, and the age of all birds is increased by one month.
2. Territories fill vacant alpha positions. If a territory lacks an alpha but has a subordinate adult (age > 12 months) of the right sex, the oldest subordinate becomes the new alpha.
3. Birds undertake scouting forays. Subordinate adults decide whether to scout for a new territory with a vacant alpha position. Birds that do scout choose randomly (with equal probability) between the two directions they can look (left or right along the row of territories). Scouting birds can explore up to five territories in their chosen direction. Of those five territories, the bird occupies the one that is closest to its starting territory and has no alpha of its sex. If no such territory exists, the bird stays at its starting territory. All birds that scout (including those that find and occupy a new territory) are then subjected to predation mortality, a stochastic event with the probability of survival 0.8.
4. Alpha females reproduce. In the twelfth month of every year, alpha females that have an alpha male in their territory produce two offspring. The offspring have their age set to zero months and their sex chosen randomly with equal probability of male and female.

Theory Development

5. Birds experience mortality. All birds are subject to stochastic mortality with a monthly survival probability of 0.99.
6. Output is produced.

Design Concepts

This discussion of design concepts may help you design alternative theories for the scouting trait.

Basic principles. This model explores the "stay-or-leave" question: when should a subordinate individual leave a group that provides safety and group success but restricts opportunities for individual success? In ecology we can assume real individuals have traits for this decision that evolved because they provide "fitness": success at reproducing. The trait we use in an ABM could explicitly consider fitness (e.g., select the behavior providing the highest expected probability of reproducing) but could instead just be a simple rule or "heuristic" that usually, but not always, increases fitness.

Emergence. The results we are interested in for theory testing are the three patterns described at the end of section 19.4.2: a successful theory will cause the model to reproduce these patterns. All the patterns emerge from the trait for scouting. The group size distribution pattern may also depend strongly on other model processes such as the reproduction and survival rates.

Adaptation. The only adaptive decision the woodhoopoes make is whether to undertake a scouting foray. You can consider several alternative traits for this decision that vary in how explicitly they represent the individuals' objective of obtaining alpha status to reproduce. You should start with "null" traits in which the decision is random or always the same. You could consider an indirectly-objective-seeking trait such as a simple rule of thumb (e.g., "scout whenever age > X"), and a trait that explicitly represents the factors that affect an individual's chance of meeting its objective.

Objectives. The subordinate birds have a clear objective: to become an alpha so they can reproduce. We also know, in this model, what processes affect the likelihood of reaching that objective. If the individual stays at its home territory, all the older birds of its sex must die for the individual to succeed to alpha. If the individual scouts, to succeed it must find a vacant alpha position and it must survive the predation risk of scouting.

Learning. The decision trait could change with the individual's experience. For example, birds could learn things on unsuccessful scouting forays that they use in subsequent decisions. (If you try learning at all, we suggest you start with simpler traits without learning.)

Prediction. The decision objective is to attain alpha status, but attain it by when? If you design a decision trait that compares the relative probability of becoming alpha for leaving vs. for staying, the trait must specify a time horizon over which that probability applies. Evaluating these probabilities would require some kind of prediction over the time horizon.

Sensing. We assume that birds know nothing about other territories and can sense whether an alpha position is open in another territory only by scouting there. However, it is reasonable to assume that a bird can sense the age and status of the others in its own group.

Collectives. The social groups are collectives: their state affects the individual birds, and the behavior of individuals determines the state of the collectives. Because the model's "territory" entities represent the social groups as well as their space, the model treats behaviors of the social groups (promoting alphas) as territory traits.

Observation. In addition to visual displays to observe individual behavior, the model's software must produce outputs that allow you to test how well it reproduces the three characteristic patterns identified in section 19.4.2. Hence, it must output the group size distribution illustrated in figure 19.2, the mean age (over all months of the entire simulation) of subordinate adult birds that do vs. do not make scouting forays, and the total number of forays made by month.

Initialization

Simulations start at January (month 1). Every territory starts with two male and two female birds, with ages chosen randomly from a uniform distribution of 1 to 24 months. The oldest of each sex becomes alpha.

Input

The model does not use any external input.

19.5 Summary and Conclusions

When you get to the theory development part of an ABM, you should finally feel like a real scientist. You have your experimental apparatus—the model—working and tested, but with a very simple trait for some key agent behavior. Now you're ready to start the inductive process of hypothesizing, testing, falsifying, and, ideally, "proving" theory for how the system's dynamics emerge from agent behavior. Even if all you do is show whether or not conventional assumptions about how agents make decisions can produce realistic behavior, you've made an important scientific contribution.

The theory development cycle of plugging alternative traits into the ABM and seeing how well the model then reproduces observed patterns requires that you define criteria for when patterns are "matched" by the model. It is important not to be too quantitative about matching patterns too soon. One reason it is better to start with just qualitative evaluation of the model is that patterns are always based on incomplete and uncertain data and may themselves be not very quantitative. More importantly, focusing on quantitatively matching a few detailed, context-specific patterns goes against a central theme of POM, which is that we best identify the basic mechanisms driving a system when we simultaneously look at a diversity of general patterns.

Another advantage of qualitative pattern-matching is that you can use it before doing all the work it often takes to find good values for model parameters via calibration—the topic we tackle in chapter 20. The ability to develop and test theory before calibration is very important because, of course, the results of calibration depend very much on the theory used in the model.

The idea of a virtual laboratory is to have "controlled" conditions, which means not changing parameter values when contrasting traits. It is therefore important to reduce uncertainty in the other submodels as much as possible, even though you have not yet calibrated the full model. For example, in the full woodhoopoe model developed by Neuert et al. (1995), parameter values were known relatively well from the long-term field study. If there is great uncertainty in the parameters for one of the traits being tested as potential theory, then the approach chosen by Janssen et al. (2009) can be a good idea: run the ABM with many combinations of values for the trait's parameters and see how many combinations reproduce the patterns. You should have more confidence in traits that reproduce the patterns robustly over wide ranges of parameter values. And you need to make good use of the trick you learned in section 12.3: testing, calibrating, and exploring submodels such as behavior traits separately, before putting them in the ABM.

Finally, we remind you that POM in general, and pattern-oriented theory development in particular, is not our invention but just the way science should be, and long has been, done. Theory development corresponds to "strong inference" and is used by many agent-based modelers. Usually, though, the process is not documented and communicated as theoretical science, so other scientists cannot learn whether the traits developed are general, reliable, and reusable. The point of POM is therefore to use multiple patterns systematically and explicitly; to contrast alternative traits for key behaviors, at a minimum comparing the trait selected for the ABM to null models; and to communicate this process as an important general finding that could be useful to others studying similar systems and problems.

19.6 Exercises

1. Select several publications that describe ABMs and review how the authors developed the "theory" used as the agents' key adaptive traits. You can select publications on the list we provide as on-line materials for chapter 3, or others that you are interested in. Identify the most important adaptive trait(s), and determine—if you can—how those traits were designed or chosen. What evidence is provided that the traits are useful—that they adequately represent the mechanisms that seem important for the system and problem being modeled? Was any effort made to try alternative traits, or to show that the selected traits work better than a null model? (Be forewarned: the answer often will be "no." We hope this situation will change once you start publishing your own work!)

2. For the simplified woodhoopoe model, develop theory for the subordinate birds' stay-or-leave decision—whether to undertake a scouting foray on each time step. Program the model from the description in section 19.4.3, and test your program thoroughly. (If you are extremely busy or lazy, you can download an implementation via a link on this book's web site.) Note that the downloadable implementation includes no sliders or inputs to change parameter values. Please leave it this way and focus only on modeling the scouting decision.

   Hypothesize several alternative traits using the information and hints provided in section 19.4. Include several null models that assume birds behave nonadaptively (e.g., randomly, or always the same way). Test and explore more elaborate traits independently before putting them in the full model, seeing how they behave in the various situations the birds might experience, and how their behavior depends on parameter values or detailed assumptions. Can you identify one or more traits that reproduce the patterns identified in section 19.4.2? What is the conceptual basis of your theory?

Parameterization and Calibration

20

Parameters are the constants in the equations and algorithms we use to represent the processes in an ABM. In our first model, which described butterfly hilltopping behavior and virtual corridors, the parameter q represented the probability that a butterfly would deliberately move uphill at a given time step (section 3.4). Other early examples include *blue-fertility* in NetLogo's Simple Birth Rates model (section 8.3) and the minimum and maximum values used to initialize patch profit and failure risk in the Business Investor model (section 10.4.1).

Parameterization is the word modelers used for the step of selecting values for a model's parameters. We have not focused on parameterization before now, mainly because few of the models we've used so far in this book are realistic representations of real systems, so we cannot say what a "good" value is. Now, however, in pattern-oriented modeling we are clearly focused on relating our models to real systems, so we must think about parameterization.

The reason that parameterization is important is, of course, that quantitative results matter. The ultimate task of agent-based modeling is to understand the organization and dynamics of complex systems. This implies understanding the relative importance of the different processes that we think are important and, therefore, include in our ABM. Changing parameter values can change the relative importance of different processes and, hence, the organization and dynamics of the model. We must use empirical information from real systems to "anchor" parameter values to reality so we can learn which processes are really most important and which dynamics are most believable.

Calibration is a special kind of parameterization in which we find good values for a few especially important parameters by seeing what parameter values cause the model to reproduce patterns observed in the real system. Calibrating an ABM is part of POM because we calibrate the model "against" (i.e., to reproduce) patterns observed in the real system. Now, however, the patterns are typically quantitative instead of qualitative and more descriptive of the whole system, not its agents and parts. Modelers typically use terms such as "calibrating a model against data" or "fitting a model to data" (calibrating its parameters so model results match observed data), and these terms are valid for what we do in this chapter.

Why do we include calibration as a part of pattern-oriented modeling separate from theory development, when they both include changing or adjusting a model until it adequately matches a set of observed patterns? The major difference is that in theory development, we are focused on one particular part of a model: its traits for agent behavior. We often test theory against qualitative patterns so that we do not need to worry yet about how closely the model reproduces observations from the real system. Calibration comes after we've identified theory for behavior and assembled the full model. Another way to think of the difference between this chapter and the rest of POM is that chapters 18 and 19 were about using patterns to reduce a model's "structural uncertainty" by finding good model designs and agent traits, and this chapter is about reducing "parameter uncertainty" by finding good parameter values. Now we are focused on quantitative results of the full model and comparing them to data from a specific real system. Sometimes a behavior trait or other submodel needs to be "calibrated" by itself; we will briefly consider this situation. But otherwise, you can assume we refer to calibration of a full model.

Model calibration serves three purposes. First, we want to force the model to match empirical observations as well as possible. When we adjust a model's parameters so that it better matches some observations from the real system, we assume that the adjusted parameters then produce more accurate results when modeling any conditions (though we talk about some potential exceptions, which we need to be careful about). When modelers talk about "calibrating a model," they generally refer to this objective of making the model fit observations and, therefore, more accurate and credible.

The second purpose of calibration is to estimate the value of parameters that we cannot evaluate directly. (In this chapter, we use the literal meaning of "evaluate": to establish the value of.) If we do not know a good value for some parameter, we can estimate a value "inversely" by adjusting it until the model best matches some observations. This is what modelers refer to when they talk about "calibrating a parameter." In some fields this type of calibration is even called "inverse modeling."

The third purpose is to test a model's structural realism: can we calibrate it to match the observations within a reasonable range? Or is there something wrong with the model so that it cannot be forced to match observations closely? (In that case, we might decide that the model is still a useful simplification even though it matches some but not all observations well.) This third purpose addresses the question of how robust the model is, which becomes our focus in chapter 23.

Learning objectives for chapter 20 are to:

- Understand the several objectives of calibration, and how parameterization and calibration of ABMs differs from calibration of more traditional models.
- Understand some fundamental calibration concepts and strategies, such as identifying good calibration variables and selecting specific criteria to measure a model's calibration.
- Develop experience by conducting several calibration exercises.

20.2 Parameterization of ABMs Is Different

For many decades, in many fields, modeling to many people has been a process of selecting one or a few simple equations with a few parameters, and then evaluating those parameters inversely by calibrating the model to empirical data. As we discuss in section 20.4, calibrating more parameters increases the uncertainty in their calibrated values. For people from this tradition, "parameterization" is equivalent to "calibration," and more parameters often means more uncertainty.

But calibration of ABMs is quite different from this traditional approach of "fitting" simple models to data. In the traditional approach, most of the information about the system is in the calibrated parameter values because the model contains very few mechanisms or processes. ABMs, though, contain more information about the system because they use more entities, state variables, and submodels. And each submodel can be parameterized and tested by itself (section 20.3). Therefore, even though ABMs typically have more equations and parameters than simple models, they are typically less reliant on calibration. Calibration of an ABM is often only a matter of "fine-tuning" a small fraction of its parameters.

Many fairly complex ABMs have proven useful and interesting with little or no calibration of the full model (in ecology, e.g., Pitt et al. 2003, Rademacher et al. 2004, Railsback et al. 2005, and Goss-Custard et al. 2006). Nevertheless, for the three purposes described in section 20.1, calibration remains an important step in the development and analysis of ABMs, especially when the model's purpose is to address specific problems of a specific real system.

20.3 Parameterizing Submodels

One of the major strategies for making ABMs reliable and credible is to develop, parameterize, and test each submodel independently and thoroughly (section 12.3). If someone says your ABM has too many parameters to calibrate reliably, you need to show how you evaluated most of those parameters before even thinking about calibrating the whole model. In this section we talk about strategies for choosing values for as many parameters as possible, especially in submodels, without falling back on calibration to evaluate them inversely.

Despite what we said in section 20.2, parameterizing *submodels* is often similar to parameterization of traditional models. In fact, one solution to parameterizing submodels is to select, as your submodel for some process, a published model of that process that has already been described, calibrated, and analyzed by someone else. (If you use a published model, you must, of course, be sure that it is appropriate for your ABM, especially its time and space scales.)

Often, though, you must make up your own submodels and evaluate their parameters. Appropriate parameter values may be available from the literature, your own data, or even via educated guesswork. One of the most important tricks for credible parameterization is to design your model so its parameters represent real quantities or rates that could, at least in principle, be measured empirically. Otherwise, you will have no choices other than guesswork and calibration to evaluate the parameters, and it will be very difficult to show that your values are "realistic."

The second important trick for credible parameterization of submodels is to show that your parameter values cause the submodel to produce reasonable results in all situations that possibly could occur. Do this as part of submodel development.

Sometimes it is necessary to calibrate a submodel to data, using methods more similar to classical model fitting and calibration than is calibration of full ABMs. If you need to calibrate a submodel, it could be wise to consult with modelers familiar with the model fitting literature.

Of course, for the first, simple versions of your models, which we encourage you to analyze early on (chapter 21), or for more theoretical models designed only to explore possible explanations, it is sufficient to use "guesstimated" parameter values. How do you do a good job of guesstimating? First, define upper and lower bounds for the parameter's value, beyond which the equation or submodel would produce nonsense. Then, think about whether the process being modeled operates on a fast or slow time scale, has a strong or moderate impact, occurs very often or rarely, varies a lot or not, etc. Doing so should help you define a reasonable range for the parameter, from which you can simply select the middle, or try several values. And, again, analyze the submodel and make sure that it "works" with your guesstimated values before putting it in the model.

Before moving on to calibrate the full ABM, you should force yourself to set the value of every submodel parameter as best you can. This is especially important if there are more than just one or two submodels. Then, during calibration you can determine which if any of these parameters to reevaluate.

20.4 Calibration Concepts and Strategies

Now we discuss calibration of a full ABM to improve and document its accuracy and to evaluate its most uncertain parameters. The basic idea of calibration is to execute a model many times, each time using different values of its parameters, and then analyze the results to see which parameter values caused the model to best reproduce some patterns observed in the real system. In any field of science where modeling is common, there is an extensive and important literature on model calibration and model fitting, addressing such topics as mathematical measures for how well model results fit an observed data set (or, which of several alternative models is most likely correct, given the data set), calculating "best" parameter values, and estimating the uncertainty in the calibrated model. For example, books important in ecology (but useful to anyone) include Burnham and Anderson (2002), Haefner (2005), and Hilborn and Mangel (1997). This literature is most relevant to models simpler than ABMs, but some of its basic concepts are very important to us.

One important concept is that there is only a limited amount of information in the observed patterns and data we calibrate a model to, so when we use the same information to calibrate more parameters, our estimate of each parameter's value tends to become less certain. A second, related concept is that of *overfitting*: by fine-tuning many parameters we can cause a model to match a few patterns very closely, while actually making the model less accurate in general. The data sets we use to calibrate models always have their own errors and uncertainties, and always represent one unique set of conditions. Consequently, if we try too hard to make the model reproduce one data set, we are likely to choose parameter values that are not as good at representing all the other conditions.

There are well-known ways of dealing with these concepts. First, we need to understand that there is a cost to calibrating more instead of fewer parameters. If we decide to adjust three instead of two parameters, we are likely to fit the calibration data better but we need to realize that we are also reducing the certainty in our parameter estimates and increasing the potential for overfitting. Second, we need to use, if possible, techniques such as validation via secondary predictions (section 18.3.4) and robustness analysis (chapter 23) to assess how well the calibrated model represents the system in general.

Here are some techniques for doing calibration effectively, presented as a sequence of six steps. (These steps assume that you have already parameterized your submodels as much as possible.)

20.4.1 Identify a Few Good Calibration Parameters

The first step is to identify a very small number of parameters to evaluate via calibration. These should be parameters that are both especially uncertain and important, and each calibration parameter should have relatively independent effects on the model.

What makes a parameter especially uncertain? One definition of "uncertain" is that we do not know the correct value accurately, only that it might fall within some broad range. But a parameter can also be uncertain because we use it to represent something that is actually variable and complex, because we do not want that variability and complexity in the model. Hence, there is no single correct value of the parameter, and it might even be unmeasurable (e.g., the coefficients of a line relating the propensity of a banking executive to buy mortgage-backed securities to how many college-age children the executive has).

And how do we know which parameters are especially important to a model? This question is only answered convincingly by sensitivity analysis, which we study in chapter 23; and it is often prudent to reconsider model calibration after conducting a sensitivity analysis. But even before the sensitivity analysis systematically examines the importance of parameters, we usually have a good idea about which parameters strongly affect model results. You can always run simple preliminary sensitivity experiments, varying the especially uncertain parameters to see how they affect results. A parameter that is highly uncertain but has little effect on results should not be used for calibration.

Finally, you should avoid having two or more calibration parameters that have similar effects. For example, imagine a business model with two uncertain parameters, one for the probability of failure due to bankruptcy and one for the probability of a business closing voluntarily (e.g., the owner retires). If you are calibrating the model against data on the number of businesses, it could be very hard to distinguish the effects of these two parameters: many combinations of them could make the model reproduce the data equally well. (You could, however, calibrate one of the parameters while assuming the other is a constant percentage of the first; e.g., calibrate probability of bankruptcy while assuming probability of voluntary closure is always 30% of bankruptcy.)

20.4.2 Choose Categorical vs. Best-Fit Calibration

There are two general ways to calibrate a model. In what we call a *categorical* calibration you search for parameter values that produce model results within a category or range you defined as acceptably close to the data. For example, you could determine from the calibration criteria (section 20.4.4) that the model is adequately calibrated if the mean number of agents is between 120 and 150 and if their mean size is between 210 and 240. You could then run the model with many combinations of parameter values and see which parameter combinations produce results within the acceptable category. In a *best-fit* calibration, you search for one set of parameter values that cause the model to best match some exact criteria—essentially, an optimization. In this case, you would try to find the parameter values that cause model results to be as close as possible to, for example, mean number of agents = 135 and mean agent size = 225.

Which approach is best? Best-fit calibration has the advantage of producing a simple, single set of calibrated parameter values instead of a whole set of acceptable values. And you can always find a "best" set of parameter values, even if the fit they produce between model and data is not very impressive. However, to do this you must be able to identify just one measure of model fit that the "best" parameter values optimize: you cannot optimize several criteria at once. Categorical calibration requires less interpretation of model results because you are not trying to find absolutely best parameter values, and makes more sense if the calibration data are themselves uncertain. But it is always possible that categorical calibration does not solve the problem of telling you what parameter values to use for future model runs because no parameter combinations produce results within the acceptable ranges.

20.4.3 Decide Whether and How to Use Time-Series Calibration

ABMs simulate how a system changes over time, and we often have data collected over time; it seems natural, then, to calibrate the model's ability to reproduce patterns of change over time. If our model's purpose includes representing how results change over time (e.g., how long does it take the system to recover from some perturbation? How is the system affected by changes in its environment over time?), then it usually does make sense to use time-series calibration. But some ABMs (e.g., the woodhoopoe model of section 19.4.3) are intended to explain long-term average conditions, so they intentionally do not contain all the processes that cause the real system to change over time and use no input data to represent how the agents' environment changes over time. In such cases, time-series calibration may not be useful or necessary.

When we have a time series of observations but choose not to calibrate the model's ability to reproduce changes over time, we can instead calibrate the model by its ability to reproduce statistical characteristics of the time series, such as the mean number of agents. We often also want to calibrate some measure of variability over time such as the standard deviation in the time-series data: does our model produce the same degree of variation around the mean as we see in the observed patterns?

When we do choose to use time-series calibration, the question is how to quantify the fit between model results and observations over time. Certainly, we want the mean and standard deviation in model results over time to be close to those in the observed time series. But these measures do not reflect trends over time. Here are some other relatively simple measures of model fit to time series:

- Maximum error: the maximum absolute difference between model results and observations, over all the times in the series. The best calibration would produce the smallest maximum error.
- Mean squared error: the mean, over all times in the series, of the square of the difference between model results and observations. Squaring the differences makes sure that negative errors do not offset positive errors, and emphasizes large errors more than small ones.
- The number of points in the time series where model results are acceptably close to observations. The modeler must decide what range around the observed values is close enough to consider a successful calibration.

Figure 20.1 illustrates these measures. Assume that this figure shows results of a model that is being calibrated with data on the number of agents alive or active at the end of a year,

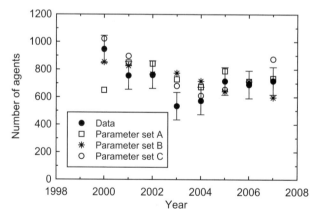

Figure 20.1

Example time-series calibration. The black dots show observed data (number of agents) collected annually from 2000 to 2007, with error bars showing the +/− 100 range considered acceptable for calibration. The other symbols represent model results for three different sets of parameter values. The data have a trend of decreasing from 2000 to 2003, then increasing gradually until 2007. Parameter set A produces results that have no clear trend over the eight years; set B has a generally downward trend, and set C has a minimum at 2004.

Table 20.1 Measures of Model Fit to Calibration Data for Example Results in Figure 20.1

| | Parameter set | | |
Measure of model fit	A	B	C
Difference in mean	36	24*	62
Difference in standard deviation	53	38	17*
Maximum error	296	240	158*
Mean squared error	19712	13976	9699*
Number of years with results within 100 units of data	5	5	5

*Lowest value, indicating best fit of model to data.

observed for eight years from 2000 to 2007. The figure shows these data plus results from three model runs using three sets of parameter values. Let us also assume that the model can be considered adequately accurate (well calibrated) if its results are within 100 agents of the observed data, each year. Can you tell from the figure which parameter set is best? In some years (e.g., 2001, 2003) none of the parameter sets produced results close to the data. For the first few years, parameter set B seems to work well, but in the last few years, set A seems to be best. The measures of model fit discussed above are evaluated for these results in table 20.1. These numbers indicate that parameter set C produced results best fitting the observations for three of the five measures, even though its mean value was farthest from the data. There was no difference among the parameter sets for one measure, the number of years in which model results were within the calibration range of 100 from the data.

20.4.4 Identify Calibration Criteria

This step is to identify and define the quantitative patterns that will be used to calibrate the model. This sounds simple, but actually is often the most challenging part of calibration. What we have to do is identify some key patterns that we calibrate the model to reproduce, and define those patterns—and the model results that we will compare them to—with sufficient quantitative precision so we can say mathematically whether the pattern is met or how close the model results are to the pattern. There are often unexpected complications.

First, we typically need to calibrate an ABM against patterns in the different kinds of results that will be used in the model's final application. Is the model going to be used to explain the number, size, and location of agents? If so, then it is best to calibrate it against patterns in all these three kinds of results if we can.

Second, we need to make sure that observed patterns and model results used for calibration actually represent the same system characteristics. There are a number of important considerations:

- The observations need to be from a system that has the same basic mechanisms that the model does, and collected under conditions similar to the ones represented in the model.
- Observations and model results need to measure the same things at the same times, as exactly as possible. If the observations are the number of agents in the system, counted in November of each year, then they should be compared to model results representing November, even if the model produces output for every month or day.
- The observations should have been made at spatial and temporal resolutions compatible with the model's time step and grid size. Patterns observed at an hourly or daily scale may not be comparable to results of an ABM that operates at a monthly or annual time step, for example.

- Special care must be taken using measures of variability as calibration criteria. There are many kinds and causes of variability and it is not legitimate to compare one kind of variability to a different kind. If, for example, you want to calibrate against the standard deviation in number of agents reported by a field study, make sure you know whether that standard deviation represents the variation among replicate experiments, or variation over time in one experiment, or uncertainty in the method used to count agents.

Third, we need some idea of how reliable and accurate the patterns are. All observations have errors and uncertainties, and we often must calibrate a model using data that we know could be inaccurate by 10, 20, or 50%. Using uncertain data is OK and usually unavoidable: when the calibration patterns are more uncertain, we don't worry so much about matching them exactly, and we recognize that calibrated parameter values are less certain. But we need to have at least some idea how accurate or certain the observations are, so we know how much information they really contain.

Finally, we must specify how we will compare the observed patterns to model results, to determine which model runs produced results that are calibrated adequately or "best." Some methods for quantitative comparison of results to observations are mentioned above in sections 20.4.2 and 20.4.3. Of particular concern is how to calibrate several different kinds of model results at once: if we want to calibrate a model to reproduce the number, size, and wealth of agents, how do we decide between a set of parameters that reproduces number and wealth well but not size, and a parameter set that reproduces size and wealth well but not the number of agents? Do we treat each kind of model result as equally important (even though it is quite possible that no combination of parameters will produce good results for all three criteria) or do we prioritize them?

At the end of this step of defining calibration criteria, we should have a specific algorithm for quantifying how well a set of model results reproduces the selected observations. The algorithm should detail exactly what model results (e.g., "the mean number of agents, evaluated on November 15 of each simulated year except for the first five years") are to be calibrated against exactly what observed numbers. And it should specify how the comparisons are made and then used to rate how well a model run matches the observations.

Calibration, like all of modeling, is often a cycle that gets repeated. If we find no parameter values that cause our model to meet the calibration criteria—or if broad ranges of parameter values meet the criteria—then we might come back and revise the criteria. We discuss this decision in section 20.4.6.

20.4.5 Design and Conduct Simulation Experiments

Now that the calibration criteria are carefully defined, we just need to run the model and find out what parameter values best meet them. It is tempting to just start trying parameter values and searching heuristically for good ones, but that approach is inefficient and very likely to miss good parameter combinations. Instead, we need to design and conduct a simulation experiment much like the other ones we've done already. This calibration experiment executes the model many times, using combinations of values over the feasible range of all parameters. The results of this experiment will tell us what ranges of parameter values produce results that meet the calibration criteria.

The first step in setting up the simulation experiment is to select values for the non-calibration parameters and input data (if any) that represent the conditions (the same time period, environment, etc.) under which calibration patterns were observed.

Then we need to identify the range of values considered feasible for each parameter, and decide how many values to use within that range. (Modelers refer to this as defining the *parameter space*: if we think of each calibration parameter as a dimension in space, then we are defining the area or volume of this parameter space where we will look for calibrated values.) Then, setting up the experiment is very easy if we use NetLogo's BehaviorSpace. For example, in the Experiment dialog field (section 8.3) where you say which variables to vary, you can enter:

```
["failure-risk" [0.0 0.01 0.2]]
["monthly-sales-std-deviation" [0 50 500]]
```

which automatically will run your model for 231 "points" in parameter space, with `failure-risk` over its feasible range of 0.0 to 0.2 and `monthly-sales-std-deviation` over its range of zero to 500. (It is usually good to include values that bound the range of feasibility; risk and the standard deviation in sales cannot be negative and are almost certainly greater than zero, but we don't know how close to zero they might be nor how the model behaves when their values are near zero.) If your model takes a long time to execute, so 231 runs are infeasible, you can start with fewer parameter combinations and then conduct a second experiment that "zooms in on" the most promising area.

A final consideration in designing the calibration experiment is how to deal with variability in results. If a model is highly stochastic, the random variation in its results can make calibration seem more challenging. What do you do if, for example, you want to calibrate the model so that the mean number of agents alive over a simulation is 103, when the model produces results that differ by 5, 10, or 20 each time you run it? It seems natural to run several replicates of each parameter combination in your calibration experiment, then average their output to get a more precise result. However, we cannot really calibrate a model to higher precision than justified by its stochasticity. Instead of estimating mean model results over a number of replicates, we recommend running more parameter combinations and analyzing them by plotting model results against parameter values. Use regression (or just your eye) to identify parameter values that produce results falling in the desired range.

20.4.6 Analyze Calibration Experiment Results

Ideally, the final step of calibration would be to analyze the results of all the model runs in the calibration experiment and identify the parameter values that meet the calibration criteria, select the final values of the calibrated parameters, and document how closely the calibrated model fit the criteria. If your ABM does meet all the calibration criteria for one or more combinations of parameter values, then you can in fact complete this step and move on to the kinds of analysis we address in part IV.

However, it is not at all unusual to learn from the calibration experiment that no combinations of parameter values cause the model to meet all of the criteria. Often, one set of parameter values will reproduce most of the criteria, but the remaining criteria are met only under different parameter values; or some calibration criteria are never met. What should you do if you cannot meet all the calibration criteria at once?

First, you should once again try very hard to find mistakes in everything from your submodels, software, and input data to the analysis of your calibration experiment. If you try but find no such mistakes, then it is very likely that your model is too simple, or too simple in the wrong ways, to reproduce the observed patterns you chose as calibration criteria. You could consider going back to the theory development stage and seeing if you can improve the traits for agent behavior, and you could consider adding processes or behaviors that were left out the first time. But keep in mind that there are costs to adding complexity to your model, especially if it is not very clear what change needs to be made.

It is very common for good modelers to keep their model simple instead of adding stuff to it until the model can reproduce all the calibration criteria. Keep in mind the overfitting issue: it can be risky to try too hard to make the model reproduce a limited set of observations. It may make sense to revise your calibration criteria so they are not as restrictive. If you choose not to revise the model to make it fit more of the calibration criteria, simply document your calibration results and the extent to which the model does not meet some criteria under your "best" parameter values, and your decision not to revise the model. Then, when you use the model to solve problems, keep in mind which results are less certain as indicated by the calibration experiment.

20.5 Example: Calibration of the Woodhoopoe Model

The Woodhoopoe model described in section 19.4.3 has few parameters, but some of them are particularly difficult to evaluate even with the extensive field studies that du Plessis (1992) conducted to support the original model. For this example, we will calibrate the model to estimate values for two parameters using several calibration criteria of the kind that could be available from a field study. (The "data" here are actually made up.)

We use the model exactly as described in section 19.4.3, adding a simple trait for how subordinate adults decide each month whether to undertake a scouting foray:

- If there are no other subordinate adults of the same sex in the territory, then do not scout.
- Otherwise, decide whether to scout via a random Bernoulli trial with probability of scouting equal to the parameter *scout-prob*.

For this exercise, let us assume that this simple trait was tested in a theory development cycle and found to reproduce the characteristic patterns of the woodhoopoe population when *scout-prob* was assumed to have a value around 0.3. (This version is available in the chapter 20 section of this book's web site.)

Now let us follow the steps described in sections 20.4.1–6.

Identify calibration parameters. We assume that two parameters are good for calibration because they are particularly uncertain and expected to have strong effects on results. First is *survival-prob*, the monthly survival probability for all birds. Second is *scout-prob*; the theory development cycle told us (we are assuming here) that *scout-prob* likely has a value around 0.3, but we do not have a certain estimate because scouting events are hard to detect in the field.

Choose categorical vs. best-fit calibration. Instead of finding the "best" set of parameter values, we will look for values that produce results within an acceptable range of the calibration criteria. We choose categorical calibration because the data available for calibration are somewhat uncertain.

Decide whether to use time-series calibration. The woodhoopoe model represents a "stable" system with no inputs or processes that change over time; only stochastic events cause the population to change over time. Hence, trying to calibrate its changes over time does not seem necessary or likely to succeed. Instead of time-series calibration, we will calibrate it against data averaged over time.

Identify calibration criteria. We will use three patterns as calibration criteria, as if they were observations from the field study. The "mean abundance" criterion is that the long-term mean number of woodhoopoes (including sub-adults) is in the range of 115 to 135. The "variation" criterion is that the standard deviation from year to year in the annual number of birds is in the range of 10 to 15 birds. The "vacancy" criterion is that the average percentage of territories that lack one or both alphas is in the range of 15–30%. All the criteria are assumed to be from data collected in November (month 11) of each year of the field study (just before breeding). Hence, the model will be calibrated by comparing these patterns to its results for average and standard deviation in number of woodhoopoes, and the number of territories lacking an alpha, in November only, and excluding results from the first two "warm-up" years of the simulation (section 19.4.3).

Design and conduct simulation experiments. Now we set up BehaviorSpace to run an experiment that simulates many combinations of the calibration parameters and produces output corresponding to our calibration criteria. Creating the parameter combinations is extremely easy: we just put statements like these in BehaviorSpace's "Vary variables as follows" field:

- `["scout-prob" [0 0.05 0.5]]`
- `["survival-prob" [0.95 0.005 1.0]]`

BehaviorSpace now will create and run 121 simulations with `scout-prob` and `survival-prob` varied in all combinations over these wide ranges of feasible values (if `survival-prob` is less than 0.95, the population rapidly dies out).

To get output comparable to the calibration criteria, we need to write reporters in the BehaviorSpace dialog to produce output that we can analyze and compare to our criteria. Because the criteria include means and standard deviation among years, we must obtain output for each step. In addition to the number of woodhoopoes, we need the number of territories lacking one or both alphas. And because we need to exclude the first two years and examine only month 11, we need to output the year and month. We can get these outputs with the reporters shown in figure 20.2.

Analyze results. BehaviorSpace produces output (using its "Table" format) similar to table 20.2. Now we need to import the file to spreadsheet or statistical software for analysis. (The "Pivot-Table" and "PivotChart" facilities of Excel are good for the kind of analysis we do here.) We can calculate the fraction of territories lacking an alpha by simply dividing the values in the rightmost column by the total number of territories, 25. Then we need to filter the results and analyze only those from month 11 of years 3–22 (20 years after the two warm-up years). (PivotTable's "Report Filter" makes this easy.) When we do this analysis in a spreadsheet, we find that there were no combinations of *scout-prob* and *survival-prob* where all three criteria were met. Does that mean the model is a failure?

Let's take a closer look at the calibration results. Contour plots are a good way to see how each model result varied with the two calibration parameters; it would be good to plot contours of how each of the three outputs varied with *scout-prob* and *survival-prob*. Figure 20.3 is similar to a contour plot, but shows which of the three calibration criteria are met for each point we simulated in the parameter space. We can see from it that the abundance criterion depended almost entirely on *survival-prob*: it was met almost always, and almost only, when *survival-prob* was 0.98. The vacancy criterion, though, was only met when *survival-prob* was 0.975 (and *scout-prob* was between 0.25 and 0.4). The variation criterion was met consistently when *survival-prob* was 0.96 or less, but also at some parameter combinations near where the abundance and vacancy criteria were met. Now the results do not look so bad: in the region around *scout-prob* = 0.25–0.4 and *survival-prob* = 0.975–0.98 there are locations close to where all three criteria were met—for example, around *scout-prob* = 0.225 and *survival-prob* = 0.9775. You could do a second BehaviorSpace experiment focused on this area to see if there are any combinations where all three criteria are indeed met. This example illustrates how calibration is often not successful immediately, but (as we discussed in section 20.4.6) that does not mean you should panic!

Figure 20.2
BehaviorSpace experiment for calibration of the woodhoopoe model.

Table 20.2 Example BehaviorSpace Output for Woodhoopoe Model Calibration

[run number]	scout-prob	survival-prob	[step]	year	month	count turtles	count patches with [count (turtles-here with [is-alpha?]) < 2]
1	0	0.95	0	1	0	100	0
1	0	0.95	1	1	1	88	5
1	0	0.95	2	1	2	83	7
1	0	0.95	3	1	3	78	6
1	0	0.95	4	1	4	74	6
1	0	0.95	5	1	5	67	8
1	0	0.95	6	1	6	63	10
1	0	0.95	7	1	7	60	10
. . .							

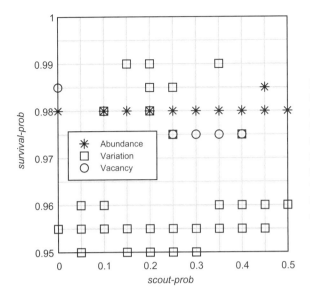

Figure 20.3
Calibration results for the woodhoopoe model example. The plot represents the parameter space: the Behavior-Space experiment ran the model once for each intersection in the grid. The symbols appear at grid locations where the parameter combination caused the model to meet one of the calibration criteria.

20.6 Summary and Conclusions

This chapter provides a very basic introduction to some modeling tasks that are fundamental to many kinds of modeling. There is a vast literature on issues such as parameterization, calibration, and model fitting, and many sophisticated methods and even software tools. (In fact, there is a separate software tool for calibrating NetLogo models; we mention it in section 24.5.)

One of the most frequent criticisms of ABMs is that "they can be calibrated to say anything." Why believe in a model if you know that users could adjust its parameters to produce whatever result they wanted? This criticism is partly a result of not understanding the difference between parameterization of ABMs and traditional, simple models. Simple equation-based models have few parameters, which can only be evaluated by calibrating the model to observed data. ABMs, in contrast, typically have submodels that each can be parameterized independently using a variety of information. So before we get to the calibration phase, we should have established values for most of our ABM's parameters.

There are, still, serious hazards in calibrating ABMs, and we need some familiarity with the literature on calibration, model fitting, and model selection and the major issues it addresses. One is the overfitting problem, which is exactly the concern about forcing a model to produce a particular result: if we adjust more parameters, we can make a model reproduce a few patterns more and more precisely, but this can make the model less and less accurate for other problems. A second important issue is that if we calibrate more parameters to the same limited data set, then the value of each parameter is less certain. How do we deal with these problems, and how do we convince critics that our modeling process is not biased to produce only the results we wanted from the start?

First, we need to evaluate as many parameters as possible directly instead of inversely via calibration. If we have to make an educated guess, we usually should do so. Avoiding ambiguous or unmeasurable parameters helps us evaluate parameters directly; the parameters of a well-designed ABM have clear meanings (e.g., the cost of a unit of production for an industry; the number of offspring an animal produces) so we can evaluate them independently. We should calibrate only a very small number of particularly uncertain parameters that have strong effects on model results.

Second, we need to be aware of the trade-offs between precision and generality in deciding what to do if we our initial attempt at calibration "fails." Because we often calibrate an ABM against several different patterns in different kinds of output, we often find that few or none of the parameter combinations cause the model to meet all the calibration criteria. When this happens, we must decide whether to revise the model, perhaps adding or modifying adaptive behaviors or trying to calibrate more parameters, even though this could make parameter values less certain and overfitting more likely. Or we might just accept our inability to calibrate the model as precisely as we wanted—perhaps loosening or ignoring some calibration criteria—as the cost of keeping the model simple.

Third, calibration, like all important parts of the modeling cycle, needs to be documented well. You will not convince critics that your calibration was unbiased unless you write down and archive how you decided which parameters to calibrate, the exact calibration criteria, the calibration experiment design and results, and the analysis. There is another important reason to document your calibration: you are very likely to repeat calibration as you iterate through the modeling cycle (e.g., by adding new submodels, applying the model to new systems, or addressing new problems with it), so documenting it the first time will save time later.

Finally, the most convincing way to show that your ABM has been calibrated without bias and produces useful general results is to analyze it thoroughly after calibration. Can you show that the model is not overfit to one particular pattern? Can you show what results emerge robustly from the mechanisms and behaviors in the model instead of depending only on the value of one or two parameters? These are, in fact, exactly the kinds of questions we explore in part IV. Now that you've learned how to formulate, program, and calibrate a model, it's time to learn how to learn from it!

20.7 Exercises

1. Section 20.5 illustrated one simulation experiment to calibrate the woodhoopoe model, but it found no parameter combinations that met all three calibration criteria. Can you find combinations that do meet all the criteria, by conducting more simulations focused on the most promising part of the parameter space? Whether or not you can, what seems to be the "best" combination of values for the parameters *survival-prob* and *scout-prob*?

2. The abundance and harvest of renewable resources (fish, trees, naive investors, etc.) are often modeled by fitting simple models to time series of data. One well-known model is:

$$N_t = N_{t-1} + rN_{t-1}[(1 - N_{t-1})/K] - H_{t-1}$$

where N_t is the number of resource units at time t, N_{t-1} is the number of units at the previous time step, K is a constant representing the number of units that would exist at "equilibrium" after a long time without harvest, r is the per-unit resource growth rate (number of units produced per existing unit, per time step), and H_{t-1} is the number of units harvested in the previous time step. To model a time series of N when we know the harvest H at each time, we need values for r, K, and N_0 (the number of units existing in the time step before our model starts). Given the time series of "observed" data for H and N in table 20.3, and the three different sets of values for these parameters in table 20.4, calculate the model results using each parameter set and then evaluate the five measures of model fit that are in table 20.1. Which parameter set seems to reproduce the data best? Can you find even better parameter values? (Real resource managers do

Table 20.3 "Observed" Data for Exercise 2

Time step (t)	Harvest (H)	Observed resource abundance (N)
1	1141	2240
2	1080	2410
3	956	2460
4	917	2630
5	825	2360
6	949	2590
7	771	2460
8	734	2690
9	628	2580
10	551	3060
11	735	2600
12	763	2460
13	802	2460
14	852	2350
15	957	2640

Table 20.4 Alternative Parameter Sets for Exercise 2

Parameter	Value for parameter set		
	1	2	3
R	1.95	1.5	2.2
K	3050	3320	2950
N_0	2250	2400	2200

this kind of analysis, but use more sophisticated measures to evaluate model fit to time-series data.)

3. In section 8.4 we explored NetLogo's Flocking model and the fish schooling research of Huth and Wissel (1992; also described in section 6.2.2 of Grimm and Railsback 2005). Can you calibrate Flocking to three patterns, observed in real fish, that Huth and Wissel (1992) used to test their model? The first pattern is that mean nearest neighbor distance (NND; the mean over all fish of their distance to the nearest other fish) is 1–2 times the fish length; use the turtle size of 1.5 patches as fish length. Second, polarization of the flock (the mean over all fish of the angle between fish heading and the mean heading of all fish) is 10–20°. The third pattern is that the fish form a single school, which you can quantify as being true if the maximum distance between any pair of fish is less than 15 patches. Like Huth and Wissel, use 8 individuals and start them near each other (e.g., within the center 1/5 of the World). Let the model "warm up," for instance by running 1000 ticks, before measuring results. You can calibrate all the parameters that appear in sliders on the Interface (except "population," which remains at 8), and these parameters can vary over the ranges defined by their sliders. What do the calibration results tell you about the model? (Important hint: code to calculate the mean heading of a turtle agentset is already in Flocking.)

4. The Mushroom Hunt model used as an introduction to NetLogo in chapter 2 has three parameters that control searching behavior. These are the number of ticks after last finding an item at which the hunter switches back to broad instead of local searching, and the range of angles the hunter turns in both broad and local searching modes. (In chapter 2, we simply "hardwired" these parameters into the code.) Find the values of these parameters that minimize the time that one hunter needs to find 50 mushrooms.

Model Analysis

IV

Introduction to Part IV

21.1 Objectives of Part IV

Testing and analyzing are, as you certainly have learned by now, integral parts of agent-based modeling (and of modeling in general). By *testing* we refer to checking whether a model, or a certain submodel, is correctly implemented and does what it is supposed to do. Throughout this book we have been testing both models and their software. For example, in chapter 6 we tested whether our NetLogo program for the hilltopping behavior of butterflies really did what we wanted it to (the answer was "no"). *Analyzing* models refers to something different: trying to *understand* what a model, or submodel, does. For example, in chapter 12 we used contour plots to understand the behavior of an important submodel of the Business Investment model.

Understanding is not produced by an ABM automatically; just building a model does not solve the problem it is intended for. Instead, we need to work hard to understand why a model system behaves in a certain way under certain circumstances. The ultimate goal of modeling is to transfer our understanding of the model system to the real system. If, for example, we find a structurally realistic representation of a natural beech forest (section 18.3) and can explain how and why spatial patterns emerge in the model, we can cautiously transfer that explanation to real beech forests. Or if we understand how to control an epidemic in a model, we can deduce practical advice for real disease management. To make this transfer we must, of course, analyze the model so that we understand it.

You should have learned by now that analysis happens throughout the modeling cycle: we analyze early versions of a model and its submodels to improve their formulations and designs, and we do pattern-oriented analysis to develop theory for agent behaviors and to calibrate the model. Modeling and model analysis cannot be separated; we need to analyze and try to understand what a model does all the time. However, here in part IV we want to do three new things: turn from analyzing models "on the fly"—while still being developed and improved—to analyzing "full" models; identify efficient approaches for analyzing ABMs and develop them into a strategic framework (chapter 22); and present several formal kinds of analysis commonly conducted on models (chapter 23).

What do we mean by a "full" model? In principle, no model is ever finished, because there will always be opportunities and incentives to modify the structure, simplify one part, go into more detail in another one, etc. And, in fact, some would-be modelers can never make themselves stop tinkering with their model's design. At some point, we need to "freeze" the design (Starfield et al. 1990), accept the current version of the model, and try to understand how it behaves and works.

When should we freeze the design? Too soon in a modeling project would obviously make no sense, because early versions are too simplified. But once we feel that we've found reasonable representations of all key processes and made sure that the model behaves reasonably like the real system, we should freeze the model's design for a while and turn to analysis. It can be a good idea to give this model version a number ("FancyModel 1.0") because quite likely we will want to make changes after we better understand the first version. If you design and analyze your models thoroughly, you should typically not need more than two or three such versions before you can start using the model as the basis for your thesis, publications, or management decisions. It is common, in fact, to find that the first version of a model is already so interesting that it is not necessary to make more.

You have also learned by now that thinking up and programming a model that looks interesting is very easy and can be done within a day or two. What turns this into science are two things that both include model analyses. First, we need to establish a relation between the model system and the real world; the model does not absolutely have to represent any specific real system, but it should at least be inspired by patterns observed in reality. The Segregation model that we explore in chapter 22 is a good example; it is highly abstract but inspired by observed patterns. This task of relating models to the real world by using patterns was covered in part III. Second, we need to analyze the model thoroughly to understand what it does and also perhaps to improve it, often by simplifying it or by adding elements that turned out to be essential but missing.

Both of these tasks can be quite challenging and time consuming, and are thus usually hard to demonstrate in classes and courses. While formulating a model may require only a few days, fully analyzing it can take months and, for more complex models, even years! A few lucky people are born model analyzers—scientific Sherlock Holmeses—and know how to do it intuitively. The rest of us have to learn how to analyze ABMs the hard way: by doing it, testing all kinds of heuristics, learning what approaches work under what circumstances, etc. Documentation of these critical steps in the modeling cycle is essential for establishing your model's usefulness (Schmolke et al. 2010), and not always trivial. Chapters 22 and 23 summarize the experience of many modelers; but you need to try it yourself. You should also make it a habit, as you read scientific publications, to pay special attention to how models are analyzed. After all, this is the homestretch where all your hard work finally pays off.

21.2 Overview of Part IV

In chapter 22, "Analyzing and Understanding ABMs," we describe general strategies for analyzing ABMs. Many of these strategies are specific to ABMs and take advantage of their structural richness and realism. The most important message of chapter 22 is that, to understand ABMs, we need to perform controlled simulation experiments (Peck 2004) driven by hypotheses. Like experimenters in the laboratory or field, we ask "what-if" questions, but unlike them we can manipulate the experimental world as much as we want. For example, how would you, in the real world, test whether spatial heterogeneity among the agents is essential to explain a certain phenomenon or not? In an ABM, this is easy: just make all agents sense the same global (instead of local) information and see what happens! Likewise, we can simulate much longer time spans and bigger spaces, and conduct many more replicates, than we can in real-world experiments. The problem is that this ease of manipulating our model world also makes it easy to get lost in myriad alternative assumptions and scenarios. Therefore, chapter 22 describes important approaches for keeping our focus. We also briefly discuss the use of statistical methods for model analysis.

In chapter 23, "Sensitivity, Uncertainty, and Robustness Analysis," we turn to more specific and traditional methods for analyzing models. In sensitivity and uncertainty analysis, we explore how model results, and the conclusions we draw from a model, depend on parameter values. In robustness analysis, we explore how robust model results are to drastic changes in parameters, or even to changes in model structure. This kind of analysis helps us understand and communicate what is really essential for generating certain patterns and system behaviors. These techniques are widely applied to simulation models and sometimes viewed as mandatory parts of model analysis.

Our last chapter (24, "Where to Go from Here") provides advice and guidance as you move on from this book to a career, perhaps even in agent-based modeling. We mention a number of technologies and tools for dealing with challenges that are likely to arise as your modeling ambitions increase. More importantly, we suggest some activities and projects to maintain and build your momentum as you evolve from a student to a practitioner of agent-based modeling.

Analyzing and Understanding ABMs

22.1 Introduction

Imagine that you just installed new software for manipulating digital photos. Like most of us, you probably don't start by reading the user manual but instead by just using the program. However, the new software might use terminology and a philosophy that you are not familiar with. What would you do? You would try it out! You would try one command or tool at a time and see how your photo changes. Once you got an idea how basic tools work, you would start using them in combination. You would base your attempts on your current, preliminary understanding: "Now, to change the color saturation of foreground objects, I probably should first use tool A, then B, while keeping the tool C set to such and such"; and then you would try it out. Sometimes your preliminary understanding, or hypotheses, would be correct and sometimes not, which tells you that you do not yet fully understand how the different tools work and interact.

What you actually would have done is to try to understand your software by performing *experiments*. The Latin verb *experiri* means just this: to try out! And you would intuitively have performed "controlled" experiments: varying only one factor at a time and checking how it affects the "results," your photo. By doing so, you would be establishing causal relationships: understanding how the results are affected by each factor.

Once we have built a new ABM, even a preliminary version of one, it becomes another piece of software that we want to understand. What results does the model produce, under what conditions? How do results change when parameters, input data, or initial conditions change? And most importantly, *why* does the model produce the results it does? What is the model trying to tell us about how it works, and how the real system works? These are the questions we want to answer once we have a working model, so we experiment with it.

To turn experimentation into a scientific method, we make our experiments reproducible: by completely describing the model just as empirical scientists describe the materials and methods used in a laboratory or field study; by precisely documenting all the parameter values, input data, and initial conditions we use; and by documenting and analyzing the results of our experiments.

This should all sound familiar to you because we have been using controlled simulation experiments throughout this course to design, test, and calibrate models. (In fact, this would be a good time to review the discussion of strong inference in section 19.2.) Controlled simulation experiments are also key to analyzing and understanding what models do. "Controlled" does not, however, imply that a simple protocol for analyzing models exists. Rather, how we analyze a model still depends on the model, the question and system it addresses, our experience, and the problem-solving heuristics we know and prefer. Heuristics, or rules of thumb, for problem solving are characterized by the fact that they are often useful, but not always: we simply have to try them. Using heuristics does not mean that modeling is unscientific: heuristics are the basis of any creative research.

Learning objectives for this chapter are to:

- Understand the purpose and goals of analyzing full ABMs, including both "finished" models and preliminary versions of models that you plan to develop further.
- Learn and try ten "heuristics"—techniques or tricks that are often useful—for analyzing ABMs.
- Become familiar with common ways that statistical analysis is used to understand ABMs.

22.2 Example Analysis: The Segregation Model

22.2.1 Model Description

The Segregation model in the Social Science section of NetLogo's Models Library was inspired by a simple model by the Nobel laureate Thomas Schelling (Schelling 1971, 2006). Following is an ODD description of this model.

Purpose

The model addresses segregation of households in cities: why do members of different groups (e.g., racial, ethnic, religious) tend to occupy different neighborhoods? The model explores the relationship between segregation patterns and the tolerance of individuals for unlike neighbors.

Entities, State Variables, and Scales

The model entities include turtles that represent households, and patches that represent houses. Households are characterized by their location (which patch they occupy) and their color, which represents the group they belong to, either blue or red. Households also have a state variable *happy?*, a boolean variable set to false if the household has more unlike neighbors than it tolerates. The grid cells make up a square of 51×51 cells, with no depiction of roads or other spaces between them. The space is toroidal. The length of a time step is unspecified but represents the time in which a household would decide whether to move. The number of time steps in a model run is an emergent outcome: the model runs until all households are happy and, therefore, stop moving.

Process Overview and Scheduling

The following actions are executed, in this order, once per time step.

- If all households are happy (*happy?* is true) then the model stops.
- The households that are not happy (*happy?* is false) execute the submodel "move." The order in which these households execute is randomly shuffled each time step.
- All households update their *happy?* variable (see submodel "update").
- Outputs for system-level results are updated.

Design Concepts

The *basic principle* of Segregation is the question of whether strong individual behaviors are necessary to produce striking system patterns—does the presence of strong segregation mean that households are highly intolerant—or can such strong patterns emerge in part from the system's structure? The key outcomes of the model are segregation patterns—especially, how strongly segregated the entire system is; these outcomes *emerge* from how households respond to unlike neighbors by moving. The households' *adaptive behavior* is to decide whether to move: they move when their *objective*—to live in a neighborhood with the fraction of unlike neighbors below their intolerance threshold—is not met. The behavior does not involve learning, or prediction other than the implicit prediction that moving might lead to a neighborhood where the tolerance objective is met. Households *sense* the color of households on the eight surrounding patches. *Stochasticity* is used in two ways: to initialize the model so that it starts unsegregated, and to determine the new location of households when they move, because modeling the details of movement is unnecessary for this model. *Observations* include a visual display of which color household is on each grid cell, and two numerical results: the mean percentage (over all households) of neighbors of similar color and the percentage of unhappy households.

Initialization

A user-chosen number of households (typically 2000, out of the 2601 patches that represent houses) are initialized. They are each placed on a random empty grid cell and given a color randomly, with equal probability of red and blue. The variable *happy?* is then calculated for all households.

Input Data

The model does not use input from external models or data files.

Submodels

The submodel "move" is performed by individual households if they are unhappy. The household chooses a direction randomly from a uniform continuous distribution between 0 and 360 degrees, then moves forward a distance drawn randomly from a uniform continuous distribution of 0 to 10 grid cell widths. If there is already a household on the grid cell at this new location, the household moves again with a new random direction and distance. If the new grid cell is empty, the household moves to its center.

The submodel "update" is conducted by all households to determine whether they tolerate their neighborhood. The tolerance of households is determined by a parameter *%-similar-wanted*, which can range from 0 to 100 and applies to all households. A household's neighbors are all households on the eight surrounding patches. The household's variable *happy?* is set to false unless the number of neighbors with the household's color is greater than or equal to *%-similar-wanted* divided by 100 and multiplied by the number of neighbors.

22.2.2 Analysis

Now, let us perform experiments to understand how the model behaves.

▪ Start NetLogo and open the Segregation model.

▪ Press the `setup` and `go` buttons.

▪ With the number of turtles set to 2000 and `%-similar-wanted` set to 30%, it takes about 15 ticks until all households are happy. The average similarity of neighborhoods is about 70%— that is, turtles have on average 70% neighbors of the same kind. The View shows relatively small clusters of red and blue households, mixed with smaller clusters of empty patches.

The model thus demonstrates, in the current settings, the point Schelling (1971) wanted to make: even relatively small intolerance to unlike neighbors can lead, without any other mechanism, to segregated neighborhoods.

Do we understand the results of our first experiment? It is not entirely clear why the small intolerance of 30% gives rise to an average similarity of 70% or more. Let us see how model output changes if we chose extreme values of *%-similar-wanted*.

Heuristic: Try extreme values of parameters. When parameters are set to the extremes of their range, model outcome often is simpler to predict and understand.

■ Set `%-similar-wanted` to small values (5%) and large values (90%). Before you press the go button, try to predict the outcome of these experiments!

If intolerance is very low, most households are happy in the initial random distribution, and it takes only a few time steps until they are all happy. The average similarity among neighbors is near 50%, which might seem surprisingly high until we remember that, with only two colors, average similarity should be near 50% even when households are randomly distributed. This is an important realization already: that the final average similarity is always at least 50% no matter how low *%-similar-wanted* is.

If intolerance is very high, almost all households are unhappy at the beginning, but movement never stops and average neighbor similarity never changes much from 50%. The first result is easy to understand; the second less clear: why does the average similarity among neighbors not just increase with *%-similar-wanted*? We can only assume that there is a critical value of *%-similar-wanted* above which it becomes very unlikely that moving households ever will find a neighborhood that suits their tolerance. Let us test this hypothesis.

Heuristic: Find "tipping points" in model behavior. If a model shows qualitatively different behaviors at different extreme values of a parameter, vary the parameter to try to find a "tipping point": a parameter range where the behavior suddenly changes.

This is an extremely important heuristic because it helps us understanding different *regimes of control*: below a critical parameter value, process A may be dominant and control the behavior of the system, but above that tipping point control is taken over by process B. Identifying regimes of control and understanding how they emerge reveals a great deal about how a model works.

To put this heuristic into practice, we would usually program some quantitative outputs, and then run sensitivity experiments (as we did in figures 5.3, 8.3, 10.1, 11.1, and 11.2). However, Segregation is such a simple model that it is sufficient to simply watch the View while we change the value of *%-similar-wanted*.

■ Set `%-similar-wanted` to 90%; click `setup` and then go. Then, use the slider to slowly decrease this parameter—slowly, because it might take some time until the system responds.

Bang! Very abruptly, when `%-similar-wanted` reaches 75%, model behavior changes and segregation occurs.

■ Press `setup` and go again for `%-similar-wanted` = 75%.

Note that for this parameter setting, the system needs quite a long time for all households to become happy (about 150 ticks). Another difference from, say, *%-similar-wanted* = 30% is that the clusters of households of the same color are much larger. And a striking feature of the final distribution of households is that regions of different colors are separated by strips of empty patches. Why?

When we try to understand how such patterns emerge, we should always make sure that we observe the model entities' state variables as thoroughly and creatively as possible (the subject of chapter 9). One of the things we need to understand about a model is how it got to its current state as a consequence of what happened over time. How could we do this without visualizing the current state from many different perspectives?

Heuristic: Try different visual representations of the model entities. To better understand an ABM, we often look at different visual outputs to detect patterns and establish causal relationships.

Kornhauser et al. (2009) list many useful ways to change the View of NetLogo models to reveal hidden structures and correlations. With the Segregation model, the ability to tell happy from unhappy households would allow us to better understand why and where clusters of happy households form or dissolve.

■ Save the Segregation model under a new name, such as "my-segregation-model.nlogo".

■ In the procedure `update-turtles`, insert, at the end of the `ask turtles []` block, code like this:

```
ifelse happy?
    [set size 1.0]
    [set size 0.6]
```

Now, happy households are represented as large turtles and unhappy ones as small.

■ Repeat the experiment described above, where you start with `%-similar-wanted =` 90% and slowly decrease it as the model runs.

With high intolerance (90%), we never see any clusters of happy households. With lower `%-similar-wanted`, more and more small clusters of happy blue or red households occur but never persist. At the very sharp transition point (75%) clusters start to persist and grow slowly, with a lot of dynamics going on at their boundaries. But still, the simulation runs too fast to really understand what happens.

Heuristic: Run the model step by step. Stepping through a simulation, tick by tick, is crucial because it allows us to look at the current state of the model, predict what will happen next, and see whether we were right.

■ Add a button `step` to the Interface and make it call the `go` procedure, but with the "forever" option off.

Now, by stepping through the simulation at the critical parameter value `%-similar-wanted` = 75%, we see that clusters of happy households tend to grow. We understand that inside such clusters there cannot be any dynamics: everybody is happy, so nobody moves. And the clusters do not contain any isolated empty patches for long. Why? (We leave answering this question to you.)

What we see is that the driving dynamics are going on at the boundaries of clusters. We realize that a household's decision to move affects not only its own happiness but also that of its old and new neighbors. Movement of households can sometimes turn other happy households into unhappy ones.

Now, how do the boundaries between blue and red clusters emerge? The following heuristic helps.

Heuristic: Look for striking or strange patterns in the model output. Patterns are key to discovering important mechanisms and the internal organization of complex systems.

■ With `%-similar-wanted` set to 75%, run the model several times.

The clusters of blue and red households are quite clear, but the pattern is not yet clear enough to indicate anything particular. So, let us use the following heuristic.

Heuristic: At an interesting point in parameter space, keep the controlling parameter constant and vary other parameters. In chapter 20 we defined *parameter space* as the set of all possible values of model parameters. A point in this space corresponds to one specific set of parameter values.

This heuristic allows us to explore how our controlling parameter, or the process represented by it, is affected by other processes in the model. Trying to understand a model includes trying to understand the interactions of mutual controls.

■ With `%-similar-wanted` set to 75%, decrease the number of households, for example to 1500, then increase it to, say, 2400.

Now you can see that the pattern becomes much clearer and more striking if we increase the number of households so there are very few free patches (figure 22.1).

It is also striking that the system now usually segregates into only one cluster of each color (remember, the boundaries are toroidal). What might be the reason for this? When you see this pattern emerge, note that small clusters, with their strongly curved boundaries, usually do not survive unless they manage, by chance, to develop longer sections of their boundary that are

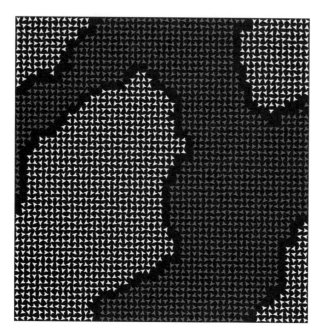

Figure 22.1
Example Segregation model results with *%-similar-wanted* = 75% and 2400 households.

more or less straight. This suggests a hypothesis: that straight boundaries create spatial configurations that make the nearby households less sensitive to the influence of single neighbors of the opposite color; there will always be enough of one's own kind to keep the proportion of unlike neighbors below the intolerance level.

This hypothesis sounds reasonable, but how could we test it? One way would be to measure (and visualize), in detail, the similarity of neighbors as perceived by households at the boundary; this would be easier if we set up specific configurations in a much smaller system of, say, 5×5 patches. Often, this is the only and best way to test our hypothesis. In this case, however, we can test our hypothesis the way physicists do: by making a prediction from it and testing the prediction. If our hypothesis that straight boundaries lead to stable results is correct, then straight boundaries should become less important when intolerance is less, because then curved boundaries can also "survive" the influence of single unlike neighbors. Hence, we can predict that when we slowly decrease `%-similar-wanted`, clusters should increase in numbers, decrease in size, and have more curved boundaries.

■ Study segregation patterns for `%-similar-wanted` smaller than 75%, keeping the number of households at 2400.

Our prediction turns out to be correct!

We still did not explain why for `%-similar-wanted` = 75% (and also down to about 58%), empty patches are only found between the boundaries of the red and blue clusters. We leave explaining this to you.

What have we achieved so far? Remember: by just running the model for the default parameter values provided by the Models Library, we got the result that the model was designed for: segregation that is much more intense than the intolerance of the individuals. But that result did not indicate by itself the mechanisms driving the segregation. Only by analyzing extreme parameter values—such as extreme intolerance, which this model was not at all about—did we identify and understand what seems to be a critical mechanism: the interaction of the level of intolerance with local spatial configurations. And we also found an unexpected tipping point where the same mechanisms (movement of unhappy households) that cause striking patterns of segregation suddenly make the segregation disappear.

The Segregation model is admittedly very simple and not that hard to understand. Nevertheless, we applied heuristics that are useful for analyzing any model. Note that Schelling's model of segregation—which, ironically, was only verbally formulated and never implemented by Schelling himself (Schelling 2006)—was and still is highly influential. Myriad modifications of this model have been made because real cities and households are of course much more complex; these models have explored things like larger neighborhoods, social networks, spatial barriers, housing prices, and variable intolerance levels.

Thus, the Segregation model is an exemplary simple conceptual model that reveals an important and robust mechanism and is the starting point for further important and more detailed studies. This exemplar should motivate you to strive for models that are as simple as possible while still producing the kind of system-level behaviors and patterns you want to understand and illustrate.

22.3 Additional Heuristics for Understanding ABMs

The previous section introduced several important heuristics for analyzing models by demonstrating how those heuristics help us gradually understand how patterns emerge. We could now introduce additional heuristics in the same way, but this is not necessary because we have already used these heuristics throughout this book! Therefore, we will just list the additional heuristics without demonstrating their use in detail.

Heuristic: Use several "currencies" for evaluating your simulation experiments. ABMs are rich in structure. Even the simplest ABM can include many agents and spatial units, all in different states. It is impossible to take all this information into account when analyzing a model. We therefore need to find "currencies," usually summary statistics or observations, that capture an important property of the model system in a single number. Currencies correspond to the things empirical researchers measure (or wish they could measure) in the real system. For example, in population models an obvious currency is population size: the total number (or weight, or wealth, etc.) of all agents. We therefore could analyze the time series of population size produced by the model. But even time series can contain too much detailed information, so we often look at even coarser currencies such as the mean and range of values over time. We have already used many currencies in this book as we analyzed models for different purposes (table 22.1).

Often, finding useful currencies is a nontrivial and important task of the modeling process. This is why "Observation" is included in the Design Concepts section of the ODD protocol. Usually, we try several currencies, or observations, and see how sensitive they are and how much they help us understanding the model. In most cases, one currency is not sufficient, and we need to view the model from different perspectives.

Several kinds of currencies are often used:

- Standard measures of the statistical distribution of results, whether distributions among agents of their state variables or distributions over time of system-level results. Common measures of distributions include the mean and median, the standard deviation or variance, the minimum and maximum, and whether they fit theoretical shapes (normal, exponential, power law, etc.) and, if so, the distribution coefficients.
- Characteristics of time series such as positive or negative trends, autocorrelation, or the time until the system reaches some state such as having no more agents or becoming static (e.g., until movement stops in the segregation model).
- Measures of spatial distributions and patterns such as spatial autocorrelation, fractal dimensions, and point pattern metrics.
- Measures of difference among agents, such as whether the distribution among agents of some state variable is uni- or multi-modal (e.g., is wealth distributed normally among agents or are there distinct groups of rich and poor?) and whether different kinds of agents are segregated over time or space.
- Stability properties (Grimm and Wissel 1997), which can provide insight into a system's internal mechanisms and be important for management problems. Currencies can evaluate stability properties such as how quickly the system returns to some "normal" state after a disturbance, how large the "domain of attraction" spanned by two or more state variables is (outside this domain, the system loses important characteristics, e.g., spatial segregation, in the Segregation model), and what buffer mechanisms damp the system's dynamics.

Heuristic: Analyze simplified versions of your model. Once we have chosen our first currencies, we proceed to the most important approach for analyzing models: Simplify! ABMs (and other models) can be hard to understand because so many different factors affect model behavior. Often, it is relatively easy to reduce this complexity and make understanding what mechanisms cause what results much more feasible.

Table 22.1 Currencies for Model Analysis Used in This Book

Model	Currencies	Chapter; Location
Butterfly hilltopping	• Corridor width • Final clumping of butterflies	5; Figs. 5.2, 5.3 Exercise 5.6
Marriage	• Age distribution of married people	6; Section 6.5.1
Simple Birth-Death	• Time to extinction	8; Fig. 8.3
Flocking	• Number of turtles who have flockmates • Mean number of flockmates per turtle • Mean distance between a turtle and its nearest neighbor • Standard deviation in heading	8; Figs. 8.5, 8.6,
Business Investor model	• Mean annual profit and risk of investors • Mean investor wealth • Mean final wealth of investors • Number of failures	10; Fig. 10.1 11; Figs. 11.1, 11.2
Telemarketer	• Number of telemarketers in business • Median weeks in business • Business size distribution • Total sales	13; Figs. 13.2, 13.3
Mousetrap	• Number of balls in air	14; Fig. 14.1
Wild dogs	• Frequency of extinction within 100 years	16; Fig. 16.2
Woodhoopoes	• Shape of the group size distribution • Mean number of birds • Standard deviation among years in number of birds • Mean percent of territories vacant	19; Fig. 19.2 20; Fig. 20.3

There are a number of ways typically used to simplify ABMs for analysis:

- Make the environment constant;
- Make space homogenous (all patches are the same, and constant over time);
- Reduce stochasticity, for example by initializing all agents identically or by replacing random draws with the mean value of their distribution (e.g., replace `set size 2 + random-float 10` with `set size 6`);
- Reduce the system's size by using fewer turtles and patches;
- Turn off some actions in the model's schedule (just comment them out in the `go` procedure);
- Manually create simplified initial configurations and states that allow you to check whether a certain mechanism works as you assume.

When you make these kinds of simplifications, the model version will no longer be a good representation of the system you want to understand. You create, so to say, simplified worlds that are unrealistic but easier to understand. Grimm (1999) found that many developers of ABMs did not systematically simplify their models for analysis, perhaps because of a psychological barrier: if you spent so much time developing a model that seems to represent a real system reasonably well, should you then make this model extremely unrealistic? Yes, definitely! Modeling includes as much deconstruction as construction!

Heuristic: Analyze from the bottom up. Any system that requires an agent-based modeling approach is very difficult to understand without first understanding the behavior of its parts: the agents and their behavior. It is important that we test and understand these behaviors first, before we turn to the full model. That is why we emphasize analyzing submodels independently before putting them in an ABM (section 12.3) and systematically developing theory for agent behaviors (chapter 19). Even more of these kinds of low-level analysis may be needed to understand the full model, especially if unexpected dynamics seem to emerge.

Heuristic: Explore unrealistic scenarios. The idea of this heuristic is to simulate scenarios that could never occur in reality. Why? Because a simple and direct way to see the effect of a certain process or structure on overall model behavior is to just remove it.

As an example, consider the forest model used by Rammig and Fahse (2009) to explore the recovery of a mountain forest after a storm blows many of the trees down. The model is grid-based and includes interaction among neighboring grid cells. How important are these local interactions? Would a simpler, nonspatial model with only global interaction be sufficient? To test this, Rammig and Fahse simulated a scenario that definitely cannot occur in reality: grid cells, and their trees, no longer interacted with just their eight neighbor cells but instead with eight cells chosen randomly from the entire grid (figure 22.2, right panel). The set of cells to interact with was randomly changed every time step for every grid cell.

In this way, they removed any spatial effects among neighboring cells. The unrealistic model led to unrealistically high numbers of trees, which indicates that the original model's spatial interactions are indeed important.

The analyses of how investor behavior affects double-auction markets discussed in section 19.3.2 provides an interesting contrast: models that deliberately used unrealistically simple investment behaviors were shown to produce system-level results that were not so unrealistic. The conclusion in that case was that complex agent behavior might not be the mechanism generating complex market dynamics after all; the market rules themselves might be more important than anticipated.

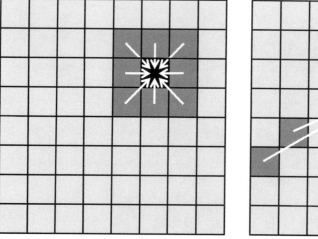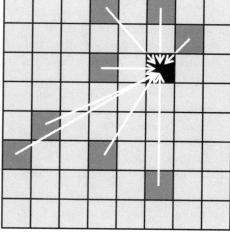

Figure 22.2
The use by Rammig and Fahse (2009) of an unrealistic scenario to analyze the importance of local interactions in a forest model. The black cell illustrates how all cells are simulated. In the normal version of the model (left panel), the state of a cell is determined by its eight neighboring cells. In the unrealistic scenario (right panel), the state of a cell is determined by eight randomly chosen cells.

The right attitude for using unrealistic scenarios is playing god: "Modelers should never hesitate to 'play god' and unrealistically manipulate individuals and their environments if it helps test hypotheses in simulation experiments" (Grimm and Railsback 2005, p. 309).

22.4 Statistics for Understanding

Many scientists automatically think of statistics when they think about analysis and understanding. The goal of statistics is to extract understanding, and perhaps infer causal relations, from a fixed and usually limited set of data. Agent-based modeling, in contrast, can produce as many "data" as we want, and offers additional ways to develop understanding and deduce mechanisms. If a simulation experiment does not provide a clear enough answer, we manipulate the model world further and run new experiments until we get clear results.

The purposes, and underlying mind-sets, of statistics and simulation modeling are thus quite different. Nevertheless, there are important uses of statistics in analyzing ABMs:

- *Summary statistics.* Aggregating model output into statistics such as the mean and standard deviation is useful, but remember that in model results, extreme or unusual values are often important clues to understanding. They should not be thrown away as "outliers" as they often are in statistical analysis of empirical data.
- *Contrasting scenarios.* Statistics can be used to detect and quantify differences between simulation scenarios, much like testing for differences among laboratory treatments. This analysis often requires arbitrary assumptions that affect results—especially, the number of replicates for each "treatment." While the same is true for laboratory and field experiments, it is easier to change these assumptions when analyzing model results.
- *Quantifying correlative relationships.* Regression methods such as *analysis of variance* and *general linearized modeling* can be used to see which inputs have strongest effects on which model outputs, and whether there are interactions among factors. The approach is simply to look for statistical relationships among inputs (here, including parameter values and initial conditions as well as time-series inputs) and outputs. This approach can be very useful, in particular if a model is complex and we initially have no idea where to expect causal relationships. For example, Kramer-Schadt et al. (2009) used several regression methods to identify model parameters and, hence, their corresponding mechanisms, that were most important for "explaining" the persistence of a wildlife epidemic. Of course, these regression methods do not directly identify causal relationships, but they can provide important clues by revealing the relative importance of different factors. As detectives trying to understand why an ABM behaves as it does, we should consider such statistical "meta-models" as a starting point, not the end, of our analysis. The meta-model points to factors that we should explore in more detail using simulation experiments.
- *Comparing model output to empirical patterns.* During calibration (chapter 20) and other times when we want to compare model results to observed patterns quantitatively, there is a wide variety of statistical approaches that can be useful. The extensive literature on model calibration, fitting of models to data, and model selection is applicable.

We go into no further detail here because many of you likely have strong backgrounds in statistics and there are many resources on this subject. In fact, one advantage of using statistics in model analysis is that other scientists are familiar with the language and techniques of statistics and will feel more "at home" reading your results. So consider statistics as part of your model analyst's toolbox, but do not rely on it too much; simulation experiments and deductive reasoning remain the most productive approach.

- Analyze from the bottom up.
- Define and test several 'currencies' for evaluating your simulation experiments.
- Try different Views (visual outputs).
- Run the model step-by-step.
- Analyze simplified versions of your model.
- Try extreme values of parameters.
- Look for striking, or strange, patterns in the model output.
- Find tipping points in model behavior.
- From an interesting point in parameter space, keep the controlling parameter constant and vary other parameters.
- Explore unrealistic scenarios.

Figure 22.3
Heuristics of model analysis.

22.5 Summary and Conclusions

To understand what an ABM does and why, we must again be detectives—as we were when searching for programming errors and doing pattern-oriented modeling. Good detectives combine reasoning and strong inference, systematic analysis, intuition, and creativity. Obviously, we cannot teach you how to reason logically, be creative, and have good intuition. You have to develop your intuition through experience. Thus, the key messages of this chapter are that once we build an ABM, or even the first time we "freeze the design," we need to try to understand what it does; and controlled simulation experiments are how to do it. Now try it!

In addition to the analysis heuristics presented here, you should try to learn from published analyses. If you read a publication that presents some exciting new insights using an ABM, ask yourself: how did they do that? The key experiments presented in a publication can give you a good idea what techniques and heuristics the authors used to obtain their insights.

We called the techniques in this chapter "heuristics" because they are often useful, but not always. Which heuristics will work depends on the model, you, and the question you are addressing. And, unfortunately, there is no clear order in which the heuristics work best. We nevertheless compiled them into a sequence (figure 22.3) in which you can try them.

We suggest you add your own heuristics to ours in figure 22.3. Intuition is important, but critical reflection on how your intuition, and heuristic approaches, work is no less important. Good science always is based on a mixture of creativity, which can include quite chaotic, unsystematic things, and rigorous methods, which can be pedantic and boring. Switch between these two attitudes, try looking at your model from different perspectives, and you will quickly learn—how to learn from ABMs!

22.6 Exercises

1. The NetLogo Segregation model can be criticized for being unrealistic because real people do not move randomly when they abandon a neighborhood; it seems more reasonable to assume instead that they will move to a neighborhood where they will (at least temporarily) be happy with their neighbors. Start with the version of Segregation you made in section 22.2 and make it more realistic by having turtles, when they move, choose an empty patch where they will be happy. (Think about how you schedule things!)

Reanalyze this version and see what results and causal relations have changed. Do you need to use different "currencies" (outputs, e.g., instead of the number of ticks until the system becomes static) to evaluate the model meaningfully? Are there different tipping points? Do different patterns emerge? Are different processes important?

2. A second way that the Segregation model is unrealistic is that it lacks neighborhoods: subareas surrounded by distinct boundaries such as large streets. Households may not care who lives across such boundaries. Add boundaries to the model; you can do this simply by, say, turning every tenth row and column of patches into a "street" where households cannot live. How does this change model behavior? Why? What happens if you vary the size of the neighborhoods?

3. Take one or more of the models we've used in this book (see table 22.1) and analyze it in a way similar to how we analyzed the Segregation model in section 22.2. (You could also use one of NetLogo's library models. Be aware that it will be more interesting to use one in which the agents actually have adaptive behavior—not all do.) Use the currencies that are already implemented, but add new ones if they help. Think about how you could change the View to get further insights. Make sure you check the heuristics in figure 22.3. Document your series of experiments, results, new assumptions, new experiments, etc.

Sensitivity, Uncertainty, and Robustness Analysis

23

If someone developed an ABM of the stock market and claimed that it explains how the market's trends and short-term variation emerge, what questions would you ask? First, because you are now trained in design concepts and pattern-oriented modeling (POM), you would ask whether the system-level patterns of variation really emerge from the traders' adaptive traits or whether they were imposed by model rules; second, you would ask whether the model only reproduces a single system-level pattern, or multiple patterns observed in real markets at both trader and market levels and at different time scales.

But even if the model passed those tests, there would be one more important set of questions to ask: does the ABM reproduce observed patterns *robustly*, or are these results *sensitive* to changes in model parameters and structure? And how *uncertain* are the model's results— would it produce the same results if different, but also plausible, parameter values were used? These questions are important because we seek robust explanations of observed phenomena. In POM, the patterns we select to characterize a system (chapter 17) are by definition general and robust; they do not depend on the idiosyncrasies of a specific system or situation. If a model reproduces those patterns only when its parameters have a very limited range of values, or only when key processes are modeled one exact way, it is very unlikely to fully capture the real mechanisms underlying these patterns.

Consider, for example, savanna ecosystems, which are defined as places where trees and grass coexist for long times, with trees covering no more than about 20% of the area. Savannas occur worldwide and make up about 10% of the land surface, but vary widely in their plant and animal species, soil types, and (arid to semi-arid) rainfall regimes. Because savannas occur under so many conditions, there must be some generic, fundamental mechanisms that produce the ecosystem called "savanna." Any model that reproduces savanna-like features non-robustly—say, for only a limited range of parameter values—is unlikely a good explanation of savanna emergence. One theory for how savannas occur is that trees and grass coexist because their roots have different lengths so they can use the limiting resource, water, from different "niches." Jeltsch et al. (1996) implemented this theory in an ABM and found that it could produce long-term coexistence of trees and grass, but only for very limited parameter ranges. They therefore discarded this theory as a viable explanation of savanna ecosystems.

Testing and documenting the sensitivity of model output to changes in parameter values is very important for two reasons. First, as in the savanna example, this kind of analysis can show how strongly the model represents real-world phenomena. Second, it helps us understand the relative importance of model processes. High sensitivity to a parameter indicates that the process linked to that parameter controls model output, and thus system behavior, more than other processes. High sensitivity to a parameter is not necessarily a bad thing; rather, sensitivity is a diagnostic tool that helps us understand our models.

Sensitivity analysis (SA) explores how sensitive a model's outputs are to changes in parameter values: if you make a small change in some particular parameter, do model results change a lot, a little, or not at all? We have already been using "sensitivity experiments" (introduced in section 8.3) to examine similar questions, but by SA we refer to a systematic analysis of all parameters. SA is routine, and more or less required, in some disciplines (e.g., ecology, engineering), but not yet common in others. We consider SA part and parcel of agent-based modeling and highly recommend including some kind of SA in your model analyses and publications. Simple, local SA is easy to perform, but often it also pays to perform more comprehensive and sophisticated analyses.

Uncertainty analysis (UA) looks at how uncertainty in parameter values affects the reliability of model results. If our model has parameters that represent, for example, the probabilities that people of four different ethnic groups will vote for each political party, we know that the model's results will be uncertain because (a) it's very hard to measure the voting tendencies of any group precisely, and tendencies change over time; (b) it's often not clear who does and does not belong to a particular ethnic group; and (c) people often vote for reasons other than what party a candidate belongs to. UA explicitly recognizes such uncertainties in parameter values and thus provides a way to see which results of a model we should or should not have much confidence in.

Robustness analysis (RA) explores the robustness of results and conclusions of a model to changes in its structure. SA often is focused on the model's response to *small* changes in parameter values; therefore, we emphasize sensitivity—whether the system responds. In contrast, RA focuses more on response to *drastic* changes in model structure; we thus emphasize robustness—whether the system does not respond. RA is not yet routine, but you should be familiar with RA and strongly consider some form of it.

We strongly recommend that you add SA, UA, and RA to your modeling toolkit. Unlike most of the topics we cover in this book, there is a vast literature on SA and UA, so once you understand the basic ideas you can look elsewhere for approaches and techniques best for your model. By performing these analyses and communicating them in your theses, presentations, and scientific publications, you clearly indicate that you are using ABMs to do science, not—as critics of ABM sometimes insinuate—as video games or "just-so stories."

Chapter 23 learning objectives are to:

- Understand and practice the straightforward method of local SA, using BehaviorSpace to set up and run the simulation experiments.
- Understand more comprehensive forms of SA: use of creative visualizations to see how model results vary with several parameters, and "global" SA.
- Understand and practice standard UA methods, again using BehaviorSpace.
- Understand what RA is and the general steps in conducting it.

Now we introduce the basic kinds of SA. To perform any kind of SA, we need:

- A "full" version of the model with its design "frozen" at least temporarily (section 21.1), documentation and testing completed, and a version number assigned to it;
- A "reference" parameter set, meaning that the model has been parameterized (chapter 20) so each parameter has a standard value;
- One or two key outputs, or "currencies," that we use as indicators of model behavior (section 22.2.2); and
- Controlled simulation conditions—initial conditions, time-series inputs, number of time steps, etc.—that are typical for how the model will be used.

23.2.1 Local Sensitivity Analysis

The objective of local SA is to understand how sensitive the model—as indicated by the currency you selected, designated here as C—is to the value of each individual parameter. Hence, we vary the parameters one at a time and examine small changes in their value. Because local SA is relatively simple and is often used to determine which parameters to use in more sophisticated analyses (and calibration; section 20.4.1), it typically is conducted for all of a model's parameters.

A very basic form of local SA repeats these steps for each parameter:

1) Identify the range to vary the parameter. To focus on local sensitivity, these ranges are usually small. Often a rule of thumb such as +/–5% is used: if the parameter's reference value is P, then the amount to vary it by (dP) is $0.05P$. The lowest parameter value analyzed is therefore $P - dP = 0.95P$ and the highest value is $P + dP = 1.05P$.

   However, it is important to think about the range of P that is even feasible—that does not produce absurd results. For example, a parameter representing the monthly probability of a company remaining in business might have a reference value of 0.989 (producing a mean time to going out of business of about 5 years). Clearly, the range of feasible values of this parameter is much less than +/–5% because small changes in monthly probabilities produce large effects over time. $0.95P$ is 0.940, producing a mean time to going out of business of less than one year; and $1.05P$ exceeds 1.0, which is the maximum value a probability can have. (In this case, you could instead think about what range of probabilities produces +/–5% change in mean time to going out of business.) If the parameter is an integer, dP will be an integer.

2) For each parameter, run the model for its reference value P, and for $P + dP$ and $P - dP$, using sufficient replicates to estimate the mean of the currency C. Determine the resulting mean currency values, C, C^+, and C^-.

3) Calculate sensitivity S as an approximation of the partial derivative of the currency with respect to the parameter: divide the change in results by the change in parameter value to estimate the rate at which the currency changes in response to the parameter. However, the value of the parameter is scaled by its reference value, to keep parameters from appearing especially important just because their reference value is small. Two values of S indicate sensitivity above and below the parameter's reference value: $S^+ = (C^+ - C)/(dP/P)$ and $S^- = (C - C^-)/(dP/P)$.

4) Compile the sensitivities for all parameters and examine them. Three kinds of parameters are of particular interest. Those with high values of S indicate processes that are especially

important in the model. Parameters with high values of S and also high uncertainty in their reference values (because there is little information from which to estimate their values) deserve special attention as calibration parameters (section 20.4.1) and as targets of empirical research to reduce their uncertainty. And parameters with low values of S indicate relatively unimportant processes that could be considered for removal from the model (and could possibly be left out of further SA and UA).

The modeling literature includes many alternatives to this basic approach. Examples include looking only at increases in P (evaluating S^+ but not S^-), using the relative instead of absolute change in results (C^+/C instead of $C^+ - C$) in the numerator of S, examining how the whole distribution of C varies with P, instead of just the mean (e.g., by looking at how the standard deviation of C changes with P), and running the model for more values of P and using linear regression of C vs. P to evaluate sensitivity.

Table 23.1 shows an example local SA, conducted on an ecological population model. The model was developed to explore the conditions under which populations of a certain grasshopper species can persist on the gravel bars of an Alpine river. This grasshopper inhabits the sparse vegetation that appears after gravel bars are destroyed and rebuilt by extreme floods, but not the denser vegetation that gradually takes over a new gravel bar. The key processes of this model are: floods that destroy gravel bars and create new unvegetated bars; plant succession, including the reestablishment of sparse vegetation on new bars and its gradual conversion to dense vegetation; extinction of the grasshopper populations of each bar due to environmental variation other than floods and to small population size ("demographic noise"); and colonization of new gravel bars by grasshoppers. The currency for this analysis is the "intrinsic mean

Table 23.1 Local Sensitivity Analysis of a Grasshopper Metapopulation Model for Alpine River Gravel Bars

Process and parameter	Meaning of parameter	Reference value	Quality of knowledge*	Sensitivity S^+
N	Number of gravel bars	30	5	2.92
Succession				
K_{max}	Maximum habitat capacity (# grasshoppers)	50	5	9.87
K_{min}	Minimum habitat capacity (# grasshoppers)	5	2–3	22.0
Shrub	Loss of habitat capacity per year (# grasshoppers)	1	4	−18.9
Delay	Time until new habitats can be colonized (years)	3	4–5	−1.54
Floods				
Flood	Probability per year of flood event	0.1	3	−3.67
Wash	Probability of being washed away during flood event	0.25	3	−0.85
Local extinction				
Extinct	Probability of extinction due to environmental noise	0.1	1	−2.07
My(K)	Probability of extinction due to demographic noise	1/K	1	
Colonization				
Migrate	Probability that a female disperses to another bar	0.1	3	3.64

Note: Parameter values were increased by dP of 5 to 20%; sensitivity S^+ was calculated as the ratio of relative change in the population's intrinsic mean time to extinction to the relative change in parameter value (after Stelter et al. 1997).

*Estimated quality of the empirical knowledge used to set the parameter value: 5 means highly certain knowledge, 1 means low certainty.

time to extinction," a measure of how likely the population is to avoid extinction over some time period (Grimm and Wissel 2004).

You do not need to understand the model to read this table: two parameters (K_{min} and *Shrub*), and thus their corresponding processes, have very strong influence on the ability of the population to persist. This sensitivity was expected for *Shrub*, because it controls how fast a gravel bar becomes unsuitable for the grasshoppers. But K_{min} represents a process (small patches of habitat that remain even after a gravel bar is overgrown) not even envisioned when the model was first constructed; the SA shows that the ability for a few grasshoppers to survive on a densely vegetated bar is critical. Because the quality of knowledge that K_{min} is based on is low, studying this process seems especially important for improving the model.

From this example you see that it is not the absolute sensitivity numbers for each parameter that are important, but the differences in sensitivities. It makes no sense to say a sensitivity of, say, 3 is high or low; it depends on how large the other sensitivities are.

Local SA as described here is simple, but sometimes also simplistic. The approach assumes that the model output C responds linearly to parameter changes, but this approximation is not always good. Restricting the analyses to small changes in parameter values makes it safer to assume the response is linear, but reduces our ability to distinguish responses to parameter values from the model's stochasticity.

A further limitation is that we vary only one parameter at a time. We can thus not capture *parameter interactions*: how the model's sensitivity to one parameter might change as other parameters change. And our local SA is valid only for the reference parameter set and the conditions simulated during the analysis—its results might not apply when we simulate some other time period, environment, etc., or recalibrate the model. Nevertheless, a local SA is always useful, often worth publishing, and a good starting point for more advanced analyses of model sensitivity and robustness.

23.2.2 Analysis of Parameter Interactions via Contour Plots

It is safe to assume that parameter interactions are potentially important in almost any ABM. How can we understand how a model's sensitivity to one parameter depends on the values of one or several other parameters? Clearly, we need to do something beyond just local SA: we need to run the model with many combinations of values, over the full range of the parameters examined (over the full "parameter space" as defined in section 20.4.5). Then we need a way to see and understand how our output currency C responded to all the parameter combinations.

We actually conducted such an analysis of parameter interactions back in section 12.3, but just for one submodel, not a full ABM. In that analysis, we executed the Business Investor model's submodel for investor utility over wide ranges of its parameters for annual profit and risk, and then used a contour plot to see how the submodel's output—utility—responded. Figure 12.3 shows, for example, that there is an interaction between these parameters. When annual profit is below about 2000, the submodel has almost no sensitivity to annual risk: the value of utility changes almost not at all as annual risk varies all the way from 0.01 to 0.1. But when annual profit is high, especially above 5000, then utility does change substantially as annual risk varies.

Contour plots are thus a useful way to examine interactions of two parameters, when all other parameters are kept constant. Because of its simplicity, this method is often very useful for understanding how two parameters, and the processes they represent, combine to affect our model currency.

Unfortunately, almost all ABMs have more, sometimes many more, than two parameters, and contour plots show only two parameters. The best we can do is select two parameters that seem most important, and then produce several contour plots of how *C* varies with them under several values of one or more other parameters. Figure 12.6 shows, like figure 12.3, how the utility submodel output varies with annual profit and annual risk, but with another submodel parameter—current wealth—at 100,000 instead of zero. Comparing these two figures shows how the parameter interactions between profit and risk also depend on current wealth. Similarly, figure 12.7 shows how these parameter interactions change with a fourth parameter, the time horizon.

Multi-panel contour figures are a way to look at model sensitivity to several parameters at once. Figure 23.1 provides an example, generated from the Wild Dog model described in section 16.4. The two parameters of most interest in this model are the frequency with which groups of dispersing dogs meet each other (at which time they often form a new pack) and

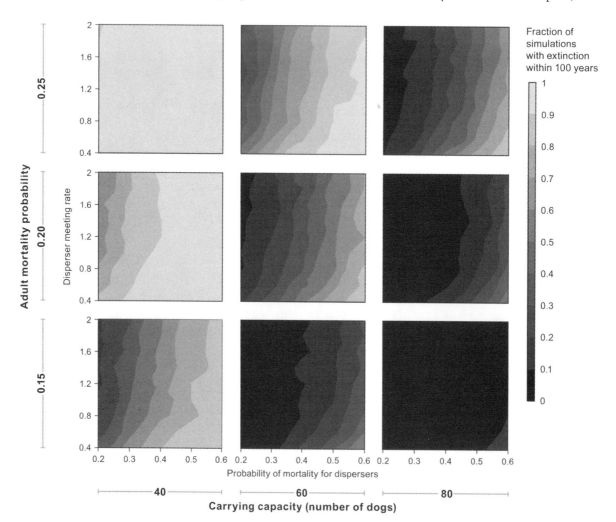

Figure 23.1
Contour plots arranged to visualize sensitivity of the Wild Dog model to four parameters. Each of the nine panels is a contour plot showing how the model currency (fraction of simulations in which the dog population went extinct within 100 years) varies with the annual mortality probability for disperser dogs (x-axis) and the mean number of times disperser groups meet per year (y-axis). Each row of panels includes subplots from model runs with the parameter for habitat carrying capacity set to 40, 60, and 80. Each column includes subplots from runs with the adult mortality probability set to 0.15, 0.2, and 0.25.

the mortality probability for these dispersing dogs. Hence, it makes sense to contour how the model's currency—here, the fraction of 500 model runs in which the population went extinct within 100 years—responds as these two parameters are varied over wide ranges. However, we need to know whether this two-parameter SA changes when we change other parameters such as the mortality probability for adult dogs that are not dispersing and the carrying capacity—the number of dogs that the nature reserve occupied by the dogs can support. We can do this by simply creating separate contour plots for the nine combinations of low, reference, and high values of adult mortality and carrying capacity (figure 23.1).

Multi-panel figures can be relatively complex if their subplots are each complex. They should be designed so that the major conclusions they support can be grasped in less than a minute. (For example, color bands usually let people see key results much more quickly than contour lines do.) If figures can be designed this way, they provide a very concise and clear summary of a model's sensitivity to up to four parameters. If it is not possible, the figure would just say that "it depends," which is not very useful because we would have expected that anyway.

23.2.3 Global Sensitivity Analysis

Seeing the limitations of local SA, contour plots, and multi-panel figures, we might ask: why not explore sensitivity globally? We could vary all parameters over their full ranges, and look at several different currencies to achieve a comprehensive understanding of our model. Unfortunately, this "brute force" kind of model analysis often does not work. First, just running the model for each parameter combination can be infeasible. Only ten values for each of five parameters, for example, is 100,000 (10^5) parameter combinations to execute and analyze. Global SA is even less feasible if multiple model runs or a complex analysis are needed to determine the currency value. Figure 23.1 required 500 model runs for each parameter combination to estimate its currency, so its 729 combinations of four parameters required 364,500 runs—feasible for such a simple model but not for models with many agents, large spaces, or complex behaviors. Second, even if we could run all these simulations and evaluate, say, five different currencies, how could we interpret the results to obtain understanding? We would almost certainly drown in a sea of numbers!

Nevertheless, global SA is widely used in simulation modeling and can be feasible and useful for ABMs. For any but the simplest ABMs, though, global SA is feasible and worthwhile only if you follow formalized methods from the SA literature (which we do not attempt to repeat here). These methods provide two critical benefits: ways of selecting a subset of all the parameter combinations (called *subsampling*) to greatly reduce the number of model runs necessary, and systematic approaches for interpreting the results into meaningful analyses. The analysis approaches often include statistical (e.g., regression) modeling of how the ABM's currency varies with parameter values. However, be aware that some of the traditional sensitivity analysis literature was developed for completely deterministic models and may not apply directly to stochastic ABMs. Some potentially useful examples of methods and applications are Campolongo et al. (2007), Kleijnen and Sargent (2000), Kramer-Schadt et al. (2009), and Saltelli et al. (2008a, b).

23.3 Uncertainty Analysis

The techniques of UA are similar to those of SA, but UA has a different objective: to understand how the uncertainty in parameter values and the model's sensitivity to parameters interact to cause uncertainty in model results. If we know that the value of each parameter is uncertain—because it simplifies into one number a process that in reality is not so simple or constant, or because it represents something that has not been measured precisely—and each parameter has different effects on model results, then how uncertain are the model results?

Traditional UA, like SA, is the subject of a large literature that we do not repeat here. The basic approach is as follows:

1. *Identify the parameters to include in the analyses.* In general it is desirable to include any parameters that are known to be uncertain and especially those that the model is more sensitive to. But the computational and analysis effort increases sharply with the number of parameters analyzed, so compromises may be necessary. It may be feasible only to include the parameters to which the model is most sensitive.

2. *For each parameter included, define a distribution of its values that describes the uncertainty it is believed to have.* "Distribution" here means the kind of distribution discussed in section 15.3.2: the type (continuous vs. discrete), shape (e.g., uniform, normal, lognormal), and parameters (e.g., minimum and maximum; mean and standard deviation) of a stochastic function from which values will be drawn. The distribution can be thought of as (a) a range that the parameter's "true" value is believed to lie in, or, (b) for parameters representing processes that are extremely hard to measure or variable, a range of values that would be obtained if you tried to measure the parameter's value multiple times.

3. *Run the model many times, each time drawing new random values of the parameters from their distributions.* Each such run therefore has a unique combination of values drawn from the parameter space.

4. *Analyze the distribution of model results produced by all the random parameter combinations.*

This kind of UA seems at first much like traditional global sensitivity analysis (section 23.2.3) but there is an important difference: for UA, the parameter values must not only cover the range of a parameter's values but also reproduce their expected probability distribution. Usually this means representing parameters using a distribution, such as the normal distribution, in which values near the standard value are more likely than those farther away. (In fact, parameters are often assumed to follow a "triangular distribution" that limits the range of values but makes values near the mean more likely.)

The Wild Dog model is good for demonstrating UA because it depends on a few probability parameters that are no doubt uncertain and because it executes quickly. Let's follow the above four steps for this model.

1. Identify parameters to include. The original Wild Dog model (section 16.4) uses four probabilities to model mortality of dogs of different social status (pups, yearlings, etc.). It also uses two probabilities to model whether single subordinates and groups of subordinates leave their pack as dispersers. All these probabilities were estimated from field observations. These probabilities no doubt are uncertain because of the limited number of observations and from the strong possibility that mortality and dispersal are not constant but vary with factors that are not in the model. So we will analyze model uncertainty due to uncertainty in these six parameters.

2. Identify distributions for parameter values. Sometimes distributions of parameter values can be estimated more or less rigorously from data, but often they are simply estimates based on scientific judgment. Let's assume that we have reason to believe that the four mortality probability parameters we're examining follow a normal distribution with mean equal to the values given in section 16.4. For example, instead of saying that the probability of a disperser dog dying in a year is 0.44, we say that the "true" value of this parameter is unknown but its distribution has a mean of 0.44 and standard deviation of 0.025. (This means we assume the "true" value is 95% likely to fall in the range of 0.44 ± 0.05—because about 95% of values drawn from a normal distribution fall within two standard deviations above and below the mean—and is more likely to be in the middle of this range.) Similarly, let's assume the probability of a single subordinate dispersing follows a normal distribution with mean of 0.5 and standard deviation of 0.05; and the probability of a group of subordinates dispersing (instead of being 1.0) follows a normal distribution of 0.95 with standard deviation of 0.05. However, of course, these parameter values cannot be greater than 1.0 or less than 0.0.

3. Run the model, drawing random parameter values from distributions. Now we need to alter our NetLogo program to draw values of the six parameters from their random distributions each time it runs. You might think we could simply replace these statements (for example) in the mortality procedure:

```
if status = "disperser"
  [ if random-bernoulli 0.44 [die] ]

if status = "yearling"
  [ if random-bernoulli 0.25 [die] ]
...
```

with these:

```
if status = "disperser"
  [ if random-bernoulli (random-normal 0.44 0.05) [die] ]

if status = "yearling"
  [ if random-bernoulli (random-normal 0.25 0.05) [die] ]
...
```

But that change would be seriously in error: we do not want a new random value of the mortality probability each of the many times it is used during a model run.

(Drawing a new random parameter value every time the parameter is used is rarely if ever a good idea. In the above example, doing so should have no effect on model results. But in other situations, such as when the parameter distribution is nonsymmetric, drawing new parameter values each time could have effects that are difficult to predict and understand. On the other hand, one common way to represent variability in the agents' environment is to draw new values once each tick, or once each simulated season or year, etc., for parameters representing the environment; see section 15.2.1.)

Here, we want one new parameter value drawn randomly at the start of a model run and used throughout the run. So we need to create global variables for each of the six probability parameters, and set their value in `setup` like this:

```
to setup

  ...

  while [disperser-mort <= 0.0 or disperser-mort > 1.0]
    [set disperser-mort random-normal 0.44 0.025]

  while [yearling-mort <= 0.0 or yearling-mort > 1.0]
    [set yearling-mort random-normal 0.25 0.025]
```

```
while [pup-mort <= 0.0 or pup-mort > 1.0]
  [set pup-mort random-normal 0.12 0.025]

while [adult-mort <= 0.0 or adult-mort > 1.0]
  [set adult-mort random-normal 0.2 0.025]

while [single-dispersal-prob <= 0.0 or
       single-dispersal-prob > 1.0]
  [set single-dispersal-prob random-normal 0.5 0.05]

while [group-dispersal-prob <= 0.0]
[
  set group-dispersal-prob random-normal 0.95 0.05
  if group-dispersal-prob > 1.0
    [set group-dispersal-prob 1.0]
]
```

Note that these statements include the kind of defensive programming discussed in the "Continuous Distributions" of section 15.3.2: a `while` loop causes a new random parameter value to be drawn if the first is outside the 0–1 range of possible probability values. The final parameter, `group-dispersal-prob` (the probability of 2 or more subordinates dispersing), is an exception: we believe that 1.0 is a very likely value of this parameter, so if we randomly draw a value of greater than 1.0, we just set it to 1.0 instead of drawing a new value.

Now, there is one more complication with the Wild Dog model: the result we are interested in is not just a single model run, but the frequency, *out of 500 model runs*, that the population goes extinct. That means we need to use the same randomly drawn parameter values for 500 model runs, then draw new parameter values and do 500 more model runs, etc. This seems like a big problem until we remember NetLogo's primitives `with-local-randomness` and `random-seed`. The solution is very simple: we put all the above statements that set the parameter values within a `with-local-randomness` statement, where the random number seed is controlled by BehaviorSpace:

```
with-local-randomness
[
    random-seed param-rand-seed

    while [disperser-mort <= 0.0 or disperser-mort > 1.0]
      [set disperser-mort random-normal 0.44 0.05]

...

  while [group-dispersal-prob <= 0.0]
  [
    set group-dispersal-prob random-normal 0.95 0.05
    if group-dispersal-prob > 1.0
      [set group-dispersal-prob 1.0]
  ]

]
```

Here, the local random number seed `param-rand-seed` is a global variable defined in a slider on the Interface. As long as `param-rand-seed` does not change, the parameters will be given exactly the same values; but each time BehaviorSpace changes `param-rand-seed` we get a new set of parameter values. Now we can set up an experiment that does 500 replicates of each of 1000 random sets of parameter values, by setting BehaviorSpace to vary `param-rand-seed` over 1000 values (e.g., `["param-rand-seed" [1 1 1000]]`) and setting the number of repetitions to 500. If we include the probability parameters such as `disperser-mort` in BehaviorSpace's outputs, we can verify that our code did indeed draw new values for them every 500 runs and that those values followed the distributions we specified.

4. Analyze the distribution of model results. Now we look at how the values of the model result—frequency of extinction in 500 model runs—changed when we added uncertainty to the six parameter values (figure 23.2).

You are no doubt asking yourself what this UA means. When we get such a wide distribution of results from what seem like relatively narrow uncertainty ranges for the probability parameters, does that mean that the model is so uncertain that we cannot learn anything from it? The

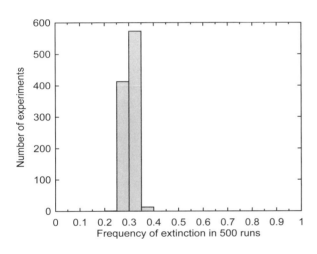

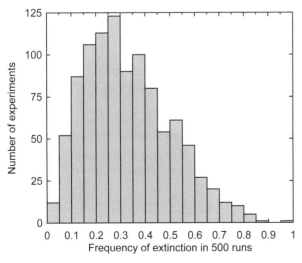

Figure 23.2
Uncertainty analysis results for the Wild Dog model; the histograms show how many simulation experiments out of 1000 (y-axis) produced extinction frequency results within the ranges shown on the x-axis. With the standard parameter values (upper panel), the frequency of extinction was between 0.25 and 0.35 in almost all of the 1000 simulation experiments. With probability parameter values drawn from random distributions (lower panel), results varied widely over the 1000 experiments, with extinction frequencies between 0.05 and 0.6 common.

UA clearly means that we should be cautious with any absolute predictions from the model, such as that the wild dog population has only a 40–50% chance of going extinct within 100 years.

On the other hand, if we want to use the model to compare alternative management scenarios, perhaps the UA means nothing because we're only interested in the simulated *differences* among scenarios, not the absolute results. In fact, we are often not that interested in a model's absolute results (in part because we know they are uncertain) but are more interested in the differences among scenarios. In the Wild Dog case, we were (in section 16.4.3) interested in whether changing the rate at which disperser groups meet each other affects the population's frequency of extinction. If we consider uncertainty in the six mortality and dispersal probability parameters, does the model still clearly show the relationship between this meeting rate and extinction frequency shown in figure 16.2? We can find out by running the UA for five scenarios: `disperser-meeting-rate` equal to 0.2, 0.6, 1.0, 1.4, and 1.8. The BehaviorSpace experiment can be changed to:

```
["param-rand-seed" [1 1 1000]]
["disperser-meeting-rate" 0.2 0.6 1.0 1.4 1.8]
```

Now, how does the model rank the five `disperser-meeting-rate` scenarios under 1000 different sets of probability parameter values? (The model is run with all five values of `disperser-meeting-rate`, for each random combination of the mortality and dispersal parameter values.) It turns out that the relationship shown in figure 16.2 is quite robust to parameter uncertainty. We found the model to predict that extinction frequency is highest at `disperser-meeting-rate` = 0.2 under all 1000 parameter combinations, lowest at `disperser-meeting-rate` = 1.8 under 77% of the parameter combinations, and intermediate at the intermediate `disperser-meeting-rate` value of 1.0 under 94% of combinations. Extinction frequency always decreased as `disperser-meeting-rate` increased under 71% of parameter combinations, even though the differences in results among `disperser-meeting-rate` values of 1.0, 1.4, and 1.8 are quite small.

This result is quite typical (but not universal!) for simulation models: parameter uncertainty can cause high uncertainty in absolute results, but other important results—especially, the model's ranking of various scenarios—can be much less affected by parameter uncertainty. Simulation models with important yet uncertain parameter values can therefore still be very useful when we use them to make *relative* predictions, such as which of several alternative scenarios is better or worse than the others. We make (and test, via UA) the assumption that errors due to parameter uncertainty apply equally to the different scenarios and hence do not affect relative predictions.

23.4 Robustness Analysis

The term *robustness analysis* (RA) is not yet well established in the simulation literature. Here, we use the meaning of Levins (1966), as summarized by Weisberg (2006): "Robustness analysis can show us 'whether a result depends on the essentials of the model or on the details of the simplifying assumptions.'" This is accomplished by studying a number of similar, but distinct models of the same phenomenon.

> [I]f these models, despite their different assumptions, lead to similar results, we have what we can call a robust theorem that is relatively free of the details of the model. Hence, our truth is the intersection of independent lies. (Levins 1966, 423)

RA has the same goal as modeling in general and POM in particular: we seek robust explanations of observed phenomena. If a model's ability to reproduce characteristic patterns of a real system is very sensitive to its details, it likely does not capture the real mechanisms driving the real system.

Thus, once we have "frozen" the design of our ABM, performed software and submodel tests, compared model output to a set of observed patterns, and done SA, it is time to create "distinct models of the same phenomenon" by creating different versions of our ABM. We can use two of the heuristics for analyzing ABMs presented in chapter 22: "Analyze simplified versions of your model" and "Explore unrealistic scenarios"; but we might also look at more complex versions of our model.

Unlike SA and UA, RA lacks a literature of established and formalized methods and instead requires more creativity. We cannot provide a straightforward recipe for RA, only some general ideas and examples, and these general steps:

- Start with a well-tested model version that reproduces, and therefore potentially explains, some patterns observed in real systems.
- Decide which element(s) of the model you want to modify. This could be the way you represent, or initialize, model entities: e.g., whether the environment is homogeneous or heterogeneous in space or time, or whether all agents are different or the same for some state variable. Or you can represent processes in different ways: for example, by using simplified, or more complex, objectives for your agents' adaptive behavior. It is rarely practical to analyze all parts of an ABM, so focus on the parts that seem most likely—from your understanding of the real system and the people who study it—to be criticized as unrealistic, too simple, or too complex.
- Test whether the modified model can still reproduce the observed patterns.

This approach may sound familiar to you. In fact, it is quite similar to how we contrast alternative submodels in theory development (chapter 19). In pattern-oriented theory development, we focus on testing alternative submodels, often to identify good models of agent behavior. In RA, we focus on testing alternative versions of the entire ABM to see whether conclusions from it are robust or an accident of its details.

Because RA cannot be boiled down to a simple recipe, and it resembles analysis methods we have already covered, we explain it further via a case study. This case study is different because—uncharacteristically for this book—it looks at what happens when you make a model more complex instead of simpler.

23.4.1 Example: Robustness Analysis of the Breeding Synchrony Model

You know this model (Jovani and Grimm 2008) if you did exercise 11 of chapter 5. It is about colonies of seabirds, which can comprise many thousands of individual birds. In many species and regions of the world, breeding in such colonies is highly synchronized: birds start laying eggs at more or less the same time, even if they vary greatly in other ways, such as when they arrived at the breeding ground, or how much energy they have stored. Environmental factors (e.g., the phase of the moon) seem not to trigger this synchrony because colonies of the same species in the same region often start breeding at different times.

The model of Jovani and Grimm (2008) seeks an explanation for how breeding synchrony occurs, exploring the possibility that it emerges from simple behaviors of the birds. They assume that it is risky for a bird to start laying eggs in a "stressful neighborhood" where neighbor birds are still aggressively competing for mates and for sites and materials for nests and, hence, are likely to destroy eggs, kill chicks, or steal nest material. In contrast, a quiet neighborhood where most neighbors are already calmly sitting on eggs is a safe place to start breeding your-self. Thus, the central concept of this very simple ABM is its state variable characterizing in-dividual birds: stress level. A nesting bird's stress level reflects both how the bird is stressed by the activity in its neighborhood and how much, in turn, a bird's activity stresses its neighbors. Here is the full ODD model description.

Purpose

The purpose of this model is to explain how local and global breeding synchrony emerges in seabird colonies.

Entities, State Variables, and Scales

The model includes one kind of entity: the female birds that each occupy a stationary nest site. A female is characterized by its own stress level (OSL, in arbitrary units), whether it has started breeding (laying eggs), and the coordinates of its nest site. Each cell (patch) of a 100×100 square grid is occupied by one female and nest. Boundaries are wrapped. One time step cor-responds to one day. Simulations run for 200 days or until all birds start breeding.

Process Overview and Scheduling

Every time step, the stress level of each bird is updated as a function of its OSL and that of its eight neighbors. If a female's OSL falls below a threshold of 10 (1/10 of the minimum initial OSL), she lays eggs and her stress level is set to zero until the end of the simulation. OSL is updated synchronously: the new OSL is based on stress levels of neighbors at the end of the previous time step, so the order in which birds update does not matter.

Design Concepts

A *basic principle* addressed with this model is synchronization in space or time. Synchroniza-tion is an important mechanisms of pattern formation in complex systems and therefore often key to understanding a system's internal organization.

Breeding synchrony at both the local and colony level *emerges* from the *interaction* of neigh-boring birds. Birds *adapt* their stress level to that of their neighbors: if all neighbors are stressed and show stressful behavior, a bird's stress level might increase, leading to a delay of breeding. Birds can *sense* the stress level of their eight neighbors. *Stochasticity* is used only to initialize stress levels, with a wide distribution of stress levels used to test the power of the model's syn-chronizing mechanism. For *observation*, we look at the distribution of laying dates and the spa-tial distribution of laying dates, to detect clusters of birds that start breeding on the same day.

Initialization

The birds are initialized by drawing their value of OSL randomly from a uniform distribution of 100 to 300 stress units.

Input Data

There is no external input to the model.

Submodels

The only submodel is the update of OSL. Each day, each bird's OSL is updated according to its OSL of the previous day, the mean OSL of its eight neighbors at the end of the previous day (*meanNSL*), and a linear stress decay:

$$OSL_t = [(1 - NR)\ OSL_{t-1}] + (NR \times meanNSL_{t-1}) - SD$$

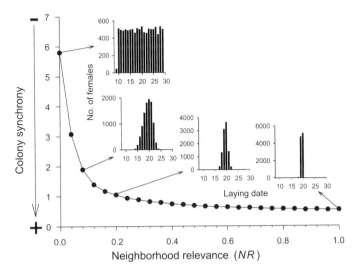

Figure 23.3
Sensitivity of breeding synchrony to *NR*, the relevance given to the mean stress level of the eight neighbor birds. The currency used to quantify synchrony (y-axis) is the standard deviation of the dates on which birds laid their eggs; histograms of these dates are inset for *NR* = 0, 0.08, 0.2, and 1.0 (after Jovani and Grimm 2008).

Neighborhood relevance *NR* (from 0 to 1) is the weight given to the mean stress level of the neighbors, *meanNSL*. If *NR* = 0 there is no interaction between neighbors at all; if *NR* = 1 a bird's *OSL* is completely determined by its neighbors' stress levels. *SD* (= 10 stress units) is the stress decay due to time: the model assumes a linear trend to calm down over time because the importance of laying eggs increases as the season progresses.

A NetLogo program of this model is available through this book's web site. The main results are summarized in figure 23.3, which analyzes the model's sensitivity to *NR*: a relatively low influence of the neighbors' stress levels (*NR* of about 0.2) is sufficient to synchronize breeding to a degree observed in real colonies. Jovani and Grimm (2008) then conducted RA; their model makes many simplifying assumptions, so the RA examines whether these assumptions make breeding synchronization unrealistically likely. Perhaps more realistic assumptions might make synchronization harder to achieve, which would cast doubt on the result that relatively weak local interactions can strongly synchronize even large colonies.

In the Online Appendix of their publication, Jovani and Grimm (2008) list the following model versions which they tested to analyze the robustness of their model. They focused on mechanisms that might destroy synchronization.

■ The stress level term that is not determined by the *OSL* of birds and their neighbors, *SD*, certainly is not constant. Therefore, for a percentage *p* of the birds, the term "–*SD*" was replaced by "–*SD* + *noise*," where *noise* was a random number between *SD* and 2*SD; *p* was varied between 0 and 25%. Hence, the stress levels for some birds temporarily increased instead of decreased, without the influence of their neighbors.

■ Synchrony in the full model is obviously a phenomenon that spreads quickly over the colony: a cluster of breeding birds creates a "front" of neighbors with stress level zero. These fronts then run through the colony, merge, and quickly lead to global synchrony. What if there were physical barriers to this spreading mechanism, such as rocks, ridges, or other places where nesting is impossible? To simulate such barriers, up to 50% of the grid cells were made unavailable for nesting.

■ What if birds "arrive" at the colony (are initialized) at different times, instead of all at once? This possibility was simulated by starting model runs with only some of (or even zero) nest sites occupied. Then, each day 10% of the remaining free nest sites were occupied. Birds were initialized with a stress level between 100 and 200, instead of 100–300; this makes synchronization a bit easier, but was considered more realistic in this scenario.

■ A classic theory of breeding synchrony predicts higher synchrony in larger colonies because birds sense breeding behavior from a larger number of individuals. How does colony size affect synchrony in this model based only on local interactions? To answer this, colonies of 9, 100, and 1024 nest sites were simulated.

The results of these analyses all supported the robustness of the original model. Interestingly, R. Jovani conducted one more RA after the model was published: What if birds do not take into account the average stress level of all neighbor birds, but randomly select only one neighbor and take that neighbor's stress level into account? We leave answering this question to the exercises.

23.5 Summary and Conclusions

Sensitivity, uncertainty, and robustness analysis have the common purposes of improving our understanding of a model, determining how its results arise and what processes in it are (and are not) important, and, ultimately, helping determine how believable its results are. However, these kinds of analyses are quite different in their implementation.

SA and UA are well established in simulation modeling, and a number of such analyses of ABMs have been published. (However, considering that SA and UA are almost mandatory in some fields, it is surprising how many ABMs in other fields have been published without them.) Standard techniques for local and global SA and for UA are well established, although variations and improvements continue to appear. Formalized, standard approaches are especially important for global SA and UA because trying to understand the effects of all parameters *and* all possible interactions among parameters is likely to be very messy or completely intractable without a formalized method.

Even with formalized methods, SA and UA require judgment and thought. They require decisions such as selecting ranges and distributions of values for each parameter and, in interpreting results, determining what sensitivity values indicate parameters of special concern (a decision that could also consider how well supported the parameters' reference values are). Even the question of what is a parameter is not always clear-cut. For example, NetLogo's Segregation model (see section 22.2) has agents that examine their eight surrounding neighbors and move if the fraction of neighbors that share their color is less than a variable *%-similar-wanted*. Clearly, *%-similar-wanted* is a parameter, but is the number of neighbors examined also a parameter that should be analyzed? One solution to this example question is to treat *%-similar-wanted* as a parameter in SA and UA, and then look at how the number of neighbors examined affects results as part of the RA.

Analyzing robustness of a model's key results to its detailed assumptions is less formalized and more creative than SA. There is less literature and far fewer examples of RA than of SA. This means you can think of approaches that are particularly meaningful for your model and the system and problem it addresses. The analyses of (non-ABM) management models by Drechsler et al. (1998, 2003) provide a good example. The models (like many ABMs) were designed to evaluate and compare management alternatives. Hence, the authors used the models to rank sets of alternative management actions by how beneficial the actions were predicted to be, then examined how robust these rankings were among versions of the model. (These examples only varied the models' parameter values, but they could also have varied model assumptions and submodels.)

A final point of this chapter is that doing SA, UA, and RA can require significant time. There is almost no end to how much of this kind of analysis you can do, so a trade-off must be made. In almost all model-based research, a local SA is the minimum necessary for a sufficient understanding of the model. If a model will be widely used, or applied to important decisions or to support particularly important research conclusions, then more comprehensive analysis is justified. Often modelers can get a feel for how robust their model is as they develop and test it. However, it is always risky to ignore the possible need for SA, UA, and RA; the wise modeler makes sure that time and resources for this kind of analysis remain after the model is developed.

23.6 Exercises

1. Conduct a local SA for one of the models you have already programmed and studied in this course. What are the model's parameters, their reference values, and their ranges of feasible values? (Remember that some important parameters might not already be global variables in your NetLogo program.) What values do you use for each parameter in the analysis runs? What currency or currencies do you analyze the sensitivity of? How do you calculate sensitivity? Which parameters is the model most and least sensitive to? If you had to calibrate the model using methods such as those in section 20.4, which appear to be good calibration parameters?

2. Figure 23.1 illustrates how the Wild Dog model responds to four of its parameters. Write an interpretation of this figure: a paragraph or two that summarizes what the figure means for how the model is used and for how wild dogs are managed. Remember that the management goal for wild dogs is to reduce the likelihood that their local populations go extinct. Do you see any clear evidence of parameter interactions?

3. The Breeding Synchrony model described in section 23.4.1 is very easy to program in NetLogo. (One slightly tricky part is making sure birds update their value of *OSL* using the *OSL* value of their neighbors from the preceding tick; use separate variables for the old and new values of *OSL*, and schedule a separate action that sets the old value to the new value after all birds have calculated their new value.) Using either your own code or (if you are extremely busy or tired) an existing implementation, conduct SA and RA on this model. Draw a contour plot of how the primary currency—standard deviation in egg-laying dates—varies with the parameters *NR* and *SD*. Then program some of the "un-simplifications" listed at the end of section 23.4.1 and see how they change this contour plot. What conclusions do you draw from these analyses?

4. At the end of section 23.4.1 we mention one further robustness experiment for the Breeding Synchrony model: examining robustness to the number of neighbors it assumes a bird considers in calculating its new stress level. Conduct this exercise. Is the model's main result unchanged if birds consider only half their neighbors, or only one neighbor? Does it matter if they always consider the same neighbors, neighbors chosen randomly each day, or only the neighbor with the highest stress level?

5. Conduct a local SA on the Virus on a Network model in the Networks section of Net-Logo's Models Library. This model was designed to represent spread of computer viruses, but is structurally similar to many models of disease epidemics. Let's treat it specifically as a model of the spread of an infectious but nonlethal human disease. Treat the "nodes" as people or households that can become infected; increase the number of nodes to 1000

(you will have to edit the slider). Assume that the disease is a virus that conveys resistance, so *gain-resistance-chance* is always 100%. (Hence, the number of resistant nodes at the end of a run equals the number that were ever infected.) Assume that all infected nodes recover, so *recovery-chance* is always 100%. Set *average-node-degree* (the average number of links per node) to a more realistic value of 15. The parameter *virus-check-frequency* can be thought of as a measure of how long a node remains infected before becoming resistant; set it to 10 ticks. Make no other changes to parameter values.

For the SA, first identify one or two meaningful model results or currencies. Then determine the sensitivity indices S^- and S^+ for these currencies using the methods in section 23.2.1. Determine sensitivity to the parameters *average-node-degree*, *initial-outbreak-size*, *virus-spread-chance*, and *virus-check-frequency*. Document all your methods. To which of these parameters are the currencies most sensitive? What do your results say about ways to control disease outbreaks? Do you notice any limitations of the local SA?

6. Conduct an uncertainty analysis of the Virus on a Network model as modified for exercise 5. Use the methods illustrated in section 23.3, for the same parameters analyzed in exercise 5. Identify what seem like reasonable distributions for each parameter. (What would make that easier?) How important is parameter uncertainty if the model is to be used to predict the *relative* benefits of these alternative interventions: treating infected patients to reduce their recovery time, keeping people away from each other (e.g., by keeping children home from school and adults home from work), and encouraging measures such as hand washing that reduce the probability of virus transmission when people do meet? Which of these interventions seems most beneficial for controlling disease outbreaks?

7. How robust are your conclusions from exercises 5 and 6 to the details of the Virus on a Network model? Identify some of the model's assumptions and modify them to make them simpler, or more realistic, or just to try another way to represent some process. (Some examples: You could make processes constant instead of stochastic. Many epidemic models assume that people are not just linked locally, but also have a few long-distance links such as associates they travel to see.) Does the modified model still support the same conclusions about which interventions are most beneficial?

8. For the Wild Dog model uncertainty analysis example in section 23.3, we set values for the uncertain parameters in the `setup` procedure, using a series of `random-normal` statements. Why not, instead, simply use BehaviorSpace to create a sequence of model runs with all possible combinations of many parameter values, like this?

```
["disperser-mort"        [0.39 0.025 0.49]]
["yearling-mort"         [0.20 0.025 0.30]]
["pup-mort"              [0.07 0.025 0.17]]
["adult-mort"            [0.15 0.025 0.25]]
["single-dispersal-prob" [0.40 0.05 0.60]]
["group-dispersal-prob"  [0.85 0.05 1.0]]
```

What advantages and disadvantages would this approach have, compared to the method used in section 23.3?

Where to Go from Here

24.1 Introduction

This book is an introduction to agent-based modeling. From it, you should have learned the basic principles of modeling in general and agent-based modeling in particular. You should also now be familiar with one specific platform for agent-based modeling, NetLogo. If you took your time, worked through the exercises, and discussed your questions with colleagues or instructors, or in a user forum, you are ready to do science using agent-based models.

However, "introduction" means that you have been introduced but you are not yet a professional modeler. After finishing this course, you are on your own and have to decide whether and how you use ABMs in your career. We do not expect everyone who uses this book to produce scientific ABMs in the future. If you do not, you still learned an important tool, know that you can return to it in the future, and have a good idea when it might be wise to collaborate with modelers. And you will be much better equipped to understand and evaluate agent-based research by other scientists. Learning ABMs is like learning statistics: you might not use it all the time, but it is good to learn its principles and know when it should be used, and to sniff out its misuse.

For those of you who are hooked by the agent-based approach and see that you need it to solve the problems you are interested in, we wrote this chapter. (We also wrote our first book, Grimm and Railsback 2005, for you, even if you are not an ecologist; it provides more detailed discussion of the concepts introduced in this book and many more examples.) You will very likely need to deal with the topics of this chapter as you continue to use—or teach—agent-based modeling.

Chapter 24's learning objectives are to:

- Learn strategies for keeping up your momentum until you are ready to start the first model of your own, and for making your first modeling projects successful.
- Briefly look at several general approaches to the most exciting and important question in agent-based research: how to model agent behaviors.
- Become aware of the many software tools that link with NetLogo; it is very likely that some will be useful to you.
- Learn ways to improve the execution speed of NetLogo programs.
- Become aware of software platforms other than NetLogo that may be better for some projects, and some ideas for preparing yourself to use them.

Developing a new model from scratch requires time and dedication, especially if you are a beginner in agent-based modeling. Therefore, the best way to keep the momentum you now have is to take existing ABMs, reimplement them, perform your own simulation experiments, ask new questions, modify the model, and progress through the modeling cycle we described back in chapter 1.

Look for ABMs in your field that you consider particularly interesting and that address questions related to your own interests. Search in the scientific literature rather than in the NetLogo Models Library; the Models Library is a wonderful collection of interesting models, but their main purpose is demonstration and teaching, not necessarily science. Or, if you are intrigued by a model in the Library, try to find related models in the literature to make sure you learn the state of the art regarding the model and the problem it addresses. The model you choose for reimplementation should not be too complex, so reimplementation is not too cumbersome. Keep in mind that NetLogo cannot deal with huge numbers of turtles and patches (see section 24.6, below).

Before you reimplement a model, rewrite its model description using the ODD protocol (unless its original description was in ODD, which is not so unusual now). Often, you will already realize at this stage that the original model description is incomplete or ambiguous. If you are lucky, you can figure out the missing information by trial and error, but you may need to contact the authors. Don't be afraid to; most authors are happy if someone is interested in their models. They may at least provide their model's software (if it was not already published), from which you can try to extract details. However, even the most helpful authors sometimes cannot help you complete the model description, for example because they lost data and files, or cannot remember which model version they used for a specific publication. (Make sure that you never find yourself in such a situation!)

Most interesting ABMs are so rich that even the most thorough analysis by their developers leaves room for you to address new questions or redesign the model for new purposes. To do this, it is most useful to take the original model and first simplify it, if possible.

As you reimplement a couple of models and invest time in testing, understanding, and modifying them, you slowly but surely will develop skill and literacy in agent-based modeling. And you will be surprised how easy it can be to achieve publishable results in little time, depending on your skills, talent, and luck. The only limiting factors are your creativity and your time. (Two examples of very simple models and NetLogo programs being the basis of publications in good journals are Jovani and Grimm 2008, and Herrera and Jovani 2010.) Always remember, of course, to give credit to the authors of the original model.

It usually is a good idea to work with one or more friends or colleagues. In addition to the obvious benefits of group work, it can make modeling more efficient when we are forced to explicitly formulate our assumptions, ideas, and hypotheses and explain them to collaborators or a supervisor.

24.3 Your First Model from Scratch

What kind of problem should you address with your first original model? First of all, keep in mind that ABMs are a tool for a certain (very broad) class of problems (chapter 1). ABMs are a means to an end. So, always ask yourself whether agent-based modeling is really the right approach for any problem you consider. Then, find a problem that is neither too complex nor too simple. Occasionally, scientists have completely new ideas and start everything from scratch, but normally we build on previous work. So make sure you have a good knowledge of the state of the art in your field and regarding the problem you are interested in. Ideally, you will already have reimplemented one or more of the important models in your field.

Use the ODD protocol to formulate your model! And then, whatever you want to model, start simple—much simpler than you think your model ultimately should be! You learned in chapter 22 that understanding is key to success in modeling. By starting simple, you develop understanding gradually and always proceed from firm ground. Naturally, you are impatient to see the "full" model you have in mind; you might consider these oversimplified null models we insist on a waste of time. But, believe us: we have seen too many beginners in modeling who got stuck in their too-complex first model version and wasted months or more before they realized that it was never going to work. Beginners are often so intent on "finishing" a "complete" model that they can't stop themselves from changing and expanding it, so they never "freeze the design," test the software, analyze a simple version, and actually learn something (such as what really needs to be in the model).

Before you start implementing your model, take time to formulate the purpose of your model, write down the model formulation, and discuss it with colleagues and, preferably, experts who have good empirical or theoretical knowledge of the system and question you are addressing. Then, just use this book as a reference as you iterate through the modeling cycle (chapter 1) of design, formulation, implementation, testing, analysis, and reformulation.

24.4 Modeling Agent Behavior

Most of this course, and even this chapter, is about the modeling process and software for ABMs. That focus is necessary to get beginners started, but an experienced modeler may consider much of it just details to get bogged down in. The real excitement in agent-based modeling—what really sets it apart from other approaches and where the basic discoveries are waiting to be made—is learning how to model agent behaviors in ways that produce realistic and useful system behavior. This is what we call Theory, and chapter 19 is about how to do it.

There are many ways of modeling the behaviors of different kinds of agent—people, organizations of people, animals, plants, etc.—and different fields of science have different traditions of modeling behavior both in ABMs and in models focused just on individuals. One of your tasks in moving forward is to learn the traditions in your field, and also to look around for other approaches that might be new and better.

Human behavior and decision-making, especially, are the subject of an immense literature and whole fields of study. In economics, the new fields of behavioral economics and behavioral finance are suddenly looking hard at how people actually make the kinds of decisions that are often represented in ABMs. The "simple heuristics" literature, for example, posits that people often make complex decisions in uncertain contexts via simple rules that usually produce good, but not optimal, results. Other work has identified whole categories of biases that affect our decision-making. There are great opportunities to use agent-based methods to link this growing understanding of behavior to the dynamics of human systems.

Another approach to modeling behavior is the "evolutionary" approaches, in which ABMs are used to "evolve" mathematical traits that cause agents to reproduce observed patterns of individual- and system-level behavior (e.g., Strand et al. 2002). Behavior is often represented via neural networks, which use a number of parameters to link decision outcomes to one or more "inputs" that the agent senses (but see Giske et al. 2003 for an interesting variation). Within the ABM, individuals with successful traits survive and reproduce, passing on their traits. Techniques based on biological evolution (genetic crossover and mutation) eventually can produce traits that are successful at whatever challenges are posed to them in the ABM (there are several books on these techniques, e.g., Mitchell 1998). Whether the evolved trait is a good, general model of behavior in other contexts can then be tested.

While we believe that agent-based modelers need to focus strongly on developing and testing theory for agent behavior, we offer one important caution. Always keep in mind that ABMs are *system* models: agent traits should only be as complex and realistic as necessary to capture the essential effects of behavior on the system. It is extremely easy to get so focused on modeling decisions that you forget that you do not need—or even want—a perfect model of what individuals do. Our experience has been that the most successful ABMs capture the essence, but just the barest essence, of the most important agent behaviors.

24.5 ABM Gadgets

A great deal of work has been done by very clever people to make certain kinds of ABMs, or certain kinds of processes in ABMs, easier to program and use. We have not focused on these specialized tools in this book, but you should be aware of them as you move on because you are very likely to need some of them. Beginning modelers often think that it's easier to build something from scratch than to learn how to use someone else's tool, but that is a naive attitude. We cannot always find (or understand) existing software for everything we want to do, but remember that (a) learning to use an existing tool may take time, but it is almost always quicker than building it yourself; (b) the kinds of tools we point to here tend to be widely used and tested, and hence less likely to contain mistakes; and (c) the farther you wander from your own area of expertise, the more likely it is that somebody else knows how to do it better than you do.

Here are some technologies that are readily available and often useful to NetLogo users. Many of these are NetLogo "extensions," which are just specialized sets of primitives that you can easily install and use; look at the Extensions section of the NetLogo Users Guide. There are more extensions than we mention here, so take a look at the User Manual and the Extensions link on the NetLogo web site. (Ignore the part about writing your own extensions for now.) Many of these extensions are contributed by users like you, and their number is always growing.

- *Geographical Information Systems (GIS).* Many ABMs address real or realistic landscapes, and the standard tool for modeling and analyzing landscapes (including cities) is GIS. NetLogo has a GIS extension that makes it very easy to import data developed in GIS to a model, and to export results from a NetLogo model for analysis in GIS.
- *Existing models of environmental variables and processes.* Many ABMs require fairly complex information on environmental conditions that the agents are exposed to: chemical concentrations, light levels, ocean currents, air temperatures, rainfall, etc. Almost any environment variable you can think of has already been modeled many times, probably with high sophistication, by scientists and engineers in the relevant field. Be extremely hesitant to write your own submodels for such processes; at least seek out and absorb the existing models first. Talk to professors or professionals in the field to find out how the process you are interested in has already been modeled. You might be able to run an existing model and import its results to your ABM, or find an existing model that is simple enough to write into your program—which saves you work and improves your credibility.
- *Mathematical and statistical platforms with links to NetLogo.* There are ways to link at least two very popular and powerful mathematical modeling platforms with NetLogo programs. Many scientists, especially in physical sciences, are familiar with Mathematica

(www.wolfram.com), a powerful platform for mathematical and statistical analysis, modeling, data storage and manipulation, and visualization. Mathematica's NetLogo link (explained in the NetLogo User Manual) lets you execute NetLogo from within Mathematica programs and report the results back to Mathematica for analysis.

Similarly, R (www.r-project.org) is a very popular platform for statistical programming and analysis. A new optional package for R allows users to execute NetLogo programs from within an R program. In the other direction, a NetLogo R extension gives access to any R command from within NetLogo (Thiele and Grimm 2010). This extension makes R's very large library of mathematical and statistical functions work essentially as NetLogo primitives. For example, you can use R's spatial analysis packages to analyze movement of your turtles (how big of an area did each turtle use?); or you can use R to program more efficient simulation experiments, for example by using Latin hypercube subsampling to conduct parameter sensitivity analyses with fewer model runs. Links to these R tools are on this book's web site.

- *Automated calibration and parameterization.* "BehaviorSearch" (www.behaviorsearch .org) is a tool that interacts with NetLogo to find model parameter values that best meet an "objective function," which could be a calibration criterion of the kind we define in section 20.4.4. Instead of using the kind of simulation experiments we describe in chapter 20, BehaviorSearch uses sophisticated algorithms to efficiently search for good parameter combinations. While modelers need to be familiar with the basic parameterization and calibration methods we introduce in chapter 20, BehaviorSearch could be a very important alternative approach, especially for complex models.

24.6 Coping with NetLogo's Limitations

NetLogo can be quite slow if the number of agents or patches is high, or if their behavior is computationally complex. Usually your attitude can be "Who cares how slow NetLogo is to run, because I saved so much time doing the programming? Besides, my computer has nothing better to do all night." But sometimes slow execution limits our simulation experiments and our ability to learn.

If you are not sure whether NetLogo can cope with the size of a model you want to build (or reimplement; section 24.2), write a simple test program that includes about the number of patches and agents of the model and some fake procedures that correspond in complexity to the model.

Often, once you get a feel for how NetLogo operates, you can dramatically improve your code's speed with a few programming changes. One interesting characteristic of NetLogo is that its very powerful primitives that make programs short, clear, and easy to write also sometimes make them dramatically slower. In NetLogo, particularly slow program statements include those that must examine every patch or turtle (e.g., `with` and `min-one-of`, which must look at the value of some reporter for every member of an agentset) and that randomize things (e.g., `ask`, which randomly shuffles the order of the agentset before having the agents do something). The problem is not that NetLogo primitives are not well designed for speed, but that these primitives do things that inherently require a lot of computation: programming the same formulation in another language or platform would not likely improve performance much. Instead, you can sometimes use slightly different formulations that require fewer computations.

The Segregation model that we analyzed in chapter 22 implements movement of households in a way that looks unnecessarily complex and computationally ineffective: unhappy households choose a random direction for movement and move a random number of steps forward, then check whether their new patch is occupied by another household; if so, they move again and again until they finally land on an empty patch (section 22.2.1). Being a NetLogo expert, you might think: Well, why don't we instead just have the turtle move directly to an empty patch with the short statement:

```
move-to one-of patches with [not any? turtles-here]
```

If you try this, the program runs much slower. Why? Because the statement `one-of patches with` requires NetLogo to look at each of the 40,401 patches to see if there are any turtles there, then make a temporary agentset of the empty patches, and then do some randomization to select one of them. It is actually faster to have the turtles just keep moving until they find an empty patch themselves, either the way the Segregation model does or via code that has turtles keep jumping to new patches until they find a vacant one:

```
let new-patch one-of patches
while [any? turtles-on new-patch]
  [ set new-patch one-of patches ]
move-to one-of patches
```

(For us, the `move-to one-of patches with [not any? turtles-here]` statement caused the model to take 15 times longer to finish, while the `let new-patch one-of` method caused the model to finish in 10% less time than the original code—but it changes the model formulation and could produce different results.)

The best way to deal with execution speed in NetLogo is to first write your program "naturally," using primitives as much as you can to keep the code short and clear. Then when the program is tested and ready to use, determine whether it is really too slow. If so, you can then figure out why and do something about it, always checking to make sure the code still produces the same results—or at least still produces useful results, if you make what seems like a slight change in formulation.

How do you figure out what statements and primitives are making NetLogo slow? You cannot just look up how the primitives work, because their exact algorithms and code are not public. Instead, you can:

- Write test programs or conduct experiments in your model program; try different ways to program things and see if they are faster without changing the model formulation undesirably.
- Use NetLogo's profiler extension. A profiler is software that determines, while a program runs, how much time is spent executing each part of the program; profilers are available for almost any programming language. The NetLogo profiler is documented in the User Manual and is easy to use. It reports how much time the computer spent executing each *procedure* (focus on the profiler's "exclusive time" results), which means you may have to break your program into more, smaller, procedures to find out exactly what statements are using up the time. (The profiler can also report how many times a procedure is called, which can be very helpful in testing and understanding more complex models.)
- Consult the email archives of the NetLogo Users Group; questions about execution speed have been asked and answered many times. Post a new question if you do not find a similar one in the archives.
- Keep your eyes on the NetLogo web site for new tools and sources of information.

Finally, check the web site for this book for links to more complete guidance for dealing with NetLogo's limitations.

24.7 Beyond NetLogo

Some of you, as you move ahead in your careers as modelers, will eventually outgrow NetLogo: you will need to build models that just do not fit NetLogo's design. For example, you may need to represent space in ways other than square patches (figure 24.1) or make execution speed the absolutely most important software consideration.

Your choices are to program your model from scratch (rarely a good idea) or to use one of the many other software platforms for agent-based modeling. Lists and reviews of these platforms are in the literature (e.g., Nikolai and Madey 2009; Railsback et al. 2006) and on web sites and forums dedicated to agent-based modeling.

Let us warn you that no other platforms or ways to design and program ABMs are remotely as easy to use, well documented, and polished as NetLogo is. You can expect a serious increase in the effort and software knowledge required to produce working models, which could be a strong incentive to keep your models simple enough to stick with NetLogo. However, if you do think you will outgrow NetLogo, we strongly recommend taking one or two serious courses in object-oriented programming, probably using the Java language. You should look at the most recent reviews of ABM platforms and communicate with their users (e.g., by joining their online user communities) to see what kinds of people are using them and what kinds of success and problems they have. Another alternative to consider seriously is finding a collaborator

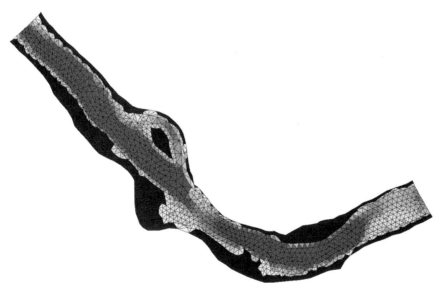

Figure 24.1
"View" of the river trout ABM of Railsback et al. (2009), which uses the Swarm platform instead of NetLogo. River habitat is depicted as irregular 3- and 4-sided cells, here shaded by water velocity. The black dots indicate cells currently occupied by trout. Users can click on cells to open "probes" similar to NetLogo's Agent Monitors. The cell geometry implementation was designed and programmed by S. Jackson using methods published in numerical recipe books (see the Programming Note on numerical recipes in section 15.3.2).

to provide the software expertise—either a paid programmer or an academic computer scientist. (Chapter 8 of Grimm and Railsback 2005 provides advice on working with software professionals.)

24.8 An Odd Farewell

The basic principle of this textbook is facilitating the emergence of agent-based modeling as a standard tool of science. This goal requires considerable adaptation within many fields: professors and instructors—and the students who someday will become professors and instructors—who have the objectives of using and teaching agent-based modeling currently have the disadvantage that there are few resources for learning this approach. Our prediction is that this situation will not remain much longer; we sense a rapidly growing interest in establishing ABMs in the standard toolboxes that young scientists are provided with.

Frequent interaction with other agent-based modelers is critical for keeping up to date in a new field like this and helping move it ahead. There is, of course, stochasticity among institutions and scientific fields in the enthusiasm for agent-based science. People who feel isolated by their interest in ABMs can, luckily, join the mostly on-line collectives of like-minded scientists. Participating in the growing number of organizations and activities such as the Open Agent Based Modeling Consortium (www.openabm.org), the Computational Social Science Society, and SwarmFest (www.swarm.org) is a very valuable way to maintain current observations of the technologies, tools, and (increasingly, we hope) theory for agent-based modeling.

The rest is, of course, just details.

References

Auyang, S. Y. 1998. *Foundations of complex-system theories in economics, evolutionary biology, and statistical physics*. Cambridge University Press, New York.

Ballerini, M., Cabibbo, N., Candelier, R., Cavagna, A., Cisbani, E., Giardina, I., Lecomte, V., Orlandi, A., Parisi, G., Procaccini, A., Viale, M., & Zdravkovic, V. 2008. Interaction ruling animal collective behavior depends on topological rather than metric distance: Evidence from a field study. *Proceedings of the National Academy of Sciences*, 105, 1232–37.

Billari, F. C., Prskawetz, A., Aparicio Diaz, B. & Fent, T. 2007. The "Wedding-Ring": An agent-based marriage model based on social interactions. *Demographic Research*, 17, 59–82.

Botkin, D. B., Janak, J. F. & Wallis, J. R. 1972. Some ecological consequences of a computer model of forest growth. *Journal of Ecology*, 60, 849–72.

Brown, D., Page, S. E., Riolo, R. & Rand, W. 2004. Agent-based and analytical modeling to evaluate the effectiveness of greenbelts. *Environmental Modelling & Software*, 19, 1097–1109.

Burnham, K. P. & Anderson, D. R. 2002. *Model selection and multimodel inference: A practical information-theoretic approach*. Springer, New York.

Camazine, S., Deneubourg, J.-L., Franks, N. R., Sneyd, J., Theraulaz, G. & Bonabeau, E. 2001. *Self-organization in biological systems*. Princeton University Press, Princeton, NJ.

Campolongo, F., Cariboni, J. & Saltelli, A. 2007. An effective screening design for sensitivity analysis of large models. *Environmental Modelling & Software*, 22, 1509–18.

Czárán, T. 1998. *Spatiotemporal models of population and community dynamics*. Chapman and Hall, London.

DeAngelis, D. & Mooij, W. 2003. In praise of mechanistically-rich models. In: *Models in ecosystem science* (ed. C. Canham, J. Cole & W. Lauenroth). Princeton University Press, Princeton, NJ, pp. 63–82.

DeAngelis, D., Rose, K. & Huston, M. 1994. Individual-oriented approaches to modeling ecological populations and communities. In: *Frontiers in mathematical biology* (ed. S. Levin). Springer, New York, pp. 390–410.

Drechsler, M., Burgman, M. A. & Menkhorst, P. W. 1998. Uncertainty in population dynamics and its consequences for the management of the orange-bellied parrot *Neophema chrysogaster*. *Biological Conservation*, 84, 269–81.

Drechsler, M., Frank, K., Hanski, I., O'Hara, B. & Wissel, C. 2003. Ranking metapopulation extinction risk: From patterns in data to conservation management decisions. *Ecological Applications*, 990–98.

du Plessis, M. 1992. Obligate cavity-roosting as a constraint on dispersal of green (red-billed) wood-hoopoes: Consequences for philopatry and the likelihood of inbreeding. *Oecologia*, 90, 205–11.

Duffy, J. 2006. Agent-based models and human subject experiments. In: *Handbook of Computational Economics*, vol. 2: *Agent-Based Computational Economics* (ed. L. Tesfatsion & K. Judd). Elsevier, Amsterdam, pp. 949–1011.

Eisinger, D. & Thulke, H.-H. 2008. Spatial pattern formation facilitates eradication of infectious disease. *Journal of Applied Ecology*, 45, 415–23.

Eisinger, D., Thulke, H.-H., Selhorst, T. & Muller, T. 2005. Emergency vaccination of rabies under limited resources—combating or containing? *BMC Infectious Diseases*, 5, 10.

Epstein, J. & Axtell, R. 1996. *Growing artificial societies: Social science from the bottom up.* MIT Press, Cambridge, MA.

Giske, J., Mangel, M., Jakobsen, P., Huse, G., Wilcox, C. & Strand, E. 2003. Explicit trade-off rules in proximate adaptive agents. *Evolutionary Ecology Research*, 5, 835–65.

Gode, D. K. & Sunder, S. 1993. Allocative efficiency of markets with zero-intelligence traders: Market as a partial substitute for individual rationality. *Journal of Political Economy*, 101, 119–37.

Goss-Custard, J., Burton, N., Clark, N., Ferns, P., McGrorty, S., Reading, C., Rehfsch, M., Stillman, R., Townend, I., West, A. & Worrall, D. 2006. Test of a behavior-based individual-based model: Response of shorebird mortality to habitat loss. *Ecological Applications*, 16, 2215–22.

Grimm, V. 1999. Ten years of individual-based modelling in ecology: What have we learned and what could we learn in the future? *Ecological Modelling*, 115, 129–48.

Grimm, V. & Berger, U. 2003. Seeing the forest for the trees, and vice versa: Pattern-oriented ecological modelling. In: *Handbook of scaling methods in aquatic ecology: Measurement, analysis, simulation* (ed. L. Seuront & P. Strutton). CRC Press, Boca Raton, FL, pp. 411–28.

Grimm, V., Berger, U., Bastiansen, F., Eliassen, S., Ginot, V., Giske, J., Goss-Custard, J., Grand, T., Heinz, S., Huse, G., Huth, A., Jepsen, J., Jørgensen, C., Mooij, W., Müller, B., Pe'er, G., Piou, C., Railsback, S., Robbins, A., Robbins, M., Rossmanith, E., Rüger, N., Strand, E., Souissi, S., Stillman, R., Vabø, R., Visser, U. & DeAngelis, D. 2006. A standard protocol for describing individual-based and agent-based models. *Ecological Modelling*, 198, 115–26.

Grimm, V., Berger, U., DeAngelis, D., Polhill, J., Giske, J. & Railsback, S. 2010. The ODD protocol: A review and first update. *Ecological Modelling*, 221, 2760–68.

Grimm, V., Dorndorf, N., Frey-Roos, F., Wissel, C., Wyszomirski, T. & Arnold, W. 2003. Modelling the role of social behavior in the persistence of the alpine marmot *Marmota marmota*. *Oikos*, 102, 124–36.

Grimm, V., Frank, K., Jeltsch, F., Brandl, R., Uchmanski, J. & Wissel, C. 1996. Pattern-oriented modelling in population ecology. *Science of the Total Environment*, 183, 151–66.

Grimm, V. & Railsback, S. 2005. *Individual-based modeling and ecology.* Princeton University Press, Princeton, NJ.

Grimm, V., Revilla, E., Berger, U., Jeltsch, F., Mooij, W., Railsback, S., Thulke, H.-H., Weiner, J., Wiegand, T. & DeAngelis, D. 2005. Pattern-oriented modeling of agent-based complex systems: Lessons from ecology. *Science*, 310, 987–91.

Grimm, V. & Wissel, C. 1997. Babel, or the ecological stability discussions: An inventory and analysis of terminology and a guide for avoiding confusion. *Oecologia*, 109, 323–34.

Grimm, V. & Wissel, C. 2004. The intrinsic mean time to extinction: A unifying approach to analyzing persistence and viability of populations. *Oikos*, 105, 501–11.

Groves, T. 1973. Incentives in teams. *Econometrica*, 41, 617–31.

Gusset, M., Jakoby, O., Müller, M., Somers, M., Slotow, R. & Grimm, V. 2009. Dogs on the catwalk: Modelling re-introduction and translocation of endangered wild dogs in South Africa. *Biological Conservation*, 142, 2774–81.

Haefner, J. W. 2005. *Modeling biological systems: Principals and applications.* 2nd ed. Springer Publishers, New York.

Heine, B.-O., Meyer, M. & Strangfeld, O. 2005. Stylised facts and the contribution of simulation to the economic analysis of budgeting. *Journal of Artificial Societies and Social Simulation*, 8, 4, http://jasss.soc.surrey.ac.uk/8/4/4.html.

Heine, B.-O., Meyer, M. & Strangfeld, O. 2007. Das Konzept der stilisierten Fakten zur Messung und Bewertung wissenschaftlichen Fortschritts. *Die Betriebswirtschaft*, 67, 583–601.

Herrera, C. & Jovani, R. 2010. Lognormal distribution of individual lifetime fecundity: Insights from a 23-year study. *Ecology*, 91, 422–30.

Hilborn, R. & Mangel, M. 1997. *The ecological detective: Confronting models with data.* Princeton University Press, Princeton, NJ.

Holland, J. H. 1995. *Hidden order: How adaptation builds complexity.* Perseus Books, Reading, MA.

Houston, A. & McNamara, J. 1999. *Models of adaptive behaviour: An approach based on state.* Cambridge University Press, Cambridge.

Huse, G., Railsback, S. & Fernø, A. 2002. Modelling changes in migration pattern of herring: Collective behaviour and numerical domination. *Journal of Fish Biology*, 60, 571–82.

Huth, A., Drechsler, M. & Kohler, P. 2004. Multicriteria evaluation of simulated logging scenarios in a tropical rain forest. *Journal of Environmental Management*, 71, 321–33.

Huth, A. & Wissel, C. 1992. The simulation of the movement of fish schools. *Journal of Theoretical Biology*, 156, 365–85.

Janssen, M. A., Radtke, N. P. & Lee, A. 2009. Pattern-oriented modeling of commons dilemma experiments. *Adaptive Behavior*, 17, 508–23.

Jeltsch, F., Milton, S., Dean, W. & van Rooyen, N. 1996. Tree spacing and coexistence in semiarid savannas. *Journal of Ecology*, 84, 583–95.

Jeltsch, F., Müller, M. S., Grimm, V., Wissel, C. & Brandl, R. 1997. Pattern formation triggered by rare events: Lessons from the spread of rabies. *Proceedings of the Royal Society of London B*, 264, 495–503.

Jovani, R. & Grimm, V. 2008. Breeding synchrony in colonial birds: From local stress to global harmony. *Proceedings of the Royal Society of London B*, 275, 1557–63.

Kaldor, N. 1961. Capital accumulation and economic growth. In: *The Theory of Capital* (ed. F. A. Lutz, & D. C. Hague). MacMillan, London, pp. 177–222.

Kleijnen, J.P.C. & Sargent, R. G. 2000. A methodology for fitting and validating metamodels in simulation. *European Journal of Operational Research*, 120, 14–29.

Kornhauser, D., Wilensky, U. & Rand, W. 2009. Design guidelines for agent based model visualization. *Journal of Artificial Societies and Social Simulation*, 12, 1, http://jasss.soc.surrey.ac.uk/12/2/1.html.

Kramer-Schadt, S., Fernandez, N., Grimm, V. & Thulke, H.-H. 2009. Individual variation in infectiousness explains long-term disease persistence in wildlife populations. *Oikos*, 118, 199–208.

Lawson, T. 1989. Abstraction, tendencies and stylised facts: A realist approach to economic analysis. *Cambridge Journal of Economics*, 13, 59–78.

LeBaron, B. 2001. Empirical regularities from interacting long-and short-memory investors in an agent-based stock market. *IEEE Transactions on Evolutionary Computation*, 442–55.

Levins, R. 1966. The strategy of model building in population biology. *American Scientist*, 54, 421–31.

Lotka, A. J. 1925. *Elements of physical biology.* Williams and Wilkins, Baltimore.

Mangel, M. & Clark, C. 1986. Toward a unifed foraging theory. *Ecology*, 67, 1127–38.

McQuinn, I. 1997. Metapopulations and the Atlantic herring. *Reviews in Fish Biology and Fisheries*, 7, 297–329.

Meredith, D. D., Wong, K. W., Woodhead, R. W. & Wortman, R. H. 1973. *Design and planning of engineering systems.* Prentice-Hall, Englewood Cliffs, NJ.

Mitchell, M. 1998. *An introduction to genetic algorithms.* MIT Press, Cambridge, MA.

Neuert, C., du Plessis, M., Grimm, V. & Wissel, C. 1995. Welche ökologischen Faktoren bestimmen die Gruppengröße bei *Phoeniculus purpureus* (Gemeiner Baumhopf) in Südafrika? Ein individuenbasiertes Modell. *Verhandlungen der Gesellschaft für Ökologie*, 24, 145–49.

Neuert, C., Rademacher, C., Grundmann, V., Wissel, C. & Grimm, V. 2001. Struktur und Dynamik von Buchenurwäldern: Ergebnisse des regelbasierten Modells BEFORE. *Naturschutz und Landschaftsplanung*, 33, 173–83.

Nikolai, C. & Madey, G. 2009. Tools of the trade: A survey of various agent based modeling platforms. *Journal of Artificial Societies and Social Simulation*, 12, 2, http://jasss.soc.surrey.ac.uk/12/2/2.html.

Parker, D., Manson, S., Janssen, M., Hofmann, M. & Deadman, P. 2003. Multi-agent systems for the simulation of land-use and land-cover change: A review. *Annals of the Association of American Geographers*, 93, 314–37.

Partridge, B. L. 1982. The structure and function of fish schools. *Scientifc American*, 246, 90–99.

Peck, S. L. 2004. Simulation as experiment: A philosophical reassessment for biological modeling. *Trends in Ecology and Evolution*, 19, 530–34.

Pe'er, G. 2003. Spatial and behavioral determinants of butterfly movement patterns in topographically complex landscapes. Ph.D. thesis, Ben-Gurion University of the Negev.

Pe'er, G., Heinz, S. & Frank, K. 2006. Connectivity in heterogeneous landscapes: Analyzing the effect of topography. *Landscape Ecology*, 21, 47–61.

Pe'er, G., Saltz, D. & Frank, K. 2005. Virtual corridors for conservation management. *Conservation Biology*, 19, 1997–2003.

Pe'er, G., Saltz, D., Thulke, H.-H. & Motro, U. 2004. Response to topography in a hilltopping butterfly and implications for modelling nonrandom dispersal. *Animal Behaviour*, 68, 825–39.

Peirce, S., Van Gieson, E. & Skalak, T. 2004. Multicellular simulation predicts microvascular patterning and in silico tissue assembly. *FASEB Journal*, 18, 731–33.

Pitt, W., Box, P. & Knowlton, F. 2003. An individual-based model of canid populations: Modelling territoriality and social structure. *Ecological Modelling*, 166, 109–21.

Platt, J. R. 1964. Strong inference. *Science*, 146, 347–52.

Polhill, J., Brown, D. & Grimm, V. 2008. Using the ODD protocol for describing three agent-based social simulation models of land use change. *Journal of Artificial Societies and Social Simulation*, 11, 3, http://jasss.soc.surrey.ac.uk/11/2/3.html.

Rademacher, C., Neuert, C., Grundmann, V., Wissel, C. & Grimm, V. 2001. Was charakterisiert Buchenurwälder? Untersuchungen der Altersstruktur des Kronendachs und der räumlichen Verteilung der Baumriesen mit Hilfe des Simulationsmodells BEFORE. *Forstwissenschaftliches Centralblatt*, 120, 288–302.

Rademacher, C., Neuert, C., Grundmann, V., Wissel, C. & Grimm, V. 2004. Reconstructing spatiotemporal dynamics of Central European natural beech forests: The rule-based forest model BEFORE. *Forest Ecology and Management*, 194, 349–68.

Rademacher, C. & Winter, S. 2003. Totholz im Buchen-Urwald: Generische Vorhersagen des Simulationsmodells BEFORE-CWD zur Menge, räumlichen Verteilung und Verfügbarkeit. *Forstwissenschaftliches Centralblatt*, 122, 337–57.

Railsback, S. & Harvey, B. 2002. Analysis of habitat-selection rules using an individual-based model. *Ecology*, 83, 1817–30.

Railsback, S., Harvey, B., Hayes, J. & LaGory, K. 2005. Tests of a theory for diel variation in salmonid feeding activity and habitat use. *Ecology*, 86, 947–59.

Railsback, S., Lamberson, R., Harvey, B. & Duffy, W. 1999. Movement rules for individual-based models of stream fish. *Ecological Modelling*, 123, 73–89.

Railsback, S. F., Harvey, B. C., Jackson, S. K. & Lamberson, R. H. 2009. inSTREAM: The individual-based stream trout research and environmental assessment model. Tech. Rep. PSW-GTR-218, USDA Forest Service, Pacifc Southwest Research Station.

Railsback, S. F., Lytinen, S. L. & Jackson, S. K. 2006. Agent-based simulation platforms: Review and development recommendations. *Simulation*, 82, 609–23.

Rammig, A. & Fahse, L. 2009. Simulating forest succession after blowdown events: The crucial role of space for a realistic management. *Ecological Modelling*, 220, 3555–64.

Reynolds, C. 1987. Flocks, herds, and schools: A distributed behavioral model. *Computer Graphics*, 21, 25–36.

Saltelli, A., Chan, K. & Scott, E. M., eds. 2008a. *Sensitivity analysis*. Wiley-Interscience, New York.

Saltelli, A., Ratto, M., Andres, T., Campolongo, F., Cariboni, J., Gatelli, D., Saisana, M. & Tarantola, S., eds. 2008b. *Global sensitivity analysis: The primer*. Wiley-Interscience, New York.

Schelling, T. C. 1971. Dynamic models of segregation. *Journal of Mathematical Sociology*, 1, 143–86.

Schelling, T. C. 2006. Some fun, thirty-five years ago. In: *Handbook of Computational Economics*, vol. 2: *Agent-Based Computational Economics* (ed. L. Tesfatsion & K. Judd). Elsevier, Amsterdam, pp. 1639–44.

Schmolke, A., Thorbek, P., DeAngelis, D. L. & Grimm, V. 2010. Ecological models supporting environmental decision making: A strategy for the future. *Trends in Ecology and Evolution*, 25, 479–86.

Starfield, A. M., Smith, K. A. & Bleloch, A. L. 1990. *How to model it: Problem solving for the computer age*. McGraw-Hill, New York.

Staufer, H. B. 2008. *Contemporary Bayesian and frequentist statistical research methods for natural resource scientists*. John Wiley & Sons, Hoboken, NJ.

Stelter, C., Reich, M., Grimm, V. & Wissel, C. 1997. Modelling persistence in dynamic landscapes: Lessons from a metapopulation of the grasshopper *Bryodema tuberculata*. *Journal of Animal Ecology*, 66, 508–18.

Strand, E., Huse, G. & Giske, J. 2002. Artificial evolution of life history and behavior. *American Naturalist*, 159, 624–44.

Thiele, J. & Grimm, V. 2010. Netlogo meets R: Linking agent-based models with a toolbox for their analysis. *Environmental Modelling & Software*, 25, 972–74.

Thiery, J., D'Herbes, J. & Valentin, C. 1995. A model simulating the genesis of banded vegetation patterns in Niger. *Journal of Ecology*, 83, 497–507.

Thulke, H.-H. & Eisinger, D. 2008. The strength of 70%: After revision of a standard threshold of rabies control. In: *Towards the elimination of rabies in Eurasia* (ed. B. Dodet, A. Fooks, T. Muller, & N. Tordo), vol. 131 of Developments in Biologicals. S. Karger A.G., Basel, pp. 291–98.

Turner, M. 1993. Book review of: Remmert, H. (ed.) 1991. The mosaic-cycle concept of ecosystems. Springer, Berlin. *Journal of Vegetation Science*, 4, 575–76.

Valone, T. 2006. Are animals capable of Bayesian updating? An empirical review. *Oikos*, 112, 252–59.

Watson, J. 1968. *The double helix: A personal account of the discovery of the structure of DNA*. Atheneum, New York.

Weimerskirch, H., Pinaud, D., Pawlowski, F. & Bost, C.-A. 2007. Does prey capture induce area-restricted search? A fine-scale study using GPS in a marine predator, the wandering albatross. *American Naturalist*, 170, 734–43.

Weisberg, M. 2006. Robustness analysis. *Philosophy of Science*, 73, 730–42.

Wiegand, T., Jeltsch, F., Hanski, I. & Grimm, V. 2003. Using pattern-oriented modeling for revealing hidden information: A key for reconciling ecological theory and conservation practice. *Oikos*, 100, 209–22.

Wiegand, T., Revilla, E. & Knauer, F. 2004. Dealing with uncertainty in spatially explicit population models. *Biodiversity and Conservation*, 13, 53–78.

Wissel, C. 1992. Modelling the mosaic-cycle of a Middle European beech forest. *Ecological Modelling*, 63, 29–43.

Index

calibration and parameterization: automated, 313; categorical vs. best-fit, 259; criteria for, 261–62; definition and purposes of, 255–56; differences for ABMs, 256–57; documentation of, 268; literature on, 258; measures of model fit for, 260–61; overfitting in, 258; parameters for, 258–59, 294; statistics for, 287; strategies for, 258–64; with stochastic results, 263; of submodels, 257–58; and theory testing, 253; time-series, 259–61

`carefully`, 121–22

checklist: of design concepts, 40–41; of NetLogo elements, 30–32

`clear-all`, 22

Code Examples in NetLogo Models Library, 16–17, 66, 121, 202

collectives (model design concept), 41, 99; definition of, 209–10; as emergent properties vs. explicit entities, 210; represented via breeds, 211–12; represented via patches or links, 210

comma-separated values (.csv) files, 69; and European computers, 69, 122

Command Center, 27

comments in code, 55–57; for temporarily deactivating code, 56, 64; in version control, 62

communication of models and science, 7, 35, 254, 292

competition, 169, 170, 181, 212, 236–38

conceptual framework for agent-based modeling, 40, 97–98

contexts: of adaptive traits, 143, 243–44, 247, 312; in NetLogo programming, 17–18, 51, 55–56, 77, 80, 129, 131, 211

continuous time simulation. *See* discrete event simulation

contour plots: in calibration, 265; using Excel, 160–61; of parameter interactions in sensitivity analysis, 295–97; in submodel analysis, 160–65; using R, 161–62

Cooperation model, 170

copying of code: benefits and ethics of, 17, 121; as cause of errors, 76–77

`count`, 27, 33, 66

`create-link(s)-to`, 139, 177

`create-turtles` (crt), 23, 51–52; context started by, 18; and initialization of turtle variables, 65, 110, 139; order of creation vs. initialization in, 139

csv files. *See* comma-separated values files

currencies of model results, 284–85; in sensitivity analysis, 293–97

debugging of code, 75. *See also* software

decision-making traits. *See* adaptation and adaptive behavior

deduced variables, 38, 239

defensive programming, 199, 201, 206, 300

design concepts: as a conceptual framework, 97–98; in ODD protocol, 40–41

detective work in modeling, 76, 93, 240, 287, 288

dictionary of NetLogo primitives, 17, 22; F1 key to access, 22

discrete event simulation, 183, 188–91

`display`, 117

display. *See* Interface tab, View

`distance`, 66, 112; distance to nearest other agent, 112, 147

`distancexy`, 50–51

documentation: of analyses, 167, 231, 274; of software tests, 89–90. *See also* ODD protocol for model description

emergence (model design concept), 10, 41, 99, 101–2, 113; criteria for, 101; in Flocking model, 108–9

entities: collectives as, 209; in ODD protocol, 37–39; selection of, 233–34

environment: emergence from, 101–2, 113, 169; modeling of, 10, 17, 38, 40, 58, 188, 285, 299, 312

epidemic models, 306–7; for rabies 3–4

`error-message`, 121–22

evolutionary modeling of behavior, 311–12

Excel spreadsheet software, 122, 160–63, 265

execution speed, 313–14. *See also* speed controller for model description

`export-` primitives, 68–69, 124

extensions to NetLogo, 202, 312–13; GIS, 71, 312

extinction: in the Grasshopper Metapopulation model, 294–95; risk in Wild Dog model, 213, 221, 296, 300–302; in the Simple Birth Rates model, 102–8

`false`. *See* boolean reporters and variables

`file-` primitives, 70, 85–86, 121–23, 185

files: input, 70; output, 85–86, 120–23; problems with due to multiple processors and Behavior-Space, 107; reasons for output, 67–68, 120; for software testing, 84–85, 86–87. *See also* BehaviorSpace; `export-` primitives

`filter`, 180

fitness, 40, 43, 143, 212, 249. *See also* objectives

Flocking model: analysis using BehaviorSpace, 110–13; calibration exercise with, 269; collectives in, 210; emergence in, 108–9, 113; related scientific models, 109, 246–47

model: definition of, 4–7
model fit measures, 260–61
Model Settings dialog, 18, 49–50, 70, 78, 81
modeling cycle, 7–9, 75, 115, 245, 273–74, 310–11; TRACE format for documenting, 89, 231
modeling practices, 11, 58, 253, 264, 274, 288. *See also* programming practices
Models Library of NetLogo: 16–17, 121, 188, 202, 310; limitations of, 61
modular code, 136
monitors. *See* Agent Monitors
Mousetrap model, 188–92
move-to, 17, 26, 54, 89, 314
multi-agent systems, xii
multi-criteria assessment of models, 228
Mushroom Hunt model: as calibration exercise, 270; as example modeling problem, 5–7; as introductory programming example, 18–32; as modeling cycle example, 7–9; as searching behavior example, 150–51
myself, 112, 131–32, 137, 146–47, 171

n-of, 21–22
nearest other agent, 112, 147
neighbors, 17, 64, 78
NetLogo: educators' forum, xiv; extensions, 202, 312–13; influence on model and research design, 30; installation, 16; introduction to, 16–18; limitations of and alternatives to, 313–15; mini-reference, 30–32; ODD protocol correspondence with, 47–48; personality and style of, 30; tab names, 15; terminology, 17; Users Group, 33; versions, xv, 15; why to use, xii
networks: in Models Library, 307; represented by links, 139–40, 210; represented by space in the Marriage model, 90–91
neural network models of behavior, 311–12
nonspatial models, 175, 213, 286
null submodels and theory, 246, 248, 249, 311

objectives (model design concept), 41, 99, 143; in the Business Investor model, 133, 148–53; in the Woodhoopoe model, 252
objectives of this book, xii
observation (model design concept), 41, 99; of the Butterfly model, 62–66; of NetLogo models, 116–25; to facilitate pattern-oriented modeling, 234, 252–53
observer and observer context in NetLogo, 17–18, 39, 128–29
ODD protocol for model description, 35–42; correspondence with NetLogo, 47–48; and design of models, 36, 98

—of the BEFORE beech forest model, 236–38
—of the Breeding Synchrony model, 304–5
—of the Business Investor model, 132–35
—of the Butterfly model, 42–44
—of the Marriage model, 90–91
—of the Segregation model, 278–79
—of the Telemarketer model, 171–74
—of the Wild Dog model, 213–15
—of the Woodhoopoe model, 250–53
of, 18, 131–32, 171
one-of, 54, 131, 314; to convert agentset to an individual agent, 178
or, 145–46
other, 111–12, 145–47, 170
output. *See* files, Interface tab in NetLogo; observation
overfitting in calibration, 258, 264, 267, 268

panic, when not to, 9, 18, 265. *See also* worry
parameters: documentation of, 42, 268; estimating values of, 255–63, 313; as global variables, 17, 23, 64, 65, 129; interaction among, 108–9, 295–97; model sensitivity to, 293–97; model uncertainty due to, 297–302; parameter space, 262–63, 267, 298; reference (standard) values of, 293–94, 295
patch, 70
patch-here, 65, 83, 146–47
patch-set, 135–36
patches-own, 48, 49, 129
patterns: that characterize a system and problem, 228–30; criteria for matching, 230, 234, 261–62; in software testing, 81, 89
pattern-oriented modeling, 227–31, 233–35, 243–46, 255–64
pen-down and pen-mode, 28, 54
plabel, 81, 117
plot, 68, 119–20
plots, 67–69, 119–20; export of, 68, 124. *See also* contour plots
Poisson distribution, 201–2, 207, 214–15
positive feedbacks, 186
precision, 81, 116
prediction (model design concept), 41, 99, 157–59, 165–66
predictions from models: absolute vs. relative, 302; independent or secondary, 238–39
primitives, 17; errors due to misunderstanding of, 77–78; and execution speed, 313–14
probability: Bayesian updating of, 165–66; logistic model of, 215–16; observed frequencies as, 196–97, 203–5; parameters in sensitivity analysis, 293; in software testing, 84–86; of success

as decision objective, 143, 153; theory and random distributions, 198–202

problem addressed by a model. *See* question addressed by a model

processes in a model: description of, 39–40; design of, 8–9, 39, 227–28, 233–35, 243–46, 257, 312

profiler extension of NetLogo, 314

programming practices, 47, 55–56, 58–59, 62, 65, 66, 75–76, 79–81, 84, 89–90, 93, 119–21, 199–200

pseudo-code, 39–40

pseudo-random numbers, 198

question (problem, purpose) addressed by a model, 4–7, 8, 10, 37, 227–28, 310

R statistical software, 86, 161–62, 311

rabies control models, 3–4

`random`, 25, 201

random number distributions, 198–202

random number generation and seeds, 198, 202–3; and BehaviorSpace, 300–302

`random-float`, 53–54, 199

`random-normal`, 76–77, 198–200

`random-poisson`, 79, 201

`random-seed`, 202–3, 300–301

regimes of control, 280–81

reimplementation: of existing models, 310; of submodels to test software, 86–89

`repeat`, 83, 190

replication and repetitions, 103–5, 112, 195, 197, 221, 287; in calibration, 263; and `random-seed`, 203

reporters, 17, 66, 136–37; in BehaviorSpace and Interface elements, 119, 123–24; as boolean logical conditions, 180–81

`reset-ticks`, 27, 185–86

`resize-world`, 70, 78

`right`, 25

robustness analysis, 292, 302–6

run-time errors, 78–79, 107, 110, 120, 122, 211. *See also* defensive programming

satisficing decision traits, 149–52

scale selection, 9, 38–39, 170, 257. *See also* scheduling

`scale-color`, 50–51, 81, 116–18

scenarios in simulation experiments, 103–5; contrast of, 112, 287, 302; usefulness of unrealistic, 286–87

scheduling, 24, 39–40, 99, 183–91; discrete event, 188–91; to represent hierarchies, 186–87; using

time steps, 185–86; of View and observer actions, 117, 118

science and agent-based modeling, 9, 30, 36, 61, 89, 97, 109, 113, 228, 244–46, 274, 311–12

scientific method. *See* strong inference and scientific method

scope of variables, 128–31

Segregation model, 274, 278–83, 313–14

sensing (model design concept), 41, 43, 99, 127, 132; of an agentset, 135; in networks of links, 139–40

sensitivity analysis, 259, 292, 293–97; global, 297; local, 293–95; parameter interactions in, 295–97

sensitivity experiments, 104, 112, 220–21

`set`, 21, 49

`set-current-plot`, 120

Settings, Model. *See* Model Settings dialog

`setup` procedure, 18–23; interaction of BehaviorSpace with, 106–7; and initialization element of ODD, 48; and variables on Interface, 64. *See also* initialization

`show`, 27, 56, 66

Simple Birth Rates model, 102–8

simulation experiments, 103–8, 109–12, 137–39, 148–49, 220–21; replication of, 197, 203. *See also* calibration and parameterization; sensitivity analysis; theory; uncertainty analysis

skeletons of procedures, 21, 50, 53, 58, 80

sliders, 63–64, 129; and BehaviorSpace, 105, 106

software: alternatives to NetLogo, 315; common errors in, 76–79; testing and verification of, 79–90. *See also* programming practices

`sort-on`, 187

spaces and landscapes: benefits of simplifying, 58; in the Business Investor model, 133–34; in the Butterfly model, 43, 49–51, 69–71; from imported data, 70–71; non-geographic, 10, 72, 90–91, 133; scales of, 38–39; wrapping, 22

spatial extent, 38–40, 49

speed controller on NetLogo Interface, 26, 81, 117, 186

spreadsheets: for analyzing models, 68–69; 86, 137, 265; for reimplementing submodels, 86–89, 90; to test and explore submodels, 158–65, 166–67, 216

stability properties, 284

standardization of agent-based modeling, xii, 36, 40

state variables, 9, 37–8. *See also* variables

statistical analysis: for software testing, 84–86, 93; of model results, 197, 287, 297

stochasticity (model design concept): 41, 99, 103, 195–97; analysis of via replication, 197; in

stochasticity (model design concept) (*continued*)
models of behavior, 203–5, 214–16; and
sensitivity analysis, 295, 297; in uncertainty
analysis, 298; uses of, 196–7
`stop`, 53, 190
Stopping rules and stopping models, 53, 185–86,
191–92; in BehaviorSpace, 104, 105, 106–7
strong inference and scientific method, 244,
253–54
structure and structural realism of models, 8,
37–39, 227–28, 233–35, 238–39, 256
stylized facts, 228, 239, 240
submodels: correspondence to NetLogo procedures, 48; definition of, 39; description of
in ODD protocol, 40–42; implementation
and analysis of, 159–62, 286; independent
reimplementation of, 86–89; parameterization
of, 257–58
syntax errors and checker, 20, 58, 76–77, 79–80

tabs (NetLogo), 15
Telemarketer model: analysis of, 174–75; addition
of mergers to, 176–78; collectives in, 210; with
customers remembering callers, 179–81; ODD
description of, 171–74; potential bias in due to
execution order, 186–87, 188
temporal extent, 38, 53, 191–92
testing software. *See* software
theory in agent-based models: definition of, 143,
243; development and testing of, 243–46,
311–12; examples of, 246–49
theory potentially useful in adaptive traits: Bayesian
updating of probabilities, 165–66; evolutionary modeling, 311–12; game theory, 239–40;
objective optimization in decision analysis
and behavioral ecology, 148; probability, 198;
satisficing, 149–52; simple heuristics, 311
`tick` and `ticks`, 27, 53, 185; and BehaviorSpace,
106–7
`tick-advance`, 188–90
time in models. *See* scheduling
time steps, 38, 53; consequences of using, 185–86.
See also scheduling; `tick`
`to-report`, 136–37, 180
topology of NetLogo's world, 23, 78. *See also* Model
Settings dialog
TRACE format for describing a modeling cycle, 12,
89, 231
trade-off decisions, 143–44, 148–49, 164–65
traits (models of adaptive behavior), 143. *See also*
adaptation and adaptive behavior
troubleshooting tips, 66

`true`. *See* boolean reporters and variables
`turtles-here`, 117, 130, 145; to provide members of a patch collective, 210
`turtles-on`, 77–78, 145
`turtles-own`, 25, 48, 129, 211

uncertainty: in calibration data, 262; reduction via
calibration, 256, 258–59; not quantified by
replication, 197; in sensing information, 127;
structural vs. parameter, 256; theory development considerations, 253
uncertainty analysis: definition of, 292; methods for,
297–302; of relative vs. absolute predictions,
302
underdetermined models, 233
`uphill`, 17, 54, 88–89
User Manual for NetLogo, 15–17, 27; printable version of, 29
`user-file`, 70
utility (economic objective), 41, 143; function of
Business Investor model, 133–34, 148–53

validation of models, 238–39. *See also* pattern-
oriented modeling; theory in agent-based
models
variables: of breeds, 211; built-in, 17, 24, 28, 116;
choosing (*see* structure and structural realism
of models); that contain agents and agentsets, 65; discrete vs. continuous, 26; local, 66,
130–31; scope of, 128–31; types of, 31, 49, 179;
of other objects, 131–32
version control, 62
View, 116–18; changing size of, 50; in continuous-
time models, 190–91; updating of, 117, 118,
124, 184–85. *See also* Interface tab

`wait`, 117, 189
web site for this book, xv
`while`, 70, 189–90; to catch normal distribution
outliers, 199–200, 300
Wild Dog model, 212–21; ODD description of,
213–15; uncertainty analysis of, 298–302
`with`, 117, 145–46; and execution speed, 314
`with-local-randomness`, 203, 300
`with-max` and `with-min`, 145, 146
Woodhoopoe model: calibration of, 264–67; ODD
description of, 250–53
`word`, 56, 68; to produce .csv output files, 122
world and wrapping, 23, 70, 78. *See also* Model Settings dialog
worry, why not to yet, 40, 45, 62, 109, 140, 256, 262.
See also panic

Index of Programming Notes